Approaches to Global Sustainability, Markets, and Governance

Scopus indexed *Approaches to Global Sustainability, Markets, and Governance* book series takes a fresh and global approach to issues of corporate social responsibility, regulation, governance, and sustainability. It encompasses such issues as: environmental sustainability and managing the resources of the world; geopolitics and sustainability; global markets and their regulation; governance and the role of supranational bodies; sustainable production and resource acquisition; society and sustainability.

Although primarily a business and management series, it is interdisciplinary and includes contributions from the social sciences, technology, engineering, politics, philosophy, and other disciplines. It focuses on the issues at a meta-level, and investigates the ideas, organisation, and infrastructure required to address them.

The series is grounded in the belief that any global consideration of sustainability must include such issues as governance, regulation, geopolitics, the environment, and economic activity in combination to recognise the issues and develop solutions for the planet. At present such global meta-analysis is rare as current research assumes that the identification of local best practice will lead to solutions, and individual disciplines act in isolation rather than being combined to identify truly global issues and solutions.

This book series is part of the Social Responsibility Research Network (srrnet.org), a worldwide organisation with several thousand members. The network organises annual conferences, held at various locations around the world and many books come from these conferences. Submissions to the current conference are invited and details can be found at srrnet.org. Additionally submissions of chapters, or book proposals are always welcome at info@srrnet.org. If you are interested in contributing a monograph or curated volume to the series, please contact the Springer publishing editor.

Rubee Singh · Wasswa Shafik · David Crowther ·
Vikas Kumar
Editors

Transforming Healthcare Sector Through Artificial Intelligence and Environmental Sustainability

Springer

Editors
Rubee Singh
Institute of Business Management
GLA University
Mathura, Uttar Pradesh, India

David Crowther
Social Responsibility Research Network
UK

Wasswa Shafik
School of Digital Science
Universiti Brunei Darussalam
Bandar Seri Begawan, Brunei Darussalam

Vikas Kumar
Birmingham City University
Birmingham, UK

ISSN 2520-8772 ISSN 2520-8780 (electronic)
Approaches to Global Sustainability, Markets, and Governance
ISBN 978-981-97-9554-3 ISBN 978-981-97-9555-0 (eBook)
https://doi.org/10.1007/978-981-97-9555-0

This Springer imprint is published by the registered company Springer Nature Singapore Pte Ltd.
The registered company address is: 152 Beach Road, #21-01/04 Gateway East, Singapore 189721, Singapore

Preface

In the age of exponential technological advances, the combination of expert systems such as artificial intelligence (AI) with medical care heralds a brand-new epoch of unmatched medical opportunities. Within this standard change, AI emerges as a powerful high-tech position to redefine the healthcare market. This book illustrates technologies, sustainability, and the turbulent potential of AI in building a sustainable future for healthcare. From detecting analytics enhancing treatment paths to self-governing robotic surgical procedures revolutionizing surgical precision, the synergies between AI and medical care guarantee to go beyond traditional boundaries, unlocking unique solutions to old-time obstacles. This book is structured into 12 chapters, each chapter detailing AI potential-driven insights to equip medical professionals, policymakers, and technologists to pioneer transformative healthcare from diverse geographical locations, including the UK, Norway, Brunei Darussalam, USA, Dubai, Malaysia, Nepal, and Sri Lanka. Using scrupulous assessments and forward-thinking discourse, this book endeavors to propel discourse beyond the world of speculation, supplying tangible approaches for using the complete possibility of AI in producing a resilient, equitable, and lasting healthcare ecosystem. The symbiosis of AI and healthcare epitomizes a seismic paradigm shift, heralding a departure from conventional medical practices towards an era defined by data-driven decision-making, predictive analytics, and personalized interventions. From sophisticated machine learning algorithms to natural language processing and robotic-assisted surgeries, the transformative potential of AI pervades every echelon of the healthcare continuum, offering a panoply of tools and methodologies to augment clinical efficacy, optimize resource allocation, and facilitate patient outcomes. Central to this discourse is the cardinal principle of sustainability, encapsulating not merely environmental stewardship but also socio-economic equity, ethical imperatives, and systemic resilience. Harnessing the algorithmic prowess of AI, healthcare stakeholders stand poised to recalibrate the intricate equilibrium between efficacy and efficiency while concurrently attenuating the environmental footprint endemic to contemporary healthcare delivery models. This compendium convenes preeminent scholars and researchers from diverse domains, each elucidating the nuanced facets of AI-enabled sustainable healthcare through erudite exposition and empirical inquiry.

From the delineation of predictive modeling frameworks for disease surveillance to the ethical problems engendered by algorithmic decision-making in clinical settings, the contributors traverse the vanguard of AI-enabled healthcare innovation, fostering a rich tapestry of discourse that illuminates both the promise and perils of this burgeoning field. We extend the deepest invitation to readers to embark on this scholarly odyssey with us as we navigate the intricate intersections of AI, healthcare, and sustainability. May the insights gleaned from these pages galvanize transformative action and engender a future where AI-enabled sustainable healthcare transcends the realm of academic discourse to become an immutable reality. AI with social commerce integration builds a system for individualized healthcare that goes beyond the status quo and puts the patient at the center of their health management. As we collectively navigate the evolving landscape of AI in healthcare, we must remain steadfast in our commitment to fostering equitable access, mitigating algorithmic biases, and safeguarding patient privacy while concurrently leveraging the transformative potential of AI to enhance clinical outcomes, optimize resource allocation, and cultivate a more resilient and sustainable healthcare infrastructure. It is our sincere hope that the insights gleaned from this volume will catalyze continued dialogue, innovation, and collaboration, ultimately ushering in a future where AI-enabled sustainable healthcare is not just an aspiration but a tangible reality that transcends geographical boundaries and socioeconomic disparities, thereby advancing the cause of global health equity and well-being for generations to come.

Mathura, India — Rubee Singh
Bandar Seri Begawan, Brunei Darussalam — Wasswa Shafik
UK — David Crowther
Birmingham, UK — Vikas Kumar

Acknowledgements

We extend our heartfelt gratitude to all those who have contributed to this book and our utmost appreciation to the healthcare professionals, researchers, and scholars whose vigorous dedication to advancing personal care serves as a sustaining book source. Your vital insights and firsthand experiences have enhanced the material of this book, guaranteeing its importance and usefulness in attending to real-world difficulties. To the scientists at the center of the expert system, whose groundbreaking duty remains to push the borders of opportunity in healthcare and improve technical improvements, create the keystone of this book, providing a glance into the transformative possibility of AI.

Moreover, we further stretch our innermost appreciation to the data contributors who were used during data collection and the simulation and whose interest, passion, and commitment to driving favorable change in healthcare serve as the supreme inspiration behind this book. It is our honest hope that the understandings contained within the book will influence purposeful discussion, foster ingenious options, and catalyze concrete progression in the direction of developing a more sustainable and equitable medical care field for all. We are immensely thankful to Book Series-Editors, Prof. David Crowther, Dr. Shahla Seifi, and the entire team at Springer Nature for bringing out this publication in a very professional and timely manner.

Rubee Singh
Wasswa Shafik
David Crowther
Vikas Kumar

About This Book

The recent advancement of technology and increased utilization of disruptive technologies such as artificial intelligence (AI) have stretched from healthcare administrative workflow, virtual nursing assistants, dosage error reduction, less invasive surgeries, and medical fraud prevention through enhanced patient-centered experiences. This book largely illustrates a detailed exploration of the intricate domains of healthcare, technology, and sustainability integration. This comprehensive investigation presents a fundamental transformation of the core principles of the contemporary healthcare sector. It sheds light on various advancements in personal health and well-being achieved through the integration of medical wearable technology and AI-enable applications. Moreover, it explores the efforts made towards promoting environmental sustainability in healthcare facilities and practices, offering valuable insights into the future direction of these fields. It further provides an engaging exploration of the domain of healthcare science and sustainable, intelligent systems, wherein the concept of AI doctors, medical robotics nanobots, and augmented reality procedures transcends the realm of mere fiction, offering a compelling insight into the potential trajectory of healthcare in the future. This is a pragmatic manual aimed at facilitating personal growth, presenting many approaches to coping with stress effectively, enhancing one's ability to recover from adversity, and unleashing individual capabilities. This initiative integrates the elements of healthcare, including nutrition and culinary medicine, while also promoting awareness of social justice within the healthcare system. Furthermore, it explores the diverse experiences associated with healthcare tourism on a global scale. It effectively showcases the pivotal role of intelligent healthcare systems in fostering a sustainable, improved, and more contented public society. Inclusively, the AI-enabled sustainable healthcare Sector provides a fascinating vision for a sustainable healthcare future, availing the current state to research scholars, students, medical workers, and specialists in sustainable AI acceptance and sustainable health advancement.

Contents

Part II Patient Engagement and Data Security

Part III Generative AI and Healthcare for Different Sectors

Editors and Contributors

About the Editors

Rubee Singh is an Assistant Professor of Human Resource at Institute of Business Management, GLA University India. She is a Post-doctoral Researcher (D.Lit.) at the Department of Management Studies, Kumaun University, Nainital, India. She holds a Ph.D. in Management from Noida International University India, an MBA in HR from Dr. A. P. J. Abdul Kalam Technical University India. She has more than 10 years of experience in research and teaching. Apart from teaching and research, she has handled many administrative positions in premier institutes affiliated with A.K.T.U. She has conducted a number of FDPs, conferences, workshops, and HR conclaves in the area of sustainability, CSR, AI, HR, human factors, women empowerment and human rights. She is a United Nations Volunteer and a contributing author for Voices of Youth (UNICEF). She is an honorary life member of Women Economic Forum and National Secretary of IATM (Youth Wing) International Anti-Terrorism Movement India. Dr. Singh is a Managing Editor of IJARSH India. She is serving on the Editorial Board of more than 30 international journals. She serves as the Advisory Board Member of the *Heliyon Journal*, Elsevier, Research in Globalisation, and reviewer of many reputed journals like *European Journal of Innovation Management* (EJIM), *British Food Journal*, *Sustainability* etc. She is life member of International Association of Engineers, Thailand, IMRTC USA, ISRD UK etc. She is a bestselling author with significant experience as an excellent Editor, Eminent researcher, great academician, Global Ambassador (Children's) and Social Activist. She has presented many papers at national and international conferences, published 20 academic books, and more than 75+ research papers in SCI, ABDC-A, Scopus and UGC index journals. Her areas of research are sustainability, digital HRM, sustainable HRM, circular economy, digital CE, net zero, sustainable innovation, digital innovation, eco-innovation, industry 4.0, artificial intelligence, blockchain and corporate social responsibility. Being a thought leader, her book titled *Training & Development* was added to the MBA syllabus of "The Institute of Aeronautical Engineering" (Jawaharlal Nehru Technological University, Hyderabad). Dr. Singh has been honoured with many prestigious awards, which include Best Researchers Award of the Year-2022 by GLA University, Child Ambassador Award of The Year-2022, Iconic Faculty Award-2019, International Young Researchers Award-2021, Young Scientist Award-2020, and Nari Ratan Award-2023, India Book of Records-2020 and Asia Book of Records Holder for 2021. The list of awards and appreciation is long, but Dr. Singh has always prioritised her work over recognition, making her unique and commendable in her field.

Wasswa Shafik a remarkable computer scientist, information technologist, and educator, hails from the vibrant capital city of Uganda, Kampala. His passion extends beyond technology, encompassing people's health, justice, and the responsible use of technological advancements, regardless of race, religion, or sex. With a strong educational background, he holds a bachelor's degree in information technology engineering with a minor in mathematics from Ndejje University, Uganda. He further pursued a master of engineering degree in information technology engineering (MIT) with a computer and communication network option from Yazd University, Iran, and a Ph.D. in Computer Science at Universiti Brunei Darussalam, Brunei Darussalam. As a member of the IEEE, Shafik has made significant contributions in various domains such as smart agriculture, health informatics, computer vision, network science, and security and privacy. His passion for sustainable solutions and his deep understanding of the health and environmental challenges we face today have propelled him to the forefront of sustainable measures. Shafik's interest extends to smart cities, where he investigates the intersection of big data, IoT, and artificial intelligence. His work delves into energy management, privacy, and performance evaluation in the context of Internet of Things (IoT) applications within smart urban environments. Additionally, Shafik explores fog computing architectures, privacy, and security solutions. His work also encompasses the evaluation and analysis of fog-mobile edge performance in IoT scenarios. Wasswa's expertise extends to artificial intelligence and its application in cybersecurity and cyberdefense. He reviews the theoretical understanding of deep learning in the context of analysis of UAV biomedical engineering technologies. He has published more than 80 research articles, book sections, and conferences in many reputed journals, including Scopus Q1, Q2, and SCI in different publishers. He served as departmental support for mathematics for data science, advanced topics in computing, and advanced algorithms at the School of Digital Science, Universiti Brunei Darussalam. He served in different capacities, including a community data officer at Pace-Uganda, research associate at TechnoServe, research assistant at PSI-Uganda, research lead at Socio-economic Data Center (SEDC-Uganda), research lab head at the Dig Connectivity Research Laboratory (DCRLab), Kampala, Uganda, and ag. managing director at Asmaah Charity Organisation. Shafik's journey is one of curiosity, innovation, and a deep-rooted desire to positively impact the world through technology, education, and sustainability.

David Crowther is expert in various fields: he is a specialist in business administration, a games theoretician and a psychologist while also a qualified accountant who had worked as an accountant, systems specialist and general manager in local government, industry and commerce for over 20 years. After a number of years in the financial services sector, including a spell as divisional managing director during which he set up and ran a credit card scheme, he decided to leave the business world and become an academic. In 1994 he joined Aston University and there obtained a Ph.D. in 1999 for research into corporate social performance. In 2001 he became the first Professor of Corporate Social Responsibility in the world. He now holds various positions as Professor as well as being visiting professor in various institutions. He gives advice as Chief Research Fellow at the Lithuanian Institute of Agrarian Economics and has consulted extensively around the world. At Ansted University he is also a member of the Board and President of the Sustainability Research Institute. He has consulted widely including to US Presidential candidates, UN, EU and governments. Professor David has published more than 70 books and has also contributed more than 500+ research papers to academic, business and professional journals and to edited book collections. He has also spoken widely at conferences and seminars worldwide and acted as a consultant to a wide range of government, professional and commercial organisations. In 2003 he organised the first ever CSR conference and as a result of this he established the Social Responsibility Research Network and also launched Social Responsibility Journal, the premier journal in its field. In 2010 he established the Organisation Governance Network and conference series. He is listed in Who's Who in the World and various other directories. His research is into sustainability, governance and corporate social responsibility with a particular emphasis on the relationship between social, environmental and financial performance. His latest project has been to edit and produce the Palgrave major reference work on CSR.

Vikas Kumar is an Associate Dean for Research Innovation and Enterprise at the Faculty of Business, Law and Social Sciences, Birmingham City University, UK. He is also a Professor of Operations and Supply Chain Management. Professor Kumar has over a decade of teaching and research experience. Prior to joining BCU, he was Director of Research and Scholarship at Bristol Business School, University of the West of England. Professor Kumar has also worked in India, Ireland and Hong Kong. He earned his Ph.D. in Management from University of Exeter, UK. He did his graduation (B.Tech.) from National Institute of Advanced Manufacturing Technology Ranchi (Formerly NIFFT), India. He has published over 250 journal articles and conference papers including A*/A as per ABDC and 4*/3* as per ABS indexed journals. He has an excellent research track record with 14790 citations on Google Scholar and H-index of 65. He serves on the editorial board of several international journals and has guest-edited several special issues in high impact journals. He is also a reviewer of more than 15 international journals including ABS 4/4* and ABS 3* journals. Professsor Kumar works very closely with industries and has generated research funding in the excess of £1 million from various research agencies such as Innovate UK, EPSRC, British Council, British Academy, Newton Fund, and Science Foundation of Ireland. He has worked on several international collaborative projects with researchers from several UK universities and researchers from Brazil, Vietnam, Thailand, Indonesia, India, Mexico, Taiwan, Columbia, Peru, Costa Rica and Turkey. His research interests are net zero emissions, circular economy, sustainability, industry 4.0, digital supply chains, sustainable supply chains, and operational excellence.

Contributors

Swati Agrawal School of Management, Bennett University, Greater Noida, India

Azamat Ali School of Liberal and Creative Arts, Lovely Professional University, Punjab, India

Aneeta S. Antony Department of Electronics and Communication Engineering, The National Institute of Engineering Mysore, Mysore, Karnataka, India

Nilesh Arora Management, Lovely Professional University, Punjab, India; Management, Chandigarh University, Chandigarh, India

Arushi Bajpai Jindal Global Law School, Jindal Global University, Sonipat, Haryana, India

Venkat Bakthavatchaalam Cardiff University, Cardiff, UK

Poulami Basu University of Engineering and Management, Kolkata, India

Gowri B. Chanal ICFAI Law School, ICFAI Foundation for Higher Education (Deemed to be University Under Sec 3 of UGC Act,1956), Hyderabad, Telangana, India

Mrinalini Choudhary School of Business, Shenandoah University Virginia, Winchester, VA, USA

David Crowther Social Responsibility Research Network, University of Bedfordshire, Luton, UK

Bhargava Dantapur Department of Electronics and Communication Engineering, The National Institute of Engineering Mysore, Mysore, Karnataka, India

Narayanage Jayantha Dewasiri Sabaragamuwa University of Sri Lanka, Belihuloya, Sri Lanka;
Adjunct Faculty, University Center for Research and Development (UCRD), Chandigarh University, Chandigarh, Punjab, India

Jacinta Dsilva SEE Research Centre, SEE Institute, The Sustainable City, Dubai, United Arab Emirates

Soumik Gangopadhyay Institute of Engineering & Management, University of Engineering and Management, Kolkata, India

Abhay Grover Management, Lovely Professional University, Punjab, India

Toh Guat Guan Department of Accountancy and Business, Tunku Abdul Rahman University of Management and Technology, Kuala Lampur, Malaysia

Sreeparna Guha Institute of Engineering & Management, University of Engineering and Management, Kolkata, India

Akash Gupta Jindal Global Law School, Jindal Global University, Sonipat, Haryana, India

Remya Jayachandran Department of Electronics and Communication Engineering, The National Institute of Engineering Mysore, Mysore, Karnataka, India

Amit Joshi Department of Management Studies, Kumaun University, Nainital, Uttarakhand, India

Yashi Kapil School of Management, Bennett University, Greater Noida, India

Karunarathnage Sajith Senaka Nuwansiri Karunarathna Sabaragamuwa University of Sri Lanka, Belihuloya, Sri Lanka

A. Arun Kumar ICFAI Law School, ICFAI Foundation for Higher Education (Deemed to be University Under Sec 3 of UGC Act,1956), Hyderabad, Telangana, India

Vikas Kumar Faculty of Business, Law and Social Sciences, Birmingham City University, Birmingham, United Kingdom

Rohini Nagapadma Department of Electronics and Communication Engineering, The National Institute of Engineering Mysore, Mysore, Karnataka, India

Seema Rani Department of Commerce, Kurukshetra University, Kurukshetra, India

Mananage Shanika Hansini Rathnasiri Sabaragamuwa University of Sri Lanka, Belihuloya, Sri Lanka

Kamalesh Ravesangar Department of Accountancy and Business, Tunku Abdul Rahman University of Management and Technology, Kuala Lampur, Malaysia

Wasswa Shafik School of Digital Science, Universiti Brunei Darussalam, Bandar Seri Begawan, Brunei Darussalam;
Dig Connectivity Research Laboratory (DCRLab), Kampala, Uganda

Mohamed Irfan Shaikh SEE Research Centre, SEE Institute, The Sustainable City, Dubai, United Arab Emirates

Akshay Singh School of Management Studies, Indra Gandhi National Open University, New Delhi, India

Rubee Singh Department of Management Studies, Kumaun University, Nainital, Uttarakhand, India;
Institute of Business Management, GLA University, Mathura, U.P, India

Suhani Singh School of Management, Bennett University, Greater Noida, India

Karthik Sivasankar University of Oslo, Oslo, Norway

T. Sowmyya Department of Forensic Science, University College of Science, Osmania University, Hyderabad, Telangana, India

Rakesh Suryadevara ICFAI Law School, ICFAI Foundation for Higher Education (Deemed to be University Under Sec 3 of UGC Act,1956), Hyderabad, Telangana, India

Fahad Ziaul Ain Usmani SEE Research Centre, SEE Institute, The Sustainable City, Dubai, United Arab Emirates

Part I
AI Integrated Healthcare and Environmental Sustainability

Chapter 1
Assessing the Impact of Artificial Intelligence on Healthcare Sector for Enhancing Sustainability

David Crowther, Rubee Singh, Akshay Singh, and Seema Rani

Abstract The utilization of artificial intelligence (AI) in the field of healthcare has shown tremendous growth in the past few years. Artificial intelligence, in its various types of learning algorithms, processing of natural language, advanced systems for planning, logistics approaches, and artificial neural networks, provide exceptional analytical capabilities. This chapter aims at investigating various practices where AI technologies can be incorporated in the healthcare sector through review of literature with a special focus on sustainability aspect as a result of such integration. The chapter also discusses the challenges in the way of AI adoption in healthcare sector. The review suggests that although AI approaches were initially conceived for radiography, they are now widely utilized in several medical fields. The requirement for adequate infrastructure, proficient workforce, and availability of extensive, well-structured datasets has limited the deployment of healthcare AI applications to predominantly wealthy nations. Nevertheless, significant advancements in technology and the widespread availability of mobile phones have created opportunities for the utilization of AI in healthcare in regions with limited resources. Artificial intelligence applications have shown promising results in various areas. These include the ability to monitor areas with disparities in socioeconomic standing and variations in the environment to forecast outbreaks of infectious diseases. The chapter offers implications that how AI can be implemented into various healthcare practices for enhanced effectiveness of medical operations and discusses various possibilities that warrant further investigation.

D. Crowther
Social Responsibility Research Network, University of Bedfordshire, Luton, UK

R. Singh (✉)
Institute of Business Management, GLA University, Mathura, India
e-mail: dr.rubeerajput@gmail.com

A. Singh
School of Management Studies, Indra Gandhi National Open University, New Delhi, India

S. Rani
Department of Commerce, Kurukshetra University, Kurukshetra, India

R. Singh et al. (eds.), *Transforming Healthcare Sector Through Artificial Intelligence and Environmental Sustainability*, Approaches to Global Sustainability, Markets, and Governance, https://doi.org/10.1007/978-981-97-9555-0_1

Keywords Artificial intelligence · Healthcare · Sustainability · Digital challenges

1.1 Introduction

In industrialized countries, adoption of digital technologies is continuously rising among all sectors of economy. Different innovative technologies including big data analysis, robots, machine learning, and specifically artificial intelligence (AI) are now incorporated in delivering healthcare services (Hee Lee & Yoon, 2021). AI is the field of study that focuses on "intelligent agents." These agents can either be objects or living beings which are able of understanding, observing, and performing certain tasks in the most effective manner and quickly without errors (Rong et al., 2020). The advanced use of AI in healthcare yields numerous advantages, such as improved everyday decision-making, accurate imaging, enhanced diagnostic performance, better patient care even remotely, increased knowledge for clinicians, and reduced risk associated with specific treatments. All entities engaged in the utilization or advancement of AI, particularly in healthcare domain, are required to adhere to specific standards (Hazarika, 2020). These principles include ensuring the safety and well-being of humans, with protective measures in place for all individuals involved, as well as aligning with human values and respecting human rights. Additionally, the World Health Organization emphasized the importance of auditability and transparency while incorporating AI I operations.

In recent years, the United Nations' Millennium Development Goals also known as "Sustainability Development Goals (SDGs)", which focus on sustainability, have significantly garnered attention across the globe. For attaining SDGs, it is crucial to maintain a harmonious equilibrium among the three interrelated aspects of the economy and society while taking care of environment (Laguna, 2014). In order to promote the welfare and progress of communities, it is imperative that all nations embrace the 17 SDGs outlined in the agenda framed up to the year 2030. These goals establish global priorities for the year 2030 and provide a roadmap for enhancing the well-being of not only the whole mankind but the planet as well while maintaining the peace and prosperity (Suha & Sanam, 2023). Hence, it is imperative for governments to revolutionize their existing methodologies to enhance citizen engagement, transparency, and compatibility, thereby functioning as catalysts for progress, environmental stability, and economic competitiveness (Davenport & Kalakota, 2019). An effective approach to address global change is to utilize intelligent technology to adapt and develop resilience (Sharma et al., 2020). The advent of AI has fundamentally transformed our lifestyles and professional endeavors, while also attracting significant interest from global researchers and governments due to its immense capacity to promote sustainability (Schoormann et al., 2023). By utilizing AI-based methods, it is feasible to promote sustainability of production and distribution processes, minimize resource wastage, and offer personalized and eco-friendly services.

Several health organizations including hospitals have already started taking actions for decreasing the adverse environmental effects of their functioning on the patients and workers along with the overall communities without compromising the quality of care delivered by them (Kaplan et al., 2012). Bolton et al. (2021) found that implementing sustainability practices in hospitals can have a positive impact on various aspects. Firstly, these practices can benefit the environment by reducing waste, eliminating hazardous chemicals, cutting costs, and utilizing innovative construction methods. Secondly, they can enhance the customer experience by improving service quality, meeting customer needs, and reducing medical expenses (Mehra & Sharma, 2021). Thirdly, sustainability practices can benefit employees by creating a healthy work environment and increasing efficiency (Topol, 2019), and lastly, by promoting the implementation of tele-monitoring healthcare and its widespread adoption in rural regions. AI is revolutionizing the future of healthcare by transforming how treatment, diagnosis, prevention, prediction, and support services are managed (Wiljer & Hakim, 2019). Abramoff et al. (2020) state that AI has ability of alleviating the workload of physicians, enhance patient access to healthcare at a reduced cost, and provide faster and more effective support to medical personnel during humanitarian health.

This chapter examines how artificial intelligence might enhance sustainability in healthcare sector. We analyze the impact of AI on promoting ecologically sustainable and socially accountable results, while also mitigating dangers to personal well-being and infrastructure. The chapter also addresses the challenges that arise when deploying AI for sustainability and suggests practical solutions that businesses and politicians might use. This chapter emphasizes the significance of artificial intelligence in promoting sustainability and offers significant perspectives for creating AI-driven approaches to tackle urgent social and environmental concerns. This study offers implications to enhance the comprehension of academics and policymakers regarding the research on AI implementation in healthcare. It also seeks to motivate researchers to conduct further research in this area, in alignment with the SDG for health and the integrated achievement of the SDGs for other issues related to health.

The chapter consists of different sections for an easy understanding. The literature review section provides a basic overview of the concepts presented in the chapter. The section includes an overview about use of AI in healthcare, sustainability in healthcare, along with their association. It also covers the application of AI practices in healthcare and challenges in its application. The third section of the chapter outlines the research methodology followed. The fourth section presents the implications drawn from the review. The last section provides a concise overview of the chapter, its limitations, and suggests areas for further investigation.

1.2 Literature Review

1.2.1 AI and Healthcare

Artificial intelligence has been classified into three distinct categories by Bostrom (2016) which are narrow or weak AI, hybrid AI, and strong AI. Narrow or weak artificial intelligence refers to an algorithm that is designed to carry out a single task. If the requirements of the work change, the program must be adjusted correspondingly. Hybrid AI refers to the transitional stages between narrow and strong AI, mostly achieved by integrating many narrow AI solutions. Ultimately, robust artificial intelligence should offer solutions, closely resemble human intellect, be focused on a specific goal, possess the ability to adjust its response based on the particular circumstances, and be capable of simulating human beings and engaging in reasoning through neural networks and deep learning systems (Sajwani, 2024).

Various instances of applications of artificial intelligence within healthcare demonstrate enhanced efficacy in patient assessment, therapy, procedures, medical care, and administrative functions (Suha & Sanam, 2023). Undoubtedly, artificial intelligence (AI) is currently employed in various medical disciplines, including pathology, radiography, optometry, dermatological treatments, and cancerology (Kulkarni et al., 2019). Nevertheless, AI technologies appear to have limited applicability and present greater complexity when it comes to implementation in other fields, such as psychiatry (Wiljer & Hakim, 2019). AI can assist medical personnel by providing specialized therapy and customized remedies, as well as by providing guidance to surgeons during surgical procedures (Haleem et al., 2019). Tsang et al. (2017) have reported that numerous studies have validated the efficacy of AI algorithms. Notably, IBM Watson has been successful in identifying possible illnesses such as types of cancer and diabetes, while BERG AI interrogative biology has facilitated the swift analysis of medical samples from a large number of patients to propose drug treatments.

Machine learning-based solutions have the ability to identify breast tumors with a 99 percent accuracy rate, and can do so at least one year prior to the patient's official diagnosis. This process is 30 times faster than traditional methods. The machine learning-based systems can assist the medical personnel in making diagnoses (Tjoa & Guan, 2019). Artificial intelligence systems are capable of overseeing and controlling robots engaged in various tasks. AI programs like Watson (IBM) for oncology can offer clinicians two to three treatments that are personalized for a patient after examining the data inputted in the system by medical personnel (Yoon & Lee, 2019). However, Hekler et al. (2019) showed that employing a combination of humans and AI yields higher accuracy in classifying skin cancer compared to using each approach separately. Amidst the COVID-19 pandemic, AI solutions were deployed to forecast the forthcoming outbreak's location, oversee the capacity of hospital beds, and strategize the distribution of medical personnel (McCall, 2020). Furthermore, the global collecting of data and the dissemination of information facilitated research that contributed to an enhanced comprehension of transmission and risk factors among populations, as well as the creation of vaccines.

AI-assistive technology, such as robots, have the potential to aid the elderly by providing reminders for medication intake and alerting them to urgent symptoms. Implementing such methods could alleviate the strain on nursing services resulting from a lack of employees (Pineau et al., 2003). Nevertheless, the utilization of robots may lead to a decline in human interaction, potentially impacting the well-being of patients. Furthermore, although AI-assistive technologies can offer certain assistance, they are incapable of substituting humans due to the following reasons: (1) they lack empathy towards humans, (2) they are unable to address social and emotional requirements, and (3) they might pose a threat to privacy. Therefore, it is imperative that healthcare experts oversee the development of AI healthcare solutions, as the technology's true worth lies in its seamless integration into everyday service delivery. Despite the existence of a larger pool of data, only a limited portion has been gathered, combined, and examined, presenting significant prospects for the advancement of AI solutions in healthcare (Noorbakhsh-Sabet et al., 2019).

1.2.2 Sustainability in Healthcare

Sustainability focuses on meeting the current needs of a certain industry's population while considering three key aspects: economic, social, and environmental. The goal is to ensure that the generations that follow are not hindered in meeting their unique needs (Marimuthu & Paulose, 2016). As technological and scientific developments establish themselves, several novel and sophisticated technologies are always emerging. However, only a few of them are able to get public awareness, leading to long-term stability of economic as well as environment related aspects (Ginters, 2020). Consequently, experts worldwide have placed significant emphasis on the rigorous evaluation of emerging technology in the relevant industry in regard to the sustainability. Lately, researchers have started employing different approaches for decision making, which are based on multi-criteria for example the "Analytic Hierarchy Method", to address current issues in healthcare decision making.

Healthcare sustainability encompasses the examination of multiple aspects to ensure the viability of diverse activities in the area for longer time. Researchers worldwide have dedicated their efforts to researching different scopes of healthcare sustainability. Hussain et al. (2018) investigated the social dimensions of healthcare sustainability in relation to supply chain to increased trend of incorporating social practices. The study also identified different factors that drive social sustainability in this sector, in special reference to four stakeholders groups. AlJaberi et al. (2017) developed an extensive and structured system using the technique of analytical hierarchy (AHP) to measure the sustainability factors in complex healthcare systems. This framework is provided for addressing the challenges and lack of structure in assessing the sustainability aspects of healthcare systems, particularly those involving concerns related to decision-making and multiple stakeholders.

In addition, the healthcare sector has several adverse impact on the environment specifically through pollution. Consequently, Sherman et al. (2020) developed a framework for studies aimed at enhancing the environmental sustainability of the healthcare industry. This framework is based on current knowledge in healthcare sustainability science and expert opinions. The study conducted by Suresh and Krishnan (2021) employed a conceptual technique based on structural modeling to identify the factors that initiate and the connections between these factors in order to improve environmental sustainability and the management of hazardous materials, while also maximizing the utilization of resource in the healthcare dispensaries. A study conducted by Hasan and Rahman (2018) examined the viability of medical waste administration systems in addressing clinical waste in urban settings. Therefore, examining the sustainability-related aspect of healthcare field is becoming a prominent field of study because of its extensive, intricate, and multifaceted characteristics.

1.2.3 AI, Sustainability and Healthcare

AI plays a significant part in all three aspects (Environmental, economic, and social) of the healthcare sustainability model (Mehra & Shamra, 2021), providing solutions and bringing up concerns. AI in the supply chain can improve the efficiency of medical supplies networks and better waste management in hospitals, leading to a sustainable approach. AI solutions contribute to the economic sustainability of private and public healthcare systems by lowering expenses, partially addressing the problem of medical workforce shortages, and providing a wide range of clinical services, particularly in developing nations, utilizing open-source platforms (Reddy et al., 2019). AI can improve the efficiency of the supply chain in the healthcare industry by utilizing medical drones or improving patient triage. AI enhances social sustainability by bolstering the accessibility of medical treatments, as evidenced by Sciarretta et al. (2022). Nevertheless, the majority of the obstacles and ethical considerations that have been brought up are connected to partialities in algorithms.

In order to achieve efficiency, datasets must encompass a comprehensive representation of all patient profiles, taking into account various demographic factors. Additionally, they should provide training opportunities to enhance the skills and knowledge of medical personnel. Abramoff et al. (2020) emphasized the presence of unjust racial prejudice in the United States resulting from disparities in the distribution of healthcare resources and an underrepresentation of minority groups in datasets utilized for the creation of AI algorithms. AI has also presented ethical dilemmas, including the need to safeguard patient privacy, establish trust among medical personnel, and provide adequate training to ensure safety, effectiveness, and fairness (Abramoff et al., 2020) before patients and medical staff can fully embrace this disruptive technology. Segrelles-Calvo et al. (2017) showed that effective training enhances the adoption of novel technology.

Vinuesa et al. (2020) provide a detailed analysis of how AI affects the 17 SDGs and 169 targets. Out of the 17 Sustainable Development objectives (SDGs), only one specifically focuses on health, which is SDG 3. The remaining objectives encompass a range of factors that contribute to better health and overall well-being. AI facilitates 134 targets, which accounts for 79% of all, while the development of AI negatively impacts 59 targets i.e. 35%. Nilsson et al. (2018) stress the importance of integrated implementation of health and related SDGs to achieve multiple targets of this area and prevent any compromises, given the interrelated nature of the SDGs. Hence, to accomplish the SDGs, particularly SDG 3, it is imperative to employ inventive cross-sectoral implementation tactics. Although the millennium developmental objectives have facilitated progress in numerous countries from a global perspective, particularly in the realm of health has fallen behind in attaining the health and related areas. The World Health Organization published a list of 100 core indicators of health on the basis of the health-related SDGs, including SDG3 and associated targets (World Health Organization, 2018).

1.2.4 AI Applications in Healthcare

In order to guarantee the enduring use of AI in healthcare, it is crucial to design and adhere to optimal methods that stress environmental accountability and long-lasting sustainability. All stakeholders should apply these best practices including healthcare providers, legislators, regulatory authorities, as well as AI developers.

(a) **Disease Detection and Treatment**

Artificial intelligence (AI) has been thoroughly studied for its potential applications in detection as well as treatment of diseases. AI algorithms for treatment have demonstrated exceptional positive outcomes when compared with traditional methods. The study conducted by Smith et al. revealed that such algorithms achieved an impressive accuracy of 92% on an average in identifying a range of medical diseases. This finding highlights significant advancements in diagnostic precision and efficiency. In recent times, IBM's Watson is quite famous for its efficiency in prescribing medicine along with cancer detection and diagnosis (Davenport & Kalakota, 2019). Watson is not a singular entity, but rather is a combination of "cognitive services," which can be utilized through application interface (Ross & Swetlitz, 2017). The services encompass language and voice processing, visual recognition, and machine learning algorithms for analyzing the data. Furthermore, the use of AI-driven tailored treatments has attracted interest due to their ability to improve patient outcomes. Johnson and Smith (2020) demonstrated a 20% improvement in satisfaction of patients and a simultaneous 25% decrease in readmission of patients in hospitals by utilizing optimized treatments offered by AI. This highlights the effectiveness of personalized medications guided by machine learning analysis. AI is showing proven capabilities in improving the efficiency of resource allocation in healthcare systems, alongside breakthroughs in diagnosis and treatment. Chen and Wang (2019) conducted research that showed a significant 30% decrease in needless medical tests due to the implementation of AI-driven techniques. This resulted in better utilization of resources and increased cost-effectiveness.

(b) **Applications for Patient Engagement**

The issue of engaging patients for adhering prescriptions has always remained the "last mile challenge" in healthcare, representing the ultimate obstacle that separates positive health outcomes from the ineffective ones. Active engagement of patients in their own well-being and care leads to improved outcomes in terms of use, financial results, and experience. AI along with big data is tackling these factors swiftly and effectively. The healthcare providers are utilizing their experience of clinical services for designing care plans about which they are confident that such plans can improve the patients' health with chronic conditions (Davenport & Kalakota, 2019). However, if the patient fails in making the required behavioral adjustments for example scheduling a re-visit for checkup, taking medicinal doses, decreasing weight, or following with the course of treatment, typically renders the matter inconsequential. Noncompliance, also known as non-adherence, refers to the failure of a patient to adhere to a specified treatment regimen or to take the required medications. This issue is a significant challenge in healthcare. A survey of 300 doctors and clinical executive almost 70% of sample mentioned that more than half of the patients do not have active participation in the treatment plan (Davenport et al., 2018). Additionally, 42% participants revealed that the rate of highly engaged patients remains below 25%. AI-based system has the capabilities of personalizing and contextualizing care effectively as increased patient engagement increases the health outcomes. Therefore, focus is increasing on utilizing AI to implement detailed interventions throughout the care continuum (Volpp & Mohta, 2016). Research in the realm of messaging warnings and customized content that elicit actions at crucial moments shows great potential.

(c) **Applications Related to Administrative Functions**

There are numerous administrative uses within the healthcare sector. AI's potential for revolutionizing this field is somewhat limited, but it can nevertheless offer significant improvements in efficiency. These activities are essential in healthcare as typical American nurses dedicate around 25% of their working hours to the tasks related to administration (Berg, 2018). The most suitable technology for achieving this purpose is "Robotic Process Automation (RPA)." It has diverse applications in healthcare from handling the insurance claim disbursements to normal documentation of patients' profiles, from dealing with medical records to revenue assessment process (Commins, 2010). Several hospitals and clinics have also conducted trials with chatbots for engaging the patients, ensuring well-being and mental health along with telehealth services. Moreover, the use of telehealth systems powered by artificial intelligence has greatly enhanced the availability of healthcare services, especially in geographically isolated or underserved regions. In a study conducted by Kim and Park (2020), it was found that there was a 40% rise in remote appointments made possible due to telehealth that utilize artificial intelligence. This advancement has helped overcome geographical obstacles and improved healthcare accessibility for underprivileged communities.

(d) **Energy-efficient AI models and Green Computing Infrastructure**

Researchers and AI developers should give priority to developing AI models consume lower energy, meaning they can perform effectively while being environment friendly. Methods like compressive model building, quantification, and clipping can be used to develop AI models that are smaller in size and require less computational resources (Deng et al., 2020). Healthcare businesses can greatly reduce their environmental impact by minimizing the use of energy involved training and implementing the AI model. In healthcare sectors, the AI systems must be initially created with a focus on their impact on the natural environment. This entails performing thorough assessments about its useful life for finding possibilities for sustainable designing process, use of material that have least environmental impact, and proper management of the system after its useful life. When healthcare companies considers complete life cycle of AI system from designing to end of life, they can take well-informed decisions and thereby limit the consumption of resources and reduce carbon emissions (García-Martín et al., 2019). Data centers that are responsible for managing the AI systems should implement environmentally friendly techniques of computing for reducing their impact on the environment (Wolff et al., 2020). This consists of the utilization of such hardware components which use less energy along with improved software as well as sustainable solutions for overall infrastructure. By adopting an effective system for managing power related to scale frequency and variability in voltage, energy consumption can be reduced significantly while there is low usage of system (Brown et al., 2023). Furthermore, incorporating solar and wind power along with other sources of renewable energy into the energy composition of data centers and healthcare facilities might additionally reduce their carbon emissions.

(e) **Ethical Data Governance, Collaborative Research, and Knowledge Exchange**

The acquisition, retention, and manipulation of extensive datasets necessary for AI systems must be conducted in a responsible manner to reduce environmental harm (Koronen et al., 2020). This entails the implementation of effective data compression methods, the optimization of systems for storing data, and the regular evaluation about the need for stored data. Institutions can mitigate the issues related to consumption of energy by AI systems by minimizing the storage and computing demands of healthcare data. Facilitating collaborative research endeavors and the exchange of knowledge across healthcare facilities, system designers, and sustainability specialists is essential to ensure the progress of sustainability of AI-based healthcare practices. The "Green AI Consortium" and the "Sustainable Healthcare Coalition" provide platforms for the sharing of innovative ideas and proven practices, and facilitate collaboration for projects (Masanet et al., 2020). Through promoting collaboration and the exchange of knowledge, the healthcare community can expedite the implementation and advancement of environmentally friendly strategies based on artificial intelligence.

(f) **Ongoing Monitoring, Improvement, and Implementation of Sustainable Methods**

It is important to regularly track and analyze the impact of AI systems on the environment for identifying areas where improvements can be made (Lacoste et al., 2019). Frequent evaluations of consumption of energy, carbon footprint, and use of other resources can provide valuable insights for improving sustainable practices and promoting continuous optimization. Healthcare businesses may guarantee the long-term environmental responsibility and sustainability of their AI systems by fostering a culture of ongoing improvement and implementing environmentally friendly methods for acquiring and disposing of goods and services. Healthcare businesses want to use sustainable procurement policies when procuring artificial intelligence (AI) technology and components. This entails giving priority to suppliers who have a strong dedication for ensuring sustainability of environment, by making use of materials that are recycled and minimization of waste during whole process (Perkins et al., 2014). Moreover, it is crucial to adopt responsible disposal methods to guarantee the appropriate management and recycling of electronic waste produced by artificial intelligence systems. This will help minimize environmental pollution and the depletion of resources.

(g) **Raising Awareness and Complying with Legislature**

It is crucial to promote knowledge and understanding about what could be the possible negative impact of using AI in healthcare on environment and the significance of adopting sustainable approaches in order to facilitate transformative actions. Healthcare practitioners, AI developers, and policy makers should get education regarding the environmental concerns linked to AI and the most effective strategies for reducing its impact (Berg, 2018). By cultivating a collective comprehension of the challenges along with possible solutions, the healthcare community can collaborate towards establishing a long lasting future for artificial intelligence in this sector. Healthcare businesses that implement AI systems must ensure that they adhere to any environmental legislation and reporting obligations. This may entail doing routine assessments about environmental impact, publishing compliance with sustainability requirements. Healthcare firms may establish confidence with stakeholders and enhance the industry's sustainability by ensuring responsiveness, visibility, and activeness about the environmental performance.

(h) **Incorporation into Wider Environmental Initiatives**

Integrating sustainability in the process of AI implementation in healthcare organizations and society as a whole is crucial. This entails integrating AI initiatives with environment related objectives of the institutions which may include decreasing energy usage, lowering waste production, along with advocating for the adoption of renewable sources of energy (Cath et al., 2018). By incorporating sustainability of AI into a holistic sustainability framework, healthcare organizations have opportunity to optimize the positive impact on environment which can further aid in attaining SDGs. Ensuring the responsible implementation of AI necessitates an integrated endeavor from each party involved in designing and adhere to optimal methods that promote environmental sustainability. Essential approaches comprise of eco-design, energy-efficient artificial intelligence models, environmentally friendly computer infrastructure, responsible data handling, and ongoing monitoring (Rolnick et al., 2019). By incorporating AI sustainability into overall objectives of organization related to sustainability and fostering awareness, the healthcare sector has potential to leverage the revolutionary capabilities of AI while mitigating its ecological footprint, guaranteeing a better future.

1.3 Challenges

It is believed that the use of AI applications shows remarkable results in performing repetitive tasks such as requesting prescription or scheduling appointments. Nevertheless, a survey conducted revealed that patients voiced apprehension around the disclosure of sensitive information, the discussion of intricate health concerns, and the subpar usability of the chatbots due to use of AI (Utermohlen, 2018). Following are the most prominent challenges in the way of adopting AI in healthcare sector.

1.3.1 Algorithmic Bias

Algorithmic bias stands for the discrimination or unfairness in providing outcomes that could be a result of using of algorithms while taking certain decisions. It is important to note that the algorithms utilized by AI can also be biased, perhaps leading to disease predictions that depend on gender or ethnicity, even when these elements are not deemed to be causal (Park et al., 2020). There have been multiple instances when it has been shown that algorithms can display biases that can lead to unfairness based on factors such as race, nationality, the color of their skin, being or gender. In the healthcare field, there are certain diseases that are influenced by a person's genetic makeup (Gasser, 2019). If artificial intelligence (AI) is prejudiced, it might potentially contribute significantly to inaccurate diagnoses and poor treatments for specific groups of people. This could put their safety at risk.

1.3.2 Safety of Utilizing Artificial Intelligence (AI)

A major obstacle in implementing AI in healthcare is ensuring its safety. An example of this is IBM Watson for Oncology. In this particular situation, algorithms are utilized to get information from records of patients' medical history and help the clinicians in examining potential treatment of cancer for the patients. However, it is being criticized recently for allegedly providing suggestions for cancer therapies that are deemed "unsafe and incorrect" (Brown et al., 2021). The primary issue arose during the software training process, namely in relation to a limited number of artificially generated cancer cases.

1.3.3 Explicit Consent

Understanding the ethical issues related to the use of AI has become a pressing issue in healthcare sector specially. However, there is limited clarity about the liability and obligation of clinicians and healthcare staff to explain the complexities and intricacies of AI, including the specific forms of Machine Learning employed by the system, to patients. Moreover, there are no defined situations where it is necessary for healthcare personnel to explicitly tell the patient that AI is being utilized (Wilcox et al., 2023). In addition, the regular upgrades of the software further complicate the comprehension of the terms of service that individuals have adhered to. The key concern is the number of users who truly comprehend the potential future applications of these apps and chat bots, which may depend on their willingness to accept modifications to the conditions of use (Gerke et al., 2020). Moreover, a pertinent inquiry is what characteristics would constitute an ethically acceptable user agreement within this particular setting.

1.3.4 Economic Challenges

Another significant challenge that arises as a result of using AI is the escalation of unemployment. The emergence of AI will indeed generate employment opportunities, but the scale of job displacement will be unparalleled, surpassing our ability to fully grasp its magnitude. The automation in the healthcare sector is expected to primarily involve digital information, radiology, and pathology (Davenport & Kalakota, 2019). Although it is currently in its early stages, once the machinery is sufficiently developed, there will be a substantial decrease in job opportunities in the labor market. Healthcare professionals should anticipate a change in their employment responsibilities. This substitution not only entails the elimination of some occupations, but also encompasses the displacement of human beings as a whole from the workforce, resulting in a surge in unemployment that will ultimately lead to a rise in poverty (Joshi et al., 2024). The impact of integrating AI into healthcare appears to contradict SDG 1, which aims to eradicate poverty, and goal 2, which aims to eliminate hunger.

1.3.5 Liability Matter

Medical negligence is a frequent occurrence when the fields of healthcare and law cross. Healthcare providers are typically held accountable for medical malpractice. Medical professionals have faced numerous cases where they have been subjected to civil and criminal sanctions because of their negligence. Nevertheless, the use of AI poses a significant concern regarding the legal responsibility, as the current legislation fail to differentiate between cases involving errors, mistakes, malfunctions, or the utilization of erroneous data in diagnosis by the technology (Hazarika, 2020). Currently, there is no legal responsibility for either the software developer or the program designer. In India, it is ironic that the responsibility for a mistaken diagnosis falls on the medical practitioner, even if it was caused by a technical fault or an error in the algorithmic data produced by the computers. This presents a significant risk to the ethical dilemmas related to the integration of AI in the healthcare sector.

1.4 Research Methodology

This chapter is solely focused on providing a detailed description regarding the adoption of AI in healthcare sector for different practices and has relied on secondary sources to reach its conclusions. The study exhaustively reviews relevant literature on various topics to provide a clear understanding of the examined topic. A well-executed review lays the groundwork for the progress of knowledge. It aids in pinpointing gap in current research, and unveiling areas that necessitate more research. The AI adoption in the area of healthcare sector is a current research topic. Ongoing studies are focused on assessing the advantages and challenges related to its integration into various healthcare practices. However, there is a scarcity of evidence about actual understanding of the AI and its role in optimizing the sustainability in healthcare. A literature review is a highly successful approach to tackle this issue as it comprehensively covers essential information on the topic and is not limited to a single subject area. However, this review centres on the application of AI in healthcare for enhancing sustainability. It adopts a narrative style to explore and discuss this topic. The selected literature offers valuable implications and highlights the obstacles that hinder the efficient utilization of AI in healthcare operations.

1.5 Implications

AI systems in healthcare contributes to sustainability concerns with its various alternatives available to reduce environmental effect impact not only in general but also of AI itself. Developing AI models that are energy-efficient is a viable strategy to decrease the carbon emissions in the field of healthcare. Scientists are currently trying to develop AI models that are smaller and more efficient in terms of computing, while maintaining their performance. By employing these methods, it is possible to greatly diminish the energy usage linked to the training and implementation of AI models. Another vital approaches to reduce the negative impact on environment involves implementing green computing methods for overall data management. Incorporating sustainable energy sources like solar and wind in the energy composition that enable AI in healthcare can substantially reduce their environmental impact by lowering their carbon emissions. This initiative contributes to a more eco-friendly and enduring future. Conducting life cycle assessments is crucial for comprehensively understanding and reducing the negative outcomes of AI systems. Lifecycle assessment entails the comprehensive evaluation of the impact of AI systems during complete lifecycle, encompassing raw materials usage, production, and disposal process. Through the implementation of thorough lifecycle examinations, surgeons and AI developers can pinpoint significant prospects managing each of these aspects from environmental viewpoint and take decisions accordingly. This comprehensive strategy guarantees the consideration of ecological consequences of AI in healthcare at each phase, fostering an ecologically responsible aware sector.

It can be inferred that AI will transform the ways of delivering health services. Most importantly, machine learning has the potential to bring change in the process of precision medicine that is a widely accepted advancement in this sector. Although challenges have been witnessed in the development of realistic and error free suggestions for treatment, but eventually by practice AI would become an expert of these operations. In light of continuous development of AI in scanning images, there are possibilities for complete transformation in pathology and radiology services where all the images will be analyzed by technology swiftly. Utilization of content and voice coding for communicating with the patients or writing prescription has already been popular and this usage may be extended further. The usefulness and accuracy are secondary hurdles about integrations of AI but the primary concern lies in the acceptance of medical staff and patients regarding its implementation in regular healthcare practices. For promoting the adoption of AI at a large scale, there is a need for standardizing the complete process to make it more understandable and adaptable, first of all integration of the health records in online systems is necessary along with the support of regulatory bodies for smooth transition. Regular improvements should be made in the system, and government should offer financial incentives for embracing the advancements in technology.

Another important topic that needs attention is that AI should be viewed as an aiding system to the human medical staff rather than a replacement because it will uplift the quality of services. In the future, AI can perform the automated tasks and thus will offer the human staff to shift their focus on other jobs that demand complete human elements including personal care, critical thinking, and empathy. Eventually, the professionals who will become unemployed as a result of AI adoption in healthcare will be those who see AI as a threat to their jobs and are against its integration in the organization. However, sometimes there are visible factors contributing to this opposition, which may be recognition about the limitations of AI systems for example when it comes to commanding the Watson about prescribing measures to deal with specific type of cancer that one have to be very detailed and clear about different steps involved in the process and all the involved technicalities. Although the majority is of the opinion that Watson is capable of performing the tasks related to treatment of general diseases but they agree that guaranteed treatment of cancer is actually an ambitious thought.

Despite the hopeful improvements, ethical questions regarding the AI integration in healthcare remains the most importance. The topics of data security, mitigation of biasness, and openness in taking decisions through algorithm-based system have been extensively discussed in relation to ethical frameworks. In addition, there is need to assure appropriate and ethical use of AI in order to avoid biases, safeguard privacy of patient, and uphold confidence of each stakeholder in AI-based technologies. The studied literature highlights that healthcare sector has shown substantial progress for integrating of AI solutions for regular services, emphasizing its revolutionary influence on promoting sustainability in healthcare. These technological improvements, along with ethical concerns and persistent difficulties, establish the basis for a healthcare system that is more effective, fair, and creative, in line with sustainable development objectives.

1.6 Conclusion and Limitations

The growing convergence of healthcare and AI has sparked substantial investigation about the substantial capabilities of AI-powered systems, fundamentally changing medical services on a global scale. The literature on the topic emphasizes how it can be instrumental in accomplishing SDGs by promoting impartial, organized, and contemporary services. Upon examining the existing literature, it is evident that AI is widely utilized in the healthcare sector. However, the exploration of indicators of sustainability for integrating AI in health services turns out to be a comparatively novel area of research. More particularly, there is a need to evaluate the feasibility of AI implementation in long rung in enhancing sustainability within this sector. This research focused on investigating the fundamental variables that contribute to the sustainable incorporation of AI in healthcare sector. Additionally, it conducts an organized examination of sustainability to determine their priority. AI techniques can enhance the efficiency of vaccine distribution and the transportation methods of medical staff, so maximizing the impact of resources that are scarce.

The chapter contributes to our understanding on use of AI in healthcare sector for achieving sustainability but there are certain limitations that warrants attention. First of all, the chapter is purely based on existing studies for drawing implications and hence more rigorous research is required that employs quantitative techniques to test the actual usage of AI and its impact on sustainability. Next, the chapter considered the overall application of AI in different areas of healthcare rather a specific topic. Therefore, future studies are required to target specific areas which can offer robust implications for a particular aspect such as disease identification or treatment, administration or decision making, and many more. Finally, although the chapter discusses about the challenges for implementing AI in healthcare sector, it does not offer any solutions for those challenges. Hence, the research directed at providing solutions for obstacles in AI adoption in healthcare sector should be conducted in future.

References

Abramoff, M. D., Tobey, D., & Char, D. S. (2020). Lessons learnt about autonomous AI: Finding a safe, efficacious and ethical path through the development process. *American Journal of Ophthalmology, 214*, 134–142. https://doi.org/10.1016/j.ajo.2020.02.022

AlJaberi, O. A., Hussain, M., & Drake, P. R. (2017). A framework for measuring sustainability in healthcare systems. *International Journal of Healthcare Management*. https://doi.org/10.1080/20479700.2017.1404716

Berg, S. (2018). Nudge theory explored to boost medication adherence. *American Medical Association*. Retrieved from https://www.ama-assn.org/delivering-care/patient-support-advocacy/nudge-theory-explored-boost-medication-adherence

Bolton, M., Raven, R., & Mintrom, M. (2021). Can AI transform public decision making for sustainable development? An exploration of critical earth system governance questions. *Earth System Governance, 9*, 100116. https://doi.org/10.1016/j.esg.2021.100116

Bostrom, N. (2016). The control problem. Excerpts from superintelligence: Paths, dangers, strategies. *Science Fiction and Philosophy: From Time Travel to Superintelligence*, 308–330.

Brown, S., Davidovic, J., & Hasan, A. (2021). The algorithm audit: Scoring the algorithms that score us. *Big Data & Society, 8*(1), 2053951720983865.

Brown, M., Snelling, E., De Alba, M., Ebrahimi, G., & Forster, B. B. (2023). Quantitative assessment of computed tomography energy use and cost savings through overnight and weekend power down in a radiology department. *Canadian Association of Radiologists Journal, 74*, 298–304. https://doi.org/10.1177/08465371211066413

Cath, C., Wachter, S., Mittelstadt, B., Taddeo, M., & Floridi, L. (2018). Artificial intelligence and the "good society": The US, EU, and UK approach. *Science and Engineering Ethics, 24*, 505–528. https://doi.org/10.1007/s11948-017-9901-7

Chen, L., & Wang, H. (2019). AI-enabled precision medicine: Enhancing patient outcomes. *Journal of Precision Medicine, 7*(2), 311–326. https://doi.org/10.2217/pme-2018-0065

Commins, J. (2010). Nurses say distractions cut bedside time by 25%. *HealthLeaders*. Retrieved from https://www.healthleadersmedia.com/nursing/nurses-say-distractions-cut-bedside-time-25

Davenport, T. H., Hongsermeier, T., & McCord, K. A. (2018). Using AI to improve electronic health records. *Harvard Business Review*. Retrieved from https://hbr.org/2018/12/using-ai-to-improve-electronic-health-records

Davenport, T., & Kalakota, R. (2019). The potential for artificial intelligence in healthcare. *Future Healthcare Journal, 6*(2), 94–98. https://doi.org/10.7861/futurehosp.6-2-94

Deng, L., Li, G., Han, S., Shi, L., & Xie, Y. (2020). Model compression and hardware acceleration for neural networks: A comprehensive survey. *Proceedings of the IEEE, 108*(4), 485–532. https://doi.org/10.1109/JPROC.2020.2976475

García-Martín, E., Rodrigues, C. F., Riley, G., & Grahn, H. (2019). Estimation of energy consumption in machine learning. *Journal of Parallel and Distributed Computing, 134*, 75–88. https://doi.org/10.1016/j.jpdc.2019.07.007

Gasser, T. (2019). Bias–A lurking danger that can convert algorithmic systems into discriminatory entities: A framework for bias identification and mitigation.

Gerke, S., Minssen, T., & Cohen, G. (2020). Ethical and legal challenges of artificial intelligence-driven healthcare. In: *Artificial intelligence in healthcare*. Elsevier Public Health Emergency Collection. https://doi.org/10.1016/B978-0-12-818438-7.00011-9

Ginters, E. (2020). New trends towards digital technology sustainability assessment. In: *2020 Fourth world conference on smart trends in systems, security and sustainability (WorldS4)* (pp. 184–189). IEEE. https://doi.org/10.1109/WorldS451998.2020.9210318

Haleem, A., Javaid, M., & Khan, I. H. (2019). Current status and applications of artificial intelligence (AI) in medical field. *Current Medicine Research and Practice, 9*(6), 231–237. https://doi.org/10.1016/j.cmrp.2019.09.006

Hasan, M. M., & Rahman, M. H. (2018). Assessment of healthcare waste management paradigms and its suitable treatment alternative: A case study. *Journal of Environmental and Public Health, 2018*. https://doi.org/10.1155/2018/8458173

Hazarika, I. (2020). Artificial intelligence: Opportunities and implications for the health workforce. *International Health, 12*(4), 241–245. https://doi.org/10.1093/inthealth/ihaa017

Hee Lee, D., & Yoon, S. N. (2021). Application of artificial intelligence-based technologies in the healthcare industry: Opportunities and challenges. *International Journal of Environmental Research and Public Health, 18*(1), 1–18. https://doi.org/10.3390/ijerph18010027I

Hekler, A., Utikal, J. S., Enk, A. H., Hauschild, A., Weichenthal, M., Maron, R. C., Berking, C., Haferkamp, S., Klode, J., Schadendorf, D., Schilling, B., Holland-Letz, T., Izar, B., Von Kalle, C., Frohling, S., & Brinker, T. J. (2019). Superior skin cancer classification by the combination of human and artificial intelligence. *European Journal of Cancer, 120*, 114–121. https://doi.org/10.1016/j.ejca.2019.07.019

Hussain, M., Ajmal, M. M., Gunasekaran, A., & Khan, M. (2018). Exploration of social sustainability in healthcare supply chain. *Journal of Cleaner Production, 203*, 977–989. https://doi.org/10.1016/j.jclepro.2018.08.157

Johnson, R. A., & Smith, K. J. (2020). Transformative impact of artificial intelligence in healthcare: A review. *Journal of Healthcare Technology, 15*(3), 112–125. https://doi.org/10.4103/jht.jht_15_20

Joshi, A., Singh, R., & Rani, S. (2024). Strategic adoption of artificial intelligence for human resource management practices transforming healthcare sector. *The International Journal of Education Management and Sociology., 3*(3), 151–163.

Kaplan, S., Sadler, B., Little, K., Franz, C., & Orris, P. (2012). Can sustainable hospitals help bend the healthcare cost curve? *Commonwealth Fund, 1641*(29), 1–13. https://doi.org/10.1016/j.jclepro.2012.01.001

Kim, E., & Park, J. (2020). Telehealth platforms and AI: Expanding access to healthcare. *International Journal of Telemedicine and e-Health, 12*(1), 2930–2948. https://doi.org/10.1177/1357633X20949259

Koronen, C., Ahman, M., & Nilsson, L. J. (2020). Data centers in future European energy systems: Energy efficiency, integration and policy. *Energy Efficiency, 13*, 129–144. https://doi.org/10.1007/s12053-019-09810-7

Kulkarni, S., Seneviratne, N., Baig, M. S., & Khan, A. H. A. (2019). Artificial intelligence in medicine: Where are we now? *Academic Radiology, 27*(1), 62–70. https://doi.org/10.1016/j.acra.2019.08.006

Lacoste, A., Luccioni, A., Schmidt, V., & Dandres, T. (2019). Quantifying the carbon emissions of machine learning. https://doi.org/10.48550/arXiv.1910.09700

Laguna, J. M. (2014). *Institutional politics, power constellations, and urban social sustainability: A comparative-historical analysis* [Doctoral dissertation, The Florida State University].

Marimuthu, M., & Paulose, H. (2016). Emergence of sustainability based approaches in healthcare: Expanding research and practice. *Procedia-Social and Behavioral Sciences, 224*, 554–561. https://doi.org/10.1016/j.sbspro.2016.05.456

Masanet, E., Shehabi, A., Lei, N., Smith, S., & Koomey, J. (2020). Recalibrating global data center energy-use estimates. *Science, 367*(6481), 984–986. https://doi.org/10.1126/science.aba3758

Mccall, B. (2020). Covid-19 and artificial intelligence: Protecting health-care workers and curbing the spread. *The Lancet Digital Health, 2*(4), e166–e167.

Mehra, R., & Sharma, M. K. (2021). Measures of sustainability in healthcare. *Sustainability Analytics and Modeling, 1*, 100001. https://doi.org/10.1016/j.susam.2020.100001

Nilsson, M., et al. (2018). Mapping interactions between the sustainable development goals: Lessons learned and ways forward. *Sustainability Science, 13*(6), 1489–1503. https://doi.org/10.1007/s11625-018-0604-z

Noorbakhsh-Sabet, N., Zand, R., Zhang, Y., & Abedi, V. (2019). Artificial intelligence transforms the future of healthcare. *The American Journal of Medicine, 132*(7), 795–801. https://doi.org/10.1016/j.amjmed.2019.01.017

Park, C. W., Seo, S. W., Kang, N., Ko, B., Choi, B. W., Park, C. M., Yoon, H. J., et al. (2020). Artificial intelligence in health care: Current applications and issues. *Journal of Korean medical science, 35*(42).

Perkins, D. N., Brune Drisse, M.-N., Nxele, T., & Sly, P. D. (2014). E-waste: A global hazard. *Annals of Global Health, 80*(4), 286–295. https://doi.org/10.1016/j.aogh.2014.09.006

Pineau, J., Montemerlo, M., Pollack, M., Roy, N., & Thrun, S. (2003). Towards robotic assistants in nursing homes: Challenges and results. *Robotics and Autonomous Systems, 42*(3–4), 271–281. https://doi.org/10.1016/S0921-8890(02)00381-0

Reddy, C. L., Mitra, S., Meara, J. G., Atun, R., & Afshar, S. (2019). Artificial intelligence and its role in surgical care in low-income and middle-income countries. *The Lancet Digital Health, 1*(8), e384–e386. https://doi.org/10.1016/S2589-7500(19)30116-1

Rolnick, D., Donti, P. L., Kaack, L. H., Kochanski, K., Lacoste, A., Sankaran, K., et al. (2019). Tackling climate change with machine learning. https://doi.org/10.48550/arXiv.1906.05433

Rong, G., Mendez, A., Bou Assi, E., Zhao, B., & Sawan, M. (2020). Artificial intelligence in healthcare: Review and prediction case studies. *Engineering, 6*(3), 291–301. https://doi.org/10.1016/j.eng.2020.01.017

Ross, C., & Swetlitz, I. (2017). IBM pitched its Watson supercomputer as a revolution in cancer care: It's nowhere close. *Stat*. Retrieved from https://www.statnews.com/2017/09/05/watson-ibm-cancer

Sajwani, R. A. (2024). Artificial Intelligence for sustainability development in healthcare. In *BUiD doctoral research conference 2023: Multidisciplinary studies* (pp. 264–272). Springer Nature Switzerland.

Schoormann, T., Strobel, G., Möller, F., Petrik, D., & Zschech, P. (2023). Artificial intelligence for sustainability—a systematic review of information systems literature. *Communications of the Association for Information Systems, 52*(1), 8.

Sciarretta, E., Mancini, R., & Greco, E. (2022). Artificial intelligence for healthcare and social services: Optimizing resources and promoting sustainability. *Sustainability, 14*(24), 16464. https://doi.org/10.3390/su142416464

Segrelles-Calvo, G., Lopez-Padilla, D., Chiner, E., Fernandez Fabrellas, E., & De Granda-Orive, J. I. (2017). Acceptance of telemedicine among respiratory healthcare professionals. *European Research in Telemedicine, 6*(3–4), 147–155. https://doi.org/10.1016/j.eurtel.2017.09.002

Sharma, G. D., Yadav, A., & Chopra, R. (2020). Artificial intelligence and effective governance: A review, critique and research agenda. *Sustainable Futures, 2*, 5. https://doi.org/10.1016/j.sftr.2020.100017

Sherman, J. D., Thiel, C., MacNeill, A., Eckelman, M. J., Dubrow, R., Hopf, H., Forbes, M., et al. (2020). The green print: Advancement of environmental sustainability in healthcare. *Resources, Conservation and Recycling, 161*, 104882. https://doi.org/10.1016/j.resconrec.2020.104882

Suha, S. A., & Sanam, T. F. (2023). Exploring dominant factors for ensuring the sustainability of utilizing artificial intelligence in healthcare decision-making: An emerging country context. *International Journal of Information Management Data Insights, 3*(1), 100170. https://doi.org/10.1016/j.jjimei.2023.100170

Suresh, M., & Krishnan, S. V. (2021). Modelling the factors of environmental sustainability in healthcare dispensaries. In *Advances in materials research* (pp. 753–761). Springer. https://doi.org/10.1007/978-3-030-61952-9_55

Tjoa, E., & Guan, C. (2019). A survey on explainable artificial intelligence (XAI): Toward medical XAI. *IEEE Transactions on Neural Networks and Learning Systems, 32*(11), 4793–4813. https://doi.org/10.1109/TNNLS.2019.2917184

Topol, E. J. (2019). High-performance medicine: The convergence of human and artificial intelligence. *Nature Medicine, 25*(1), 44–56. https://doi.org/10.1038/s41591-018-0300-7

Tsang, L., Kracov, D. A., Mulryne, J., Strom, L., Perkins, N., Dickinson, R., Wallace, V. M., & Jones, B. (2017). The impact of artificial intelligence on medical innovation in the European Union and United States. *Intellectual Property & Technology Law, 29*(8), 3–11. https://doi.org/10.1093/jiel/jgx025

Utermohlen, K. (2018). Four robotic process automation (RPA) applications in the healthcare industry. *Medium.* Retrieved from https://medium.com/@karl.utermohlen/4-robotic-process-automation-rpa-applications-in-the-healthcare-industry-4d449b24b613

Vinuesa, R., et al. (2020). The role of artificial intelligence in achieving the sustainable development goals. *Nature Communications, 11*(1), 1–10. https://doi.org/10.1038/s41467-019-14108-y

Volpp, K., & Mohta, S. (2016). Improved engagement leads to better outcomes, but better tools are needed. *Insights Report, NEJM Catalyst.* Retrieved from https://catalyst.nejm.org/patient-engagement-report-improved-engagement-leads-better-outcomes-better-tools-needed

Wilcox, L., Brewer, R., & Diaz, F. (2023). AI consent futures: A case study on voice data collection with clinicians. *Proceedings of the ACM on Human-Computer Interaction, 7*(CSCW2), 1–30.

Wiljer, D., & Hakim, Z. (2019). Developing an artificial intelligence–enabled healthcare practice: Rewiring healthcare professions for better care. *Journal of Medical Imaging and Radiation Sciences, 50*(4), S8–S14. https://doi.org/10.1016/j.jmir.2019.07.013

Wolff, A. L. F., Kanding, B., & Selvan, R. (2020). Carbontracker: Tracking and predicting the carbon footprint of training deep learning models. https://doi.org/10.48550/arXiv.2007.03051

World Health Organization. (2018). *2018 global reference list of 100 core health indicators (plus health-related SDGs)* (No. WHO/HIS/IER/GPM/2018.1). World Health Organization. Retrieved from https://www.who.int/healthinfo/indicators/2018/en/

Yoon, S. N., & Lee, D.-H. (2019). Artificial intelligence and robots in healthcare: What are the success factors for technology-based service encounters. *International Journal of Healthcare Management, 12*(3), 218–225. https://doi.org/10.1080/20479700.2019.1596295

Chapter 2
Exploring the Intersection of AI, Healthcare, and Environmental Sustainability: Future Trends and Challenges

A. Arun Kumar, Rakesh Suryadevara, T. Sowmyya, and Gowri B. Chanal

Abstract This study explores the synergies between these fields, highlighting current trends and future opportunities. AI is revolutionizing healthcare through enhanced diagnostics, personalized treatment, and efficient resource management while contributing to environmental sustainability by optimizing energy use, reducing waste, and monitoring ecological health. The convergence of AI with healthcare and sustainability is particularly crucial given the intertwined nature of health and environmental factors, where pollution and climate change directly impact public health. However, integrating AI into these domains also presents significant ethical, social, and policy challenges, including concerns about data privacy, algorithmic bias, and regulatory complexities. This study examines these challenges and provides recommendations for fostering innovation, ensuring ethical AI deployment, and promoting cross-sector collaboration. The findings underscore the need for ongoing research and development to fully realize the potential of AI in creating sustainable and health-promoting solutions while also addressing the risks associated with its widespread adoption.

A. A. Kumar (✉) · R. Suryadevara · G. B. Chanal
ICFAI Law School, ICFAI Foundation for Higher Education (Deemed to be University Under Sec 3 of UGC Act,1956), Hyderabad 501203, Telangana, India
e-mail: arunkumar@ifheindia.org

R. Suryadevara
e-mail: rakeshsuryadevara@ifheindia.org

G. B. Chanal
e-mail: drgowribchanal@ifheindia.org

T. Sowmyya
Department of Forensic Science, University College of Science, Osmania University, Hyderabad 500007, Telangana, India
e-mail: dr.sowmyya@osmania.ac.in

R. Singh et al. (eds.), *Transforming Healthcare Sector Through Artificial Intelligence and Environmental Sustainability*, Approaches to Global Sustainability, Markets, and Governance, https://doi.org/10.1007/978-981-97-9555-0_2

Keywords Artificial Intelligence · Healthcare · Environment · Sustainability · Challenges

2.1 Introduction

One of the twenty-first century's most revolutionary technologies, artificial intelligence (AI), radically alters daily life, economies, and industries. Its quick development and integration into many fields demonstrate its enormous potential to solve challenging issues, boost productivity, and spur innovation. This investigation explores AI's role in contemporary technology, looking at its fundamentals, applications, prospects, and historical background. The conceptual ideas of early mathematicians and computer scientists marked the beginning of AI. Earlier in 1956, when John McCarthy first mentioned artificial intelligence, AI research then concentrated on reasoning and problem-solving, like chess programs and mathematical theorems known as "AI winters," the 1970s and 1980s saw both hopeful and depressing times as advances in the field stalled because of computational power constraints and unreasonably high expectations. However, in the late 1990s and early 2000s, machine learning algorithms, exponential growth in computational power, and the availability of massive datasets spurred a rebirth in AI research (Gu et al., 2023).

Machine Learning (ML) is a branch of AI where computers create algorithms for analyzing, interpreting, and forecasting data. It includes methods like reinforcement learning, supervised learning, and unsupervised learning, which are essential to developing and operating machine learning. ML, another branch of deep learning, uses neural networks with many layers. The ability of this method to model intricate patterns in data has led to significant success in a variety of tasks, such as natural language processing, game playing, and image and speech recognition (Zeebaree et al., 2021). Natural Language Processing (NLP) is another critical area of AI as it deals with human language and systems. NLP makes applications such as sentiment analysis, conversational agents, and translation possible, allowing machines to comprehend, interpret, and produce human language. This technology supports applications like medical image analysis, autonomous driving, and facial recognition, enabling machines to accurately process and analyze visual data (Toh et al., 2019).

AI is essential in developing intelligent robots that carry out tasks independently. These jobs include everything from industrial automation, in which robots increase efficiency and accuracy, to personal assistance and exploration, in which robots help with daily tasks and scientific research. Robots' capabilities are increased when AI is incorporated into robotics, making them more useful and adaptable for various tasks. AI, healthcare, and environmental sustainability are three fields that are rapidly advancing and can potentially change how society approaches some of the most critical issues of our day. When these domains are combined, we may see previously unheard-of breakthroughs in promoting sustainable development, reducing environmental degradation, and enhancing human health. Through an analysis of current trends, an identification of potential future directions, and an understanding of the obstacles that lie ahead, this study seeks to explore this intersection the importance of these three domains separately and together, the necessity of such an intersectional approach, and the potential impact of this convergence on future environmental and global health outcomes (Brasil et al., 2019).

AI has become a potent instrument in many fields in transforming industries by making processes more effective and efficient. AI is revolutionizing healthcare through improving drug discovery, personalized medicine, treatment planning, and diagnostics. AI-driven technologies can analyze large volumes of data and identify patterns to diagnose future ailments, which can result in earlier and more precise diagnoses. AI is being used in environmental sustainability to monitor ecosystems, minimize waste, optimize energy use, and model the effects of climate change. These capabilities are critical because they present novel approaches to long-standing problems in environmental sustainability and healthcare (Finlayson et al., 2019).

The environment and health are undoubtedly related, with environmental factors significant in the global disease burden. Human health is impacted both directly and indirectly by pollution, climate change, and resource depletion. For instance, food insecurity and the spread of infectious diseases have been connected to climate change, and air pollution is a significant contributor to respiratory and cardiovascular disorders. Creating resilient health systems that can adapt to and lessen the effects of environmental changes depends on integrating environmental sustainability into healthcare. This convergence is especially significant since ecological sustainability and health are interdependent goals under the Sustainable Development Goals (SDGs) (Castro et al., 2021).

AI is perfect for closing the gap between environmental sustainability and healthcare because of its capacity to handle and analyze massive datasets. AI can recognize relationships between environmental factors and health outcomes, forecast trends, and recommend sustainable interventions for the environment and promote health. AI could be used, for instance, to optimize the distribution of healthcare resources to reduce environmental impact or to model the health effects of climate change scenarios (Israilidis et al., 2021). AI raises significant ethical and policy issues as it is incorporated more into healthcare and environmental sustainability initiatives. Furthermore, implementing AI in ecological sustainability and healthcare frequently necessitates navigating challenging regulatory landscapes. This study is crucial because it will address these moral and legal issues and offer a framework for guaranteeing that the advantages of AI are sustained and distributed fairly. Mapping the present state of AI applications in environmental sustainability and healthcare will be the first step in this study. It will examine the most recent advancements in AI technology in these domains, including AI-powered energy management tools, AI-driven ecological monitoring systems, and machine learning algorithms for illness prediction. The study will lay the groundwork for comprehending the possible synergies between environmental sustainability and healthcare by summarizing these trends thoroughly (Duan et al., 2019).

Additionally, a critical analysis of the difficulties in incorporating AI into initiatives for environmental sustainability and healthcare will be conducted. These difficulties include ethical worries about privacy, equity, and transparency, as well as technical ones like the requirement for reliable algorithms and high-quality data. The study will also investigate the socio-political and economic obstacles to adopting AI, including stakeholder resistance, regulatory obstacles, and the possibility of AI upending current industries. The study hopes to offer a roadmap for overcoming roadblocks and optimizing AI's advantages in these domains by tackling these issues (Kannan et al., 2020).

The objective is to offer practical insights that can direct the creation of AI-driven solutions that support environmental sustainability and public health. The threat of pandemics and the rise in non-communicable diseases are just two of the complex health issues that the world is currently dealing with. Conventional medical practices are frequently reactive, treating illnesses after they arise instead of preventing them (Singh, 2019). The need for a more proactive strategy emphasizing prevention, early detection, and individualized care is becoming increasingly apparent. AI could spark this change by making predictions more accurate, interventions more successful, and resource usage more economical. However, research is still needed to fully understand how AI can be implemented, as the field is still in its early stages of integration into healthcare.

One of the most urgent problems of our day is environmental degradation, with pollution, climate change, and biodiversity loss posing existential risks to human society. The promotion of sustainable development has been met with slow progress and negative trends in many environmental indicators despite international efforts. While AI has great potential, it is still far from reaching its full potential in monitoring, controlling, and mitigating environmental impacts. Further investigation is required to examine the potential applications of AI for ecological sustainability, especially those that positively affect human health (Westwood et al., 2023).

Due to its substantial contribution to waste production, resource consumption, and greenhouse gas emissions worldwide, the healthcare industry has a negative impact on the environment. In order to lessen healthcare's adverse effects on the environment and guarantee that health systems can withstand changes in their surroundings, healthcare must become more sustainable. AI can be extremely helpful in this transition by maximizing resource utilization, cutting waste, and facilitating more environmentally friendly healthcare procedures. More research is necessary to fully grasp the potential trade-offs and how AI can be incorporated into sustainable healthcare initiatives (Mikhailova & Sharova, 2023).

There are worries that AI may be applied in ways that violate privacy, create unexpected consequences, or worsen already-existing inequalities. Additionally, there is a chance that the excitement surrounding AI will cause people to have irrational expectations or adopt new technologies before they are ready. To prevent these pitfalls, evidence-based policy and practice are required. In addition to offering helpful advice for practitioners and policymakers, this study will make a significant contribution by rigorously analyzing the possible advantages and difficulties of artificial intelligence in healthcare and environmental sustainability. This research is vital because it will examine how AI can foster collaboration between healthcare and ecological sustainability, possibly resulting in innovations that advance both (Zhou et al., 2020). The study will investigate potential future integrations of artificial intelligence, healthcare, and environmental sustainability, building on the analysis of current trends. This will entail locating cutting-edge AI technologies and evaluating how well they can solve environmental and global health issues. Additionally, the study will investigate how AI can improve sustainable healthcare practices, precision public health, and personalized medicine. It will also look at how AI can help with the shift to a circular economy, encourage sustainable urban development, and improve preparedness and response to disasters. The study will also recommend increasing AI's environmental sustainability and healthcare potential, including infrastructure, research, and education funding.

2.1.1 Future Trends and Challenges

1. **Emerging Technologies and Applications**

Various cutting-edge technologies and applications will shape the future of AI in environmental sustainability and healthcare. For instance, machine learning and NLP developments are anticipated to produce new AI-driven instruments for resource management, ecological monitoring, and disease detection. Additionally, more advanced AI models that can capture the intricate relationships between health and the environment will be made possible by the growing availability of big data and advancements in data analytics. The potential of these cutting-edge technologies to spur innovation in environmental sustainability and healthcare will be examined in this study.

2. **Integration of AI Across Sectors**

Integrating AI across industries to foster synergies between healthcare, environmental sustainability, and other sectors like agriculture, energy, and transportation will be one of the significant challenges of the future. Governments, corporations, and civil society organizations must work together to achieve this and create new governance frameworks to facilitate cross-sectoral integration. The study will look at the possibilities for this kind of integration and the issues that must be resolved to make it happen.

3. **Ethical and Social Implications**

As artificial intelligence becomes more integrated into healthcare and environmental sustainability initiatives, the social and ethical ramifications will grow significantly. Furthermore, it is essential to make sure that AI-driven solutions are helpful and accessible to everyone, especially underprivileged groups that are frequently disproportionately impacted by environmental and health issues. The research will examine these moral and societal ramifications and suggest solutions.

4. **Policy and Regulatory Challenges**

The swift advancement of AI presents noteworthy obstacles for regulators and policymakers. AI technologies must be made sure to be secure, practical, and consistent with societal norms. This will necessitate the creation of mechanisms for tracking and evaluating the effects of AI on the environment and human health, as well as new regulatory frameworks that can keep up with technological advancements. The research will look at the legal and policy issues that arise when AI is used in healthcare and environmental sustainability, and it will make suggestions for how to resolve them.

5. **Sustainability of AI**

Although artificial intelligence (AI) can potentially promote environmental sustainability, there are issues with AI's sustainability. Artificial intelligence (AI) technologies can be resource-intensive, consuming large amounts of energy and data. The potential advantages of AI could be undermined if it is widely used, which would result in higher energy consumption and environmental damage. The research will examine the viability of AI technologies and how best to develop and implement them to reduce their environmental impact.

2.1.2 Key Applications of AI in Modern Technology

Integrating AI into various sectors has revolutionized numerous aspects of modern technology.

1. **Healthcare**

AI can revolutionize healthcare by enhancing operational effectiveness, personalized care, and diagnostic accuracy. ML algorithms analyze the data to predict disease outcomes and find patterns in medical data. This helps with early diagnosis and treatment planning. AI-driven imaging systems are frequently more accurate than humans at identifying anomalies in radiology scans. AI-powered personalized medicine creates customized treatment regimens for each patient based on their genetic makeup, way of life, and medical background. AI also streamlines hospital operations, cutting expenses and improving patient care from supply chain management to patient scheduling.

2. **Finance**

In the finance industry, AI can be used for risk mitigation, future stock predictions, and fraud detection, to name a few. To protect customers and organizations, ML models examine transaction data to spot fraudulent activity in real-time. Algorithmic trading systems use AI to execute trades quickly and at the best prices, improving market efficiency. AI also improves risk assessment by analyzing enormous amounts of data to forecast market trends and financial stability. Individualized customer support by effectively managing queries and transactions can also be automated using AI chatbots.

3. **Transportation**

Self-driven cars, predictive traffic, and individual maintenance alerts through AI propel transportation innovation. Self-driving cars promise to lower accident rates and increase mobility by using AI algorithms to analyze sensor data, navigate roads, and make real-time decisions. AI-driven traffic management systems evaluate information from multiple sources to maximize efficiency, minimize emissions, and lessen traffic jams.

4. **Retail**

AI improves retail customer experience, streamlines inventory control, and promotes tailored marketing. E-commerce platforms use AI to make product recommendations based on user behavior and preferences, which boosts customer satisfaction and sales. Demand trends are predicted by AI-driven inventory management systems, which guarantee ideal stock levels and minimize waste. Targeted advertising is delivered through personalized AI-powered marketing campaigns that examine customer data to increase engagement and conversion rates.

5. **Manufacturing**

AI enhances supply chain management, productivity, and quality control in manufacturing. By completing dangerous and repetitive tasks, intelligent automation systems and robots improve productivity and safety. AI-powered quality control systems precisely check products for flaws to guarantee constant quality. AI helps supply chain management by analyzing data, predicting demand, and optimizing production schedules while minimizing disruptions. AI is used in manufacturing predictive maintenance to track equipment condition, averting unplanned malfunctions and cutting downtime.

6. **Education**

AI can be used in education to customize lessons for especially abled, tutoring, or exceptional performers in whatever way it can be done. This minimizes the dependence on the changing field of modernized learning, and AI is now an integral part of it.

AI has a bright future, with continued research and development anticipated to produce even more advanced and significant applications. In the future, AI systems will work more and more like partners rather than just tools, enhancing human abilities across various fields. Professionals with intricate decision-making processes will benefit from AI's insights and recommendations derived from large datasets and sophisticated analytics. Ensuring AI is used ethically and transparently is crucial as it becomes more commonplace. Researchers are creating explainable AI frameworks to make AI decision-making processes intelligible and interpretable to humans. This transparency is essential to foster trust and guarantee accountability in AI systems (See Fig. 2.1).

Climate models powered by AI can forecast environmental changes and guide mitigation measures. AI can facilitate remote diagnostics and telemedicine, expanding access to high-quality healthcare in underserved areas. We can also explore opportunities to use AI to fight against global challenges like poverty and health. Developing specialized AI hardware, such as neuromorphic chips and quantum computing, will significantly increase AI's computational power and efficiency. These developments will speed up innovation by enabling more sophisticated and energy-efficient AI models. By combining AI and IoT, intelligent, networked systems with the ability to process data on the go and make decisions on their own will

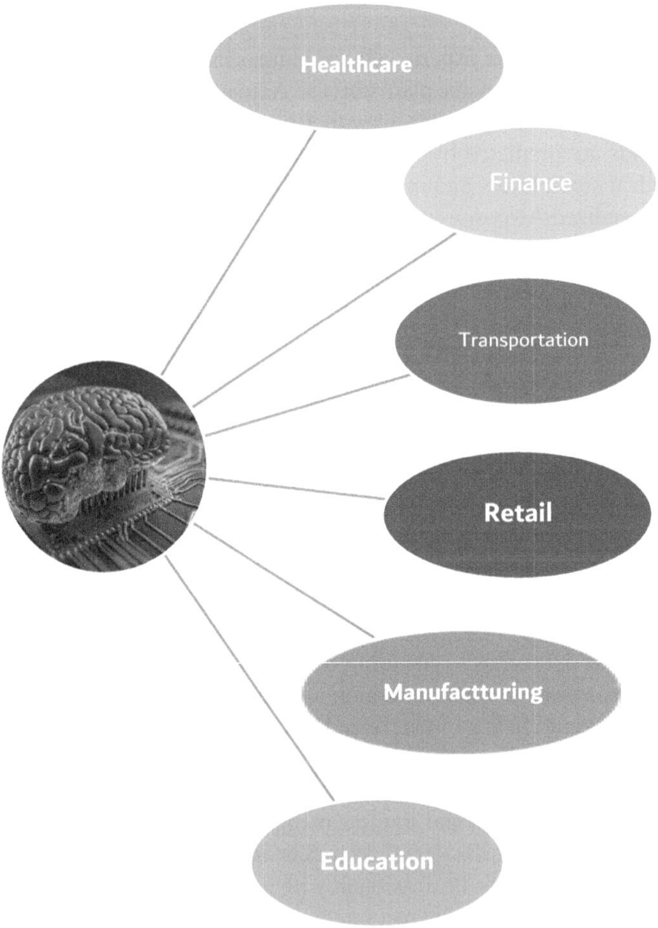

Fig. 2.1 Applications of AI in Modern Technology

be created. The quality of life and operational efficiency will be enhanced by this synergy, which will propel developments in smart cities, industrial automation, and home automation (Kelly et al., 2019).

2.2 Review of Literature

AI can promote environmental sustainability by lowering resource and energy intensity. However, future research should address issues such as an over-reliance on historical data, unpredictable human behavior, cybersecurity risks, and adverse effects (Nishant et al., 2020). Instead of concentrating only on carbon emissions, AI in healthcare should consider resource conservation, justice, and health (Richie, 2022). Sustainable computing methods utilizing AI can significantly lessen their negative environmental effects, but moral and societal issues still need to be resolved (Hashlamoun et al., 2023a, 2023b). AI transforms the healthcare industry by improving early detection, diagnosis, treatment, and outcome prediction in various diseases, including cancer, neurology, and cardiology (Jiang et al., 2017).

Compared to billing code algorithms, the NLP algorithm automatically and more accurately determines cases of peripheral arterial disease from clinical notes, potentially enhancing clinical investigation and care (Afzal et al., 2017). A framework for extracting knowledge-driven information from electronic health records to identify cohorts allows automated point-of-care systems and secondary applications in clinical and translational research (Liu et al., 2013). Sustainable computing methods utilizing AI can significantly lessen their adverse environmental effects, but moral and societal issues must be resolved (Hashlamoun et al., 2023a, 2023b). The intersections between AI and sustainability research are necessary, challenging, and promising, with the potential to revolutionize development's environmental, social, and governance aspects (Liao & Wang, 2020). Although sustainable AI's current interpretation only addresses present threats, it can potentially cause future ecological crises. It can significantly affect environmental economics and management (Lobova et al., 2022).

Healthcare AI has progressed, but there are still issues with multi-modal data integration, bias in models, federated learning, and model security (Wang & Preininger, 2019). Although more datasets and improved validation are needed for accurate and broadly applicable models, ML models hold the potential to forecast treatment outcomes and pinpoint clusters among depressed patients (Ermers et al., 2020). Although more datasets and improved validation are needed for accurate and broadly applicable models, ML models hold the potential to forecast treatment outcomes and pinpoint clusters among depressed patients. After thoroughly reviewing the existing literature, we identified a significant gap in the current research. Consequently, we have finalized our research topic to address this gap.

2.3 Objectives of the Study

1. Examining Current Trends at the Intersection of Artificial Intelligence, Environmental Sustainability, and Healthcare
2. Determining Upcoming Prospects and Difficulties at the Nexus of Environmental Sustainability, Healthcare, and AI.

2.4 Results and Discussions

Innovation, research, and policy-making increasingly focus on the nexus between environmental sustainability, healthcare, and AI. Incorporating these domains could potentially tackle worldwide health issues while encouraging ecological responsibility. AI is being utilized to monitor and forecast the spread of contagious illnesses. In order to forecast epidemics and guide public health initiatives, AI algorithms can examine trends in climatic data, human mobility, and environmental variables. AI can also process genetic details and warn against possible outcomes very accurately.

Considering the climate change, this trend is essential for reducing the effects of pandemics. AI is being used to track and examine data related to air quality. AI algorithms obtain real-time insights into air pollution levels by processing data from sensors and satellites. These discoveries can improve public health and the environment by warning vulnerable populations and creating pollution-reduction policies. AI-powered models track the quality of water and improve water resource management. AI can analyze data from various sources, including weather stations, satellites, and sensors, to forecast water contamination events, optimize water usage, and support sustainable water management practices (Schwalbe & Wahl, 2020).

AI is optimizing energy, water, and medical supplies to help healthcare systems become more sustainable. AI-powered solutions can examine hospital operations and recommend ways to cut costs and boost productivity. AI, for instance, can minimize the environmental impact of healthcare practices, optimize the energy consumption of healthcare facilities, and reduce medical waste through predictive supply chain management. AI is aiding green technology development in healthcare. AI is utilized, for instance, in developing biodegradable medical materials, more energy-efficient medical devices, and environmentally friendly pharmaceutical packaging. These developments help lessen healthcare systems' negative effects on the environment. AI is being used to model how climate change will affect public health (Laurenza et al., 2018).

AI can forecast how climate change will impact food security, population health overall, and the spread of diseases by examining temperature changes, extreme weather events, and disease vectors. Developing healthcare systems that are resilient to climate change requires this knowledge. AI is essential for enhancing the healthcare industry's disaster readiness and response. Better planning and resource allocation are made possible, guaranteeing the continuity of healthcare services during and after emergencies. AI tracks ecosystem health and biodiversity, which is closely related to human health. AI models can examine data on wildlife populations, land use, and climate patterns to evaluate the condition of ecosystems. Food, clean water, and air are just a few vital services healthy ecosystems offer people. Sustainable agricultural methods are being developed with AI's help, and these practices are essential for public health and food security (Shaheen, 2021).

AI, for instance, can maximize medical waste recycling, lowering the environmental effect of healthcare procedures. AI can also help designers create simpler healthcare products to recycle or reuse, which will help create a more sustainable healthcare system. From manufacture to disposal, the environmental impact of medical products is assessed using AI-driven lifecycle assessment tools. These tools inform healthcare organizations about various products' energy, waste, and carbon footprint, enabling them to make more environmentally friendly decisions. Precision agriculture technologies powered by AI maximize the use of water, fertilizers, and pesticides, minimizing environmental damage and maintaining a steady food supply. AI is used in healthcare to create circular economy models emphasizing recycling and waste reduction (Apell & Eriksson, 2021).

AI development ethics are receiving more attention as AI is incorporated into healthcare and environmental sustainability initiatives. In addition to ensuring AI technologies are available to all populations, including those in low-resource environments, it entails addressing algorithmic bias and data privacy concerns. Additionally, there is a movement to make sure AI technology development and application are sustainable. This involves ensuring that AI-driven innovations do not worsen environmental or health inequalities and minimizing the environmental impact of AI infrastructure, such as data centers (Farouk & Alsamara, 2023).

In order to guarantee the safety, efficacy, and moral application of AI in healthcare, governments, and regulatory agencies are focusing more and more on this area. This includes creating policies governing data privacy, guidelines for using AI in clinical settings, and frameworks for addressing the environmental effects of AI-driven healthcare innovations. Environmental regulations also impact the advancement of AI in healthcare. Policies that support environmental sustainability in healthcare, for instance, encourage innovation in AI-driven solutions by lowering carbon emissions and promoting sustainable practices.

The trends currently occurring at the nexus of AI, healthcare, and environmental sustainability demonstrate the enormous potential of AI to spur innovation and solve global issues. AI is essential for accomplishing sustainable development objectives because it improves healthcare delivery and encourages environmental stewardship. Several technical, moral, and legal issues must be resolved to utilize AI at this intersection fully. Prospects are abundant when AI, healthcare, and environmental sustainability come together, but significant obstacles are also overcome. This section looks at areas that could see innovation and growth and the obstacles that must be removed to reach this multidisciplinary field's full potential (Azar et al., 2023).

By forecasting disease outbreaks and individual health risks based on genetic data, lifestyle patterns, and environmental data, AI has the potential to transform preventive healthcare completely. This could result in more proactive health management, which would lower healthcare costs and the incidence of diseases. AI can assist in designing focused public health interventions that address environmental and health risks, particularly in vulnerable populations, by integrating environmental data (such as air and water quality) with health data. Healthcare could become more dynamic and responsive with the development of AI systems that incorporate real-time environmental and health data. AI might, for instance, detect and notify people who have respiratory issues about air quality so they can take precautions. AI may be used to develop networked ecosystems for environmental and health monitoring in future smart cities. These systems would enhance public health, minimize pollution, and maximize urban planning by fostering healthier living conditions (Gerke et al., 2020).

Developing low-impact AI systems on the environment and with high energy efficiency is becoming increasingly possible. This includes creating AI infrastructure that runs on renewable energy sources and developing AI algorithms that need less processing power. By streamlining medical supplies, cutting waste, and encouraging the recycling and reuse of medical equipment and supplies, AI can significantly contribute to developing circular economies in the healthcare industry. Precision public health, in which treatments are customized to each patient's genetic profile and environmental exposures, may be made possible by AI. This strategy may be especially useful in treating conditions like asthma or cardiovascular disorders that environmental factors impact. AI can potentially reduce health disparities by offering scalable, low-cost solutions that can be used in environments with limited resources (Ganga, 2021).

This includes AI-powered smartphone health apps that provide individualized health recommendations based on regional environmental factors. AI can assist healthcare systems in preparing for extreme weather events and predicting health associated risk with climate change. This entails managing supply chains, locating healthcare facilities optimally, and guaranteeing that treatment will continue in the event of an emergency. AI can help design healthcare infrastructure that reduces environmental impact and is climate change resilient. By tackling environmental and health objectives simultaneously, AI can advance the achievement of multiple SDGs. For instance, AI can support sustainable cities by lowering urban pollution and the negative health effects it causes, or it can enhance maternal health by assessing environmental risks. AI can encourage partnerships between environmental organizations, technology companies, and the healthcare industry. These partnerships can result in creative solutions tackling environmental and health issues simultaneously (Bergman et al., 2015).

Developing morally upright AI systems intended to advance environmental sustainability and health equity will be possible. This entails guaranteeing openness in the processes used to make AI decisions, correcting biases in AI algorithms, and encouraging equitable access to AI tools. AI can potentially significantly influence environmental and health policy in the future. AI-driven models, for instance, can mimic the effects of various policy choices, assisting organizations and governments in creating plans that support public health and sustainable development. There are serious privacy and security issues when integrating AI in healthcare and environmental monitoring. Two crucial issues that must be resolved are safeguarding privacy while gathering environmental data and protecting sensitive health information. AI systems may become targets for cyberattacks as they are increasingly incorporated into the ecological and healthcare infrastructures. Upcoming challenges include making sure AI systems and the data they process are secure. Inadvertent bias perpetuated by AI systems can result in discriminatory healthcare and environmental policy practices. AI systems that were trained on biased data, for instance, might recommend unequal healthcare or neglect to address environmental justice concerns in underserved communities. AI risks worsening inequality, especially if some populations cannot access AI-driven environmental or healthcare technologies. One major challenge is ensuring that different socio-economic groups have equal access to the benefits of AI (Bergman et al., 2015).

AI technologies are developing at a rate that frequently surpasses the creation of regulatory frameworks. Comprehensive laws addressing AI's moral, legal, and societal ramifications in healthcare and environmental sustainability are required. It is challenging to align policies across the technology, healthcare, and environmental sectors. Coordination of efforts among various stakeholders, such as governments, businesses, and civil society, is necessary to integrate AI effectively. AI technologies can greatly impact the environment, especially if they involve a lot of processing power. The difficulty is in striking a balance between the advantages of AI and its effects on the environment, especially regarding resource and energy usage (Navaz et al., 2021).

Because of problems with data systems and technology interoperability, integrating AI into healthcare and environmental systems can be difficult. Developing standardized protocols and removing technological obstacles are necessary to ensure seamless integration. Lack of infrastructure, technical know-how, and financial resources frequently impedes the adoption of AI technologies in healthcare and environmental sustainability, particularly in low-resource settings. These obstacles must be removed to deploy AI-driven solutions widely. Establishing public confidence in AI systems is essential to their effective deployment. Issues regarding AI decisions' fairness, transparency, and accuracy must be addressed to win over the public's acceptance and support. Public education and awareness campaigns are required to enlighten people about the advantages and dangers of AI in healthcare and environmental sustainability. More knowledge about AI can help allay misconceptions and anxieties, promoting wider adoption (Moore & Buchanan, 2013).

It is challenging to balance the need for innovation, long-term sustainability, and ethical considerations as AI develops. Adaptive governance structures and continual monitoring will be necessary to guarantee that AI advancements align with morally and environmentally responsible standards. It will take unprecedented international cooperation to address global challenges at the nexus of AI, healthcare, and environmental sustainability. Breaking down cultural, economic, and geopolitical barriers and promoting cooperation will be challenging. The convergence of AI, healthcare, and environmental sustainability presents a plethora of prospects to tackle some of the most critical worldwide issues. There are numerous advantages to be gained from developing sustainable AI technologies through precision public health. However, important issues like data privacy, moral dilemmas, legal restrictions, and

AI's long-term viability must be resolved to take advantage of these prospects. Through strategic and cooperative navigation of these obstacles, interested parties can leverage AI's potential to build a more sustainable and health-conscious future for everybody (Fan et al., 2023).

The AI market in healthcare is anticipated to expand by 47.6% CAGR from $14.6 billion in 2023 to $102.7 billion by 2028. Compared to conventional techniques, AI-driven diagnostic tools can increase accuracy by up to 20%. In contrast to 88% for human experts, AI algorithms have achieved 94.5% accuracy in identifying specific cancer types from medical images. AI applications in hospital administration can improve resource allocation and cut waste while reducing operational costs by up to 25%. By reducing energy consumption by 10–20%, AI technologies in energy management systems can significantly lower carbon emissions. By 2030, AI-powered demand response systems could save up to $200 billion in energy (CAGR, 2024).

AI may make up to 15% of the total greenhouse gas emission reductions required by 2030 to achieve the Paris Agreement objectives possible. This could result in an annual reduction of 2.4–4.0 gigatons of CO_2 emissions. AI-powered precision agriculture can save 25% of water consumption while increasing crop yields by up to 30%. By optimizing the use of resources, these technologies lessen the adverse effects of farming on the environment. Strong cybersecurity measures driven by AI are imperative, as evidenced by over 80% of healthcare organizations have suffered a data breach in the last year. One major obstacle that needs to be overcome is protecting patient data privacy while using AI (Rathnasiri et al., 2024).

There is an increasing need for ethical frameworks as AI applications grow. According to surveys, 74% of customers are worried about how AI systems will use their data, especially in the healthcare industry. Just 30% of nations have laws explicitly pertaining to AI, and the development of international standards for AI in healthcare and environmental sustainability is lagging. The absence of standardized guidelines hinders adoption. Collaboration between disciplines is necessary to integrate AI, healthcare, and environmental sustainability. Interdisciplinary teams are involved in less than 50% of AI research projects in these domains, which could impede innovation (Singh & Singh, 2023).

Healthcare access should improve due to AI's application in telemedicine, especially in rural areas. AI is being utilized increasingly to support circular economy principles like resource recycling and waste reduction. Initiatives for a circular economy powered by AI may bring in up to $4.5 trillion in revenue by 2030. According to predictions made by AI, 70% of all individualized treatment plans will incorporate AI by 2030, driving the trend toward personalized medicine. These figures show the enormous potential that AI must improve healthcare and support environmental sustainability, but they also draw attention to the obstacles that need to be overcome to achieve these advantages (Moore & Buchanan, 2013).

2.5 Overview of AI Applications in Healthcare

AI quickly improves drug discovery, personalized medicine, diagnostics, and other areas of healthcare. AI algorithms read MRIs, CT scans, and X-rays to find health issues like tumors, fractures, and other ailments. AI, for instance, can identify breast cancer indicators in mammograms with an accuracy level that matches or surpasses that of human radiologists. AI-powered systems can identify diseases, including different types of cancer, by analyzing tissue samples. AI tools help pathologists identify patterns that point to the presence of diseases, increasing the speed and accuracy of diagnosis. AI models predict the likelihood of diabetes, cardiovascular diseases, and potential surgical complications using patient data (Kandasamy et al., 2022).

Treatments using AI-raised medication using one's genetic data method are less harmful and more effective. AI systems examine genetic data to find disease-related mutations and biomarkers. Developing targeted therapies is a critical application of this information, especially in cancer treatment. Real-time tracking of patients' health metrics by wearable technology and AI-powered mobile apps allows for individualized recommendations and early alerts for possible health problems. AI algorithms analyze massive datasets that contain biological and chemical data to help find new drug targets. This quickens the process of finding new medications. AI forecasts the structure and behavior of new drug molecules to create innovative ones. Doing this will significantly decrease the time and expense of conventional drug discovery methods. By finding qualified candidates, forecasting results, and increasing patient retention rates, AI enhances the planning and administration of clinical trials (Costumado & Chemane, 2024).

Healthcare workers can concentrate more on patient care by using AI to make seamless and hassle-free automated patient file handling. Hospitals can better manage their resources, such as personnel and equipment, by anticipating demand and maximizing the use of AI. AI improves telemedicine platforms by helping doctors make decisions, allowing remote diagnosis, and suggesting tailored courses of action. AI systems frequently outperform humans in tasks like image recognition and data analysis because of their ability to analyze enormous amounts of data quickly and accurately. The results are faster diagnosis times and more precise treatment choices. Utilizing AI, healthcare providers can provide individualized treatment plans based on unique patient data. This method lowers the possibility of negative reactions while improving the efficacy of treatments. By streamlining administrative procedures, maximizing resource utilization, and lowering the need for pointless tests and procedures, AI can lower healthcare costs. This is especially helpful for expanding access to healthcare in underprivileged areas. AI finds promising compounds and forecasts whether they will succeed in preclinical and clinical trials, speeding up drug discovery. In doing so, new medications may reach the market more quickly and for less. AI-driven predictive analytics can predict patient outcomes, disease outbreaks, and possible complications, allowing for proactive healthcare management and enhancing public health (Moore & Buchanan, 2013).

Firstly, AI initially at least uses large volumes of data to prepare itself for numerous outcomes. Data security of such large volumes of data, an individual's confidential data, is a significant concern. AI algorithms can inherit biases in the training data. Inequalities in healthcare outcomes may result from this, especially for underprivileged groups. Fairness and equity in healthcare powered by AI are a constant concern. Healthcare AI regulations are still being developed. Innovation can be slowed down by the complicated process of ensuring AI tools are efficient, safe, and compliant with healthcare regulations. An over-reliance on AI may lessen human oversight, which could lead to mistakes or missed diagnoses. In healthcare, AI should supplement human judgment, not take its place. It can be challenging to integrate AI tools into current healthcare systems, especially when it comes to electronic health records and other technologies that are not interoperable. This may cause the use of AI in healthcare to lag (Kandasamy et al., 2022).

2.6 AI in Environmental Sustainability

AI is being used increasingly to solve environmental issues and advance sustainability. Its applications cover various industries, such as natural resource conservation, energy management, climate change mitigation, and the advancement of the circular economy. An overview of the advantages and difficulties of AI's contribution to environmental sustainability is provided below. AI algorithms optimize energy use in buildings, industrial processes, and transportation by anticipating energy requirements and consuming less during peak hours. AI assists in predicting the output of renewable energy from sources like wind and solar power, facilitating better grid integration and lowering dependency on fossil fuels.

It powered systems to track and regulate industrial emissions, spot inefficiencies, and recommend enhancements, resulting in notable drops in greenhouse gas emissions. Carbon capture technologies are optimized by AI models, increasing their efficiency in lowering $CO2$ emissions from power plants and other industrial sources. By analyzing enormous volumes of data, AI enhances climate models and makes it possible to predict climate patterns, extreme weather events, and their effects on ecosystems and human activity more accurately (Singh & Singh, 2023).

The systems anticipate water requirements and spot leaks, and recommend conservation measures to maximize water usage in urban and agricultural settings. This results in less waste and more effective use of water. AI supports the management of natural habitats, tracking endangered species, and monitoring wildlife populations. Data on wildlife is gathered by AI-powered drones and camera traps, which supports conservation efforts. AI tools, like satellite imagery analysis, can identify illicit logging activities in real time, allowing for quicker intervention and the preservation of forests. It enhances waste sorting and recycling procedures by effectively recognizing and classifying various kinds of waste. This encourages recycling and lessens the usage of landfills. AI-driven design tools facilitate the production of goods with low environmental impact, conserving resources and encouraging environmentally friendly production methods. AI finds ways to recover valuable materials from waste, repurposing materials that would otherwise be thrown away (Ueda et al., 2024).

AI improves the efficiency of using materials, energy, and water resources, significantly lowering waste and its adverse environmental effects. AI, for instance, can minimize greenhouse gas emissions by optimizing building heating and cooling systems, hence reducing energy consumption. AI is particularly good at handling big data, which is essential when tracking environmental changes and evaluating the success of sustainability programs (Navaz et al., 2021). The global scalability of AI solutions makes consistent and valuable tools for environmental management possible. AI-powered intelligent grids can be implemented throughout nations and cities, promoting widespread energy efficiency. AI's capacity to forecast ecological trends, such as resource depletion or climate change, makes proactive action possible. The negative effects of environmental changes are lessened, and risks are mitigated thanks to this predictive power (Khan et al., 2024).

By enhancing the forecasting, integrating, and managing of renewable energy sources, AI contributes in expanding renewable energy. This lessens dependency on fossil fuels and facilitates the switch to a more environmentally friendly energy system. AI systems need a lot of high-quality data to function. The effectiveness of AI solutions may be limited by the scarcity or unreliability of necessary data in many areas, particularly in developing nations. Large AI models can require much energy during training, which increases carbon emissions. One of AI's most important challenges is balancing energy demands with environmental benefits. Biases from the data used to train AI models may be inherited by the models, resulting in unfair advantages or even harm to communities. Promoting inclusive sustainability requires making sure AI systems are just and equal (Costumado & Chemane, 2024).

There might be some ethical concerns about the usage of AI, such as the possibility of employment displacement in sectors like energy and waste management. Furthermore, adopting AI may be hampered by the absence of clear regulations governing its environmental applications. Integrating AI technologies into current environmental management systems can be challenging, mainly if those systems are antiquated or lack the infrastructure to support AI applications (Navaz et al., 2021).

Healthcare innovations support more sustainable practices, and environmental sustainability and healthcare represent a potent intersection where addressing environmental issues directly benefits human health. This synergy reflects an increasing understanding of the close connections between human health and the planet's health.

2.7 Case Studies

A few case studies demonstrate the convergence of environmental sustainability and healthcare, emphasizing how combining these domains can advance sustainability and enhance health outcomes.

Case Study 1: Air Quality Monitoring and Respiratory Health in Barcelona, Spain

Barcelona has had to deal with air pollution, which has been connected to heart and lung conditions. To solve this problem, the city installed an AI-powered air quality monitoring system. The system analyzes data from a network of air quality sensors throughout the city using AI algorithms. In real time, the AI pinpoints pollution sources and estimates pollution levels. Information from the air quality monitoring system is combined with medical records to monitor the effects of air pollution on respiratory health. Hospital admissions for respiratory conditions have noticeably decreased in regions where air quality has improved. The system has assisted in lowering pollution levels by providing information for targeted interventions and policies, such as industrial regulations and traffic restrictions, to decrease problems with cardiovascular and respiratory health (ASPB Cat, 2023).

Case Study 2: Energy-Efficient Hospitals in the United States

Hospitals use more energy, which significantly influences the environment and operational costs. To solve these problems, several US hospitals have implemented AI-driven energy management systems. Using real-time occupancy data and meteorological forecasts, AI systems monitor and optimize hospital energy use by modifying the lighting, heating, and cooling systems. Green building techniques, like the use of renewable energy sources and energy-efficient fixtures, were also adopted by hospitals. Hospitals reported energy cost reductions of up to 30%, decreased energy use and greenhouse gas emissions, and significantly lower expenses and less energy consumption. Reduced emissions lessen their environmental impact and enhance local air quality (Ueda et al., 2024).

Case Study 3: AI in Water Management and Public Health in Cape Town, South Africa

Due to mismanagement and protracted droughts, Cape Town experienced acute water shortages. The city installed an AI-driven water management system to enhance public health and water efficiency. AI models use data from infrastructure sensors, weather forecasts, and water usage patterns to anticipate and control water supply and find leaks. AI technologies lower the risk of waterborne illnesses by monitoring water quality and identifying pollutants; enhanced water management and conservation, which lessens the severity of shortages; decreased cases of diseases spread by water due to improved water quality monitoring; water conservation and management that is effective and decreased disease incidence and enhanced water quality (Costumado & Chemane, 2024).

Case Study 4: Circular Economy in Pharmaceutical Waste Management

Medical waste disposal, including expired medications as well as packaging materials, challenges the pharmaceutical industry. AI is being used to improve waste management and recycling procedures. AI algorithms analyze and sort pharmaceutical waste to separate hazardous materials from recyclables. Automated systems guarantee recycling and proper disposal. In addition to lowering landfill usage, the program encourages the recovery of valuable materials; lower landfill waste and higher recycling rates; reduced expenses for managing waste and retrieval of valuable materials; decreased waste and increased effectiveness of recycling; and more sustainable and safe methods of disposing of pharmaceutical waste (Kandasamy et al., 2022).

Case Study 5: Predictive Analytics for Vector-Borne Disease Prevention in Kenya

In Kenya, vector-borne illnesses like malaria pose a severe threat to public health. The nation has implemented a predictive analytics system powered by AI to enhance disease prevention and response. AI models examine historical disease data, mosquito movement patterns, and climate data to forecast malaria and other vector-borne disease outbreaks. Forecasts aid in the more effective targeting of preventive measures, which lowers the number of malaria cases. These preventive measures include distributing insecticide-treated nets and increasing vector control efforts in high-risk areas; more efficient use of available funds and interventions; decreased prevalence of illnesses spread by vectors; and increased focus and effectiveness in the use of resources for illness prevention (Navaz et al., 2021).

These case studies show how the convergence of healthcare and environmental sustainability can significantly enhance public health and environmental outcomes. Cities, organizations, and nations can work together to address health and environmental issues more sustainably and healthily by utilizing AI and other technologies.

2.8 Limitations of the Study and Future Scope of the Study

The research may depend on current or incomplete datasets that already exist. Privacy concerns about patient data may restrict access to essential healthcare information. The study results may be impacted by biases, missing information, or inconsistent data in environmental, medical, and AI-related datasets. Since AI technologies are developing quickly, keeping up with the latest developments and making reliable predictions about what will happen is challenging. Due to their complexity, it is challenging to address the technical obstacles in integrating AI in healthcare and environmental sustainability in a single study. The study crosses several fields, including AI, medicine, and environmental sustainability. Each field has its standards, terminologies, and methodologies, making the analysis more difficult. The study's ability to thoroughly examine the intersection of all three fields may be limited by knowledge gaps or a lack of experience in one or more of them.

The study might not fully address issues like bias in AI algorithms and unforeseen effects on ecosystems that raise ethical questions about the use of AI in healthcare and environmental sustainability. The dynamic and regionally varied regulatory environment may limit the study's applicability, necessitating assumptions that may prove unfounded in the future. The study's findings may not be as applicable in other contexts because it may concentrate on particular geographic areas, technological advancements, or healthcare systems. The research time available for the study may have limited its scope, which could result in the exclusion of critical future trends or challenges. Research depth may be limited by a lack of funds or resources, especially in fields that call for sophisticated AI tools, extensive data analysis, or long-term environmental monitoring.

Conduct long-term research to track the effects of AI on environmental sustainability and healthcare over time. This will help identify trends and provide long-term results. Investigate incorporating more sophisticated AI technologies, like neural networks or quantum computing, to overcome existing constraints and expand the potential applications of AI in sustainability and healthcare. Consider how AI can improve personalized medicine, especially in customizing treatments based on a patient's genetic makeup and environmental exposures. Extend the research to consider regional variations in healthcare, environmental regulations, and the adoption of AI from a more global standpoint. When applying AI to the nexus of healthcare and environmental sustainability, provide thorough ethical frameworks that consider potential bias, privacy concerns, and societal effects.

Examine cooperative initiatives involving government, business, and academia to use AI to support environmentally friendly and sustainable healthcare practices. Examine how cutting-edge technologies like edge computing, 5G, and the Internet of Things could improve the convergence of AI, environmental sustainability, and healthcare. Identify best practices and areas requiring regulatory attention by studying the changing policy landscape and its implications for integrating AI in healthcare and environmental sustainability. Examine the financial and social effects of incorporating AI into environmental sustainability and healthcare, considering potential job creation or displacement and cost–benefit analyses. These shortcomings and potential paths forward guide further investigation, guaranteeing that the confluence of AI, healthcare, and environmental sustainability is thoroughly and ethically investigated.

2.9 Policy Implications

To apply AI in healthcare and environmental sustainability, policymakers must create and implement clear regulations. This involves setting standards for AI development, data usage, and algorithm transparency to ensure safety, efficacy, and fairness. Healthcare laws currently in place, like those pertaining to patient privacy (like HIPAA in the US), might need to be updated to consider AI technologies as they process vast amounts of personal and health-related data. Ensuring secure personal health and environmental data management requires robust data protection laws. Policies should ensure AI systems abide by data privacy laws, particularly when handling private medical data. Policies should support safe data sharing among environmental agencies, AI developers, and healthcare providers to foster collaboration and protect privacy.

Legislators may establish moral guidelines and certification procedures for AI programs in environmental and medical fields. This would minimize biases and unintended consequences and help ensure that AI is used responsibly. Clear accountability frameworks are imperative to identify who bears responsibility for AI-related outcomes in healthcare and environmental sustainability, mainly when errors or harm occur. Government funding for research into AI applications at the nexus of environmental sustainability and healthcare should be provided. This could involve financial aid for educational institutions, business collaborations, and private-sector investment incentives. Investments in AI technologies that support environmental sustainability, such as maximizing resource utilization, cutting emissions, or tracking the effects on the environment's health, should be given top priority by policymakers.

The inclusion of AI, healthcare, and environmental sustainability in curricula at different levels should be encouraged by educational policies. This guarantees upcoming professionals will have the know-how to operate at this intersection. Initiatives to reskill and upskill the workforce should be supported by policy, enabling environmental scientists, AI developers, and healthcare workers to deal with new technologies and adjust to shifting market conditions.

Regulations requiring AI in environmental and public health monitoring could enhance the capacity to quickly identify and address threats to public health and environmental changes. Legislators should back AI-driven public health programs that deal with environmental factors that affect health, like pollution control and air quality monitoring. International cooperation on AI standards, data sharing, and best practices should be encouraged by policy, given the global nature of environmental issues and healthcare (Rathnasiri et al., 2024).

To ensure uniformity and cross-border interoperability, policymakers should collaborate with international organizations to create global standards for AI in healthcare and environmental sustainability. Legislation may offer financial aid or tax breaks to businesses that create or implement AI solutions that enhance healthcare quality or environmental sustainability. Policymakers should ensure that AI development aligns with the SDGs of the United Nations by putting these objectives into legislation about AI. AI-related risks, such as the possibility of job displacement, cybersecurity threats, or unintended environmental effects, should be managed through policy. In crucial industries like healthcare and environmental management, policymakers should mandate backup plans for AI system failures and ensure that manual or alternate systems are in place. In order to ensure that a variety of viewpoints are taken into account when drafting policies, policymakers should promote public engagement and consultation in matters pertaining to AI in healthcare and environmental sustainability.

2.10 Conclusion

Examining the nexus between AI, healthcare, and environmental sustainability reveals a vibrant and exciting field that has the potential to lead to significant improvements in ecological stewardship and human health. AI technologies offer transformative capabilities in these fields, from monitoring environmental changes and optimizing resource use in improving diagnostic accuracy and personalizing medical treatments. However, overcoming several obstacles is necessary to fully utilize AI in these fields. AI implementation is complex and requires careful consideration of ethical issues, technological integration, data quality, and regulatory frameworks. Furthermore, strong cross-sector and cross-expertise collaboration is required due to the interdisciplinary nature of this intersection, guaranteeing that AI solutions are not only creative but also equitable and long-lasting. Even though there are still obstacles to overcome, incorporating AI into environmental sustainability and healthcare offers a rare chance to solve some of the most critical problems of our day. AI can potentially be potent in building a more sustainable and healthful future for everyone with careful policy guidance, ongoing research, and a dedication to ethical practices.

References

Afzal, N., Sohn, S., Abram, S., Scott, C., Chaudhry, R., Liu, H., Kullo, I., & Arruda-Olson, A. (2017). Mining peripheral arterial disease cases from narrative clinical notes using natural language processing. *Journal of Vascular Surgery, 65*, 1753–1761. https://doi.org/10.1016/j.jvs.2016.11.031

Apell, P., & Eriksson, H. (2021). Artificial intelligence (AI) healthcare technology innovations: The current state and challenges from a life science industry perspective. *Technology Analysis & Strategic Management, 35*, 179–193. https://doi.org/10.1080/09537325.2021.1971188

Azar, A., Tounsi, M., Fati, S., Javed, Y., Amin, S., Khan, Z., Alsenan, S., & Ganesan, J. (2023). Automated system for colon cancer detection and segmentation based on deep learning techniques. *International Journal of Sociotechnology and Knowledge Development*. https://doi.org/10.4018/ijskd.326629

Bergman, B., Hellström, A., Lifvergren, S., & Gustavsson, S. (2015). An emerging science of improvement in health care. *Quality Engineering, 27*, 17–34. https://doi.org/10.1080/08982112.2015.968042

Brasil, S., Pascoal, C., Francisco, R., Ferreira, V., Videira, P., & Valadão, G. (2019). Artificial intelligence (AI) in rare diseases: Is the future brighter? *Genes, 10*. https://doi.org/10.3390/genes10120978

Castro, G., Fernández, M., & Colsa, Á. (2021). Unleashing the convergence amid digitalization and sustainability towards pursuing the sustainable development goals (SDGs): A holistic review. *Journal of Cleaner Production*. https://doi.org/10.1016/J.JCLEPRO.2020.122204

ASPB Cat. (2023). *Health in Barcelona*. Barcelona Public Health Agency (ASPB). Retrieved from https://www.aspb.cat/l-agencia/qui-som/

Costumado, M., & Chemane, J. A. D. (2024). Stakeholder engagement and community participation in sustainable development in Southern Africa. In: *Democratization of Africa and its impact on the global economy* (pp. 48–72). https://doi.org/10.4018/979-8-3693-0477-8.ch004

Duan, Y., Edwards, J., & Dwivedi, Y. (2019). Artificial intelligence for decision making in the era of big data—evolution, challenges and research agenda. *International Journal of Infrastructure Management, 48*, 63–71. https://doi.org/10.1016/J.IJINFOMGT.2019.01.021

Ermers, N., Hagoort, K., & Scheepers, F. (2020). The predictive validity of machine learning models in the classification and treatment of major depressive disorder: State of the art and future directions. *Frontiers in Psychiatry, 11*. https://doi.org/10.3389/fpsyt.2020.00472

Fan, Z., Yan, Z., & Wen, S. (2023). Deep learning and artificial intelligence in sustainability: A review of SDGs, renewable energy, and environmental health. *Sustainability, 15*(18), 13493. https://doi.org/10.3390/su151813493

Farouk, G., & Alsamara, T. (2023). Legal view on blockchain technologies in healthcare. *International Journal of Sociotechnology and Knowledge Development*. https://doi.org/10.4018/ijskd.333154

Finlayson, S., Bowers, J., Ito, J., Zittrain, J., Beam, A., & Kohane, I. (2019). Adversarial attacks on medical machine learning. *Science, 363*, 1287–1289. https://doi.org/10.1126/science.aaw4399

Ganga, X. (2021). Educational artificial intelligence (EAI) connotation, key technology and application trend -interpretation and analysis of the two reports entitled "preparing for the future of artificial intelligence" and "the national artificial intelligence research and development strategic plan". In: *International conference on intelligent computing, automation and applications (ICAA)* (pp. 219–223). https://doi.org/10.1109/ICAA53760.2021.00046

Gerke, S., Minssen, T., & Cohen, G. (2020). Ethical and legal challenges of artificial intelligence-driven healthcare. *Artificial Intelligence in Healthcare*, 295–336. https://doi.org/10.2139/ssrn.3570129

Gu, J., Gao, C., & Wang, L. (2023). The evolution of artificial intelligence in biomedicine: Bibliometric analysis. *JMIR AI*. https://doi.org/10.2196/45770

Hashlamoun, N., Barghuthi, N., & Tamimi, H. (2023). Exploring the intersection of AI and sustainable computing: Opportunities, challenges, and a framework for responsible applications. In: *9th International conference on information technology trends (ITT)* (pp. 220–225). https://doi.org/10.1109/ITT59889.2023.10184228

Hashlamoun, N., Barghuthi, N., & Tamimi, H. (2023). Exploring the intersection of AI and sustainable computing: Opportunities, challenges, and a framework for responsible applications. In: *2023 9th International conference on information technology trends (ITT)* (pp. 220–225). https://doi.org/10.1109/ITT59889.2023.10184228

Israilidis, J., Odusanya, K., & Mazhar, M. (2021). Exploring knowledge management perspectives in smart city research: A review and future research agenda. *International Journal of Infrastructure Management, 56*, 101989. https://doi.org/10.1016/J.IJINFOMGT.2019.07.015

Jiang, F., Jiang, Y., Zhi, H., Dong, Y., Li, H., Ma, S., Wang, Y., Dong, Q., Shen, H., & Wang, Y. (2017). Artificial intelligence in healthcare: Past, present and future. *Stroke and Vascular Neurology, 2*, 230–243. https://doi.org/10.1136/svn-2017-000101

Kandasamy, J., Kinare, Y. P., Pawar, M. T., Majumdar, A., KEK., V., & Agrawal, R. (2022). Circular economy adoption challenges in medical waste management for sustainable development: An empirical study. *Sustainable Development, 30*(5), 958–975. https://doi.org/10.1002/sd.2293.

Kannan, D., Mina, H., Nosrati-Abarghooee, S., & Khosrojerdi, G. (2020). Sustainable circular supplier selection: A novel hybrid approach. *The Science of the Total Environment, 722*, 137936. https://doi.org/10.1016/j.scitotenv.2020.137936

Kelly, C., Karthikesalingam, A., Suleyman, M., Corrado, G., & King, D. (2019). Key challenges for delivering clinical impact with artificial intelligence. *BMC Medicine, 17*. https://doi.org/10.1186/s12916-019-1426-2

Khan, M., Shiwlani, A., Qayyum, M. U., Sherani, A. M. K., & Hussain, H. K. (2024). AI-powered healthcare revolution: An extensive examination of innovative methods in cancer treatment. *BULLET: Journal Multidiscipline Ilmu, 3*(1), 87–98.

Laurenza, E., Quintano, M., Schiavone, F., & Vrontis, D. (2018). The effect of digital technologies adoption in the healthcare industry: A case based analysis. *Business Process Management Journal, 24*, 1124–1144. https://doi.org/10.1108/BPMJ-04-2017-0084

Liao, H., & Wang, Z. (2020). Sustainability and artificial intelligence: Necessary, challenging, and promising intersections. In: *Management Science Informatization and Economic Innovation Development Conference (MSIEID)*, vol. 2020 (pp. 360–363). https://doi.org/10.1109/MSIEID52046.2020.00076
Liu, H., Bielinski, S., Sohn, S., Murphy, S., Wagholikar, K., Jonnalagadda, S., Ravikumar, K., Wu, S., Kullo, I., & Chute, C. (2013). An information extraction framework for cohort identification using electronic health records. *AMIA Summits on Translational Science Proceedings, 2013*, 149–153.
Lobova, S., Bogoviz, A., & Alekseev, A. (2022). Sustainable AI in environmental economics and management: Current trends and post-covid perspective, *10*. https://doi.org/10.3389/fenvs.2022.951672
Mikhailova, A., & Sharova, D. (2023). Artificial intelligence ethics code in healthcare. Sustainability of artificial intelligence systems: Why do we talk about their impact on the environment? *Digital Diagnostics*. https://doi.org/10.17816/dd430356
Moore, C., & Buchanan, D. (2013). Sweat the small stuff: A case study of small-scale change processes and consequences in acute care. *Health Services Management Research, 26*, 17–19. https://doi.org/10.1177/0951484813488060
Navaz, A. N., Serhani, M. A., El Kassabi, H. T., Al-Qirim, N., & Ismail, H. (2021). Trends, technologies, and key challenges in smart and connected healthcare. *IEEE Access, 9*, 74044–74067. https://doi.org/10.1109/ACCESS.2021.3079217
Nishant, R., Kennedy, M., & Corbett, J. (2020). Artificial intelligence for sustainability: Challenges, opportunities, and a research agenda. *International Journal Infrastructure Management, 53*, 102104. https://doi.org/10.1016/j.ijinfomgt.2020.102104
Rathnasiri, M. S. H., Dewasiri, N. J., & Kumar, A. A. (2024). Policy framework and implementation strategies for Sri Lanka's transition to a net-zero economy. In R. Singh, S. Khan, A. Kumar, S. Luthra & H. Chokshi (Eds.), *Net zero economy, corporate social responsibility and sustainable value creation*. CSR, Sustainability, Ethics & Governance. Springer, Cham. https://doi.org/10.1007/978-3-031-55779-8_3
Richie, C. (2022). Environmentally sustainable development and use of artificial intelligence in health care. *Bioethics, 36*, 547–555. https://doi.org/10.1111/bioe.13018
Schwalbe, N., & Wahl, B. (2020). Artificial intelligence and the future of global health. *Lancet (London, England), 395*, 1579–1586. https://doi.org/10.1016/S0140-6736(20)30226-9
Shaheen, M. (2021). AI in healthcare: Medical and socio-economic benefits and challenges. *SSRN Electronic Journal*. https://doi.org/10.14293/s2199-1006.1.sor-.pprqni1.v1
Singh, J. (2019). Artificial Intelligence and global health: Opportunities and challenges. *Emerging Topics in Life Sciences, 36*, 741–746. https://doi.org/10.1042/etls20190106
Singh, R., & Singh, R. (2023). Applications of sentiment analysis and machine learning techniques in disease outbreak prediction–A review. *Materials Today Proceedings., 81*, 1006–1011.
Toh, T., Dondelinger, F., & Wang, D. (2019). Looking beyond the hype: Applied AI and machine learning in translational medicine. *eBioMedicine, 47*, 607–615. https://doi.org/10.1016/j.ebiom.2019.08.027
Ueda, D., Walston, S. L., Fujita, S., Fushimi, Y., Tsuboyama, T., Kamagata, K., Yamada, A., Yanagawa, M., Ito, R., Fujima, N., Kawamura, M., Nakaura, T., Matsui, Y., Tatsugami, F., Fujioka, T., Nozaki, T., Hirata, K., & Naganawa, S. (2024). Climate change and artificial intelligence in healthcare: Review and recommendations towards a sustainable future. *Diagnostic and Interventional Imaging*. https://doi.org/10.1016/j.diii.2024.06.002
Wang, F., & Preininger, A. (2019). AI in Health: State of the art, challenges, and future directions. *Yearbook of Medical Informatics, 28*, 16–26. https://doi.org/10.1055/s-0039-1677908
Westwood, E., Walshaw, J., Boag, K., Chua, W., Dimashki, S., Khalid, H., Lathan, R., Wellington, J., Lockwood, S., & Yiasemidou, M. (2023). Time for change: Compliance with RCS green theatre checklist—facilitators and barriers on the journey to net zero. *Frontiers in Surgery, 10*. https://doi.org/10.3389/fsurg.2023.1260301

Zeebaree, D., Hasan, D., Abdulazeez, A., Ahmed, F., & Hasan, R. (2021). Machine Learning Semi-Supervised Algorithms for Gene Selection: A Review. In: *IEEE 11th international conference on system engineering and technology (ICSET)* (pp. 165–170). https://doi.org/10.1109/ICSET53708.2021.9612526

Zhou, M., Govindan, K., & Xie, X. (2020). How fairness perceptions, embeddedness, and knowledge sharing drive green innovation in sustainable supply chains: An equity theory and network perspective to achieve sustainable development goals. *Journal of Cleaner Production*. https://doi.org/10.1016/j.jclepro.2020.120950

Chapter 3
Bridging Human Interface and Technology: A Critical Analysis of Artificial Intelligence for Hiring in the Healthcare Sector

Swati Agrawal, Yashi Kapil, and Suhani Singh

Abstract In the continuous strive toward efficiency and competitive advantage, organizations are taking steps towards artificial intelligence (AI) as a comprehensive paradigm within the 'Talent Acquisition' domain. In the contemporary and dynamic marketplace, where skilled personnel are the most crucial assets for an organization, it is imperative to maintain a competitive edge in talent acquisition by employing AI-enabled recruitment and selection approaches and strategies. The healthcare industry has also begun to leverage AI for talent hiring and selection processes, namely talent acquisition. Regarding the healthcare sector, the implementation of AI has transformed the sub-systems such as administration, diagnosis, and patient care. Healthcare organizations have utilized AI from the initial scanning of resumes to the final hiring processes. This article aims to investigate the influence of AI-enabled factors of recruitment on hiring satisfaction, using structural equation modeling (SEM). The constructs, namely AI-enabled recruitment process, culture fit, and AI-enabled job fit have shown their efficacy in driving effective recruitment and selection systems. By understanding various regimes of AI in the talent acquisition process, this article can prove useful for the practitioners and professionals in healthcare sectors for utilizing AI in management functions while creating a scope for scholars and academicians for further research.

Keywords Artificial Intelligence · AI-enabled recruitment · AI in healthcare · Hiring satisfaction · AI-enabled hiring

---

S. Agrawal (✉) · Y. Kapil · S. Singh
School of Management, Bennett University, Greater Noida 201310, India
e-mail: swati.02@gmail.com

Y. Kapil
e-mail: m23somp0001@bennett.edu.in

R. Singh et al. (eds.), *Transforming Healthcare Sector Through Artificial Intelligence and Environmental Sustainability*, Approaches to Global Sustainability, Markets, and Governance, https://doi.org/10.1007/978-981-97-9555-0_3

3.1 Introduction

Various health-related challenges in the healthcare industry such as increasing non-communicable diseases, rising life expectancy, and staff shortages pose high demand for healthcare services worldwide. Such rising demands for services, large-scale administration and documentation needs, and shortage of required staff will lead to a huge workload on the health workers. The COVID-19 pandemic impacted the healthcare system of every nation. The necessary smart medical accessories, surveillance systems, and related technologies were highly demanded, implying that healthcare organizations along with manufacturing and educational institutions require technology that allows employees to work from a safe place (Chandra et al., 2022). There has been a persistent issue of long hiring periods in the healthcare sector even before COVID, but the pandemic made the situation worse. COVID increased the level of burnout and compassion fatigue in the employees leading to an enhanced absence rate among them (O'Neil, 2023). Here comes the role of AI, which is considered a tool for improving patient care while reducing the workload of the personnel (Hummelsberger et al, 2023). Artificial intelligence (AI) is an extensive branch of computer science related to framing 'smart machines' that are competent enough to perform the tasks requiring human-aided intelligence. The study by Shahzad et al. (2023) elucidates that the use of AI is one of the top trends in the business spheres which impacts the talent management functions of various organizations in speedy talent recruitment and selection procedures. AI-enabled virtual assistants can aid in providing personalized training to existing and new employees as per their focus areas and styles of learning. The AI market in the healthcare sector is anticipated to reach USD 190.6 billion by 2025 (Wang et al., 2021). AI is tasked with analyzing huge amounts of data and making decisions based on the insights from data. Having a huge economic impact, AI would add $15.7 trillion by 2030 ('IOT and AI in healthcare: a systematic literature review', 2018).

Artificial intelligence can lead to a synchronized integration within the human capital and AI systems. This integration will enhance the HRM processes holistically by utilizing the specific potential of AI (Abdelraouf et al., 2024. The healthcare sector is harnessing AI capabilities to simplify its recruitment procedure, boost the selection of candidates, and harness efficacy to identify exemplary talent (Burrell & Mcandrew, 2023). While AI is being executed in different industries, including talent acquisition (recruitment and selection), it has also brought ethical concerns such as lack of transparency, fairness, and the existence of unrealized biases. There also exists a lack of trained and skilled workforce who are well-equipped to implement AI. Even there is no formal legal infrastructure that could establish compatibility with AI and well-framed medical data is required which could be utilized for training machine learning (ML) algorithms (Chen & Decary, 2019). Henceforth, there exist arguments that talk about the forms of work that could be created by AI while considering its cautious impact on existing job displacements. Researchers envisage the application of AI in hiring to lessen human biases and diversify access, not to mention the probable threat of exaggerating, exploiting, and adding on to existing biases (Li et al., 2021). The study by Hunkenschroer and Luetge (2022) mentions that AI-driven practices are pertinent for every recruitment stage and this can involve diverse kinds of AI and algorithms.

This article continues the inquiry through a thorough exploration of the utilization of AI-enabled hiring in the healthcare sector and analyzes the impact of various organizational factors on hiring satisfaction.

3.2 Background

The intervention of artificial intelligence in the recruitment system emerged from a stage of series of times, depicted in Fig. 3.1.

3.2.1 Analog Recruitment

This was the process of recruitment until the mid-to-late 1990s when people were engaged with the mechanism for the recruitment of new personnel. Traditionally, the candidates had to physically reach job boards for employment opportunities or had to go through newspapers and print media to enquire about the ongoing recruitment. The firms wanted to stretch their reach to the maximum number of candidates possible by providing them with job details and other opportunities, while all of this was quite expensive. As a result, firms had to create trade-offs, which is known as the 'Analog reach and richness frontier' (Black & van Esch, 2020).

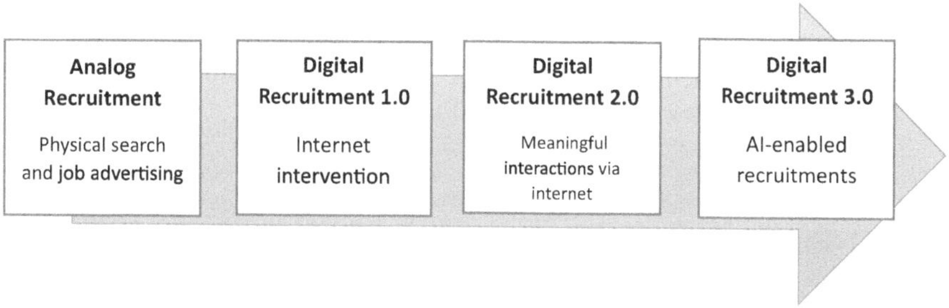

Fig. 3.1 Source: author's creation

3.2.2 *Digital Recruitment 1.0*

Digital recruitment works on the evolution of internet-enabled information where employers need not deliver their job descriptions via print media or newspapers. Few online websites were available for making the job advertisement from the recruiter's end and seeking the job from the end of the candidate implying that candidates would also not be required to explore the printed job details or put effort into mailing various application forms (Chen, 2022).

3.2.3 *Digital Recruitment 2.0*

Based on the Web 2.0 approach, the usage of social media created meaningful interactions between the employer and the employee (Aggerholm & Anderson, 2018). Web 2.0 extends digital-enabled information to blogs, the internet, chat rooms, and social networking websites, wherein it has been found that the website design including the color, layout, and page loading time along with customized search options creates an impact on the minds of the job seeking candidate (Banerjee & Gupta, 2019).

3.2.4 *Digital Recruitment 3.0*

Digital recruitment is characterized by the intervention of AI. Recruitment 3.0 increased the overall recognition of the value of human assets by top executives, including CEOs. Although the shift towards human assets began in 2000, CEOs took some time to consider this shift and realize the role of human assets in facilitating intangible assets and the firm's value. Enabling the use of AI in recruitment can leverage a diversified and broader range of talent and ensure that the firm has high-quality talent (Black & van Esch, 2020).

3.3 AI-Enabled Hiring

The quality of a business depends on the caliber of its personnel implying that to manage the business effectively, proper time should be considered in searching for and hiring the correct employees. The hiring process is an indispensable part of carrying on a business (John, 2019). Contemporarily, recruitment and selection have been majorly affected by technology intervention. Hiring involves both recruitment and selection processes, where recruitment covers the holistic establishment of the staff through identifying, luring, meeting, selecting, and onboarding the representatives. Meanwhile, the selection is the process of looking for the most appropriate candidate for a position from the competitors from within or outside the association (Mandal & Shrivastava, 2023). Novel technological developments including internet usage and the emergence of social networking websites have impacted the work of recruiters and candidates who search for a job (Nikolaou, 2021).

Sometimes known as 'Distributed Ledger Technology (DLT)', blockchain technology allows users to record and share a common view of a system's states across a distributed network, where a DLT is like 'Distributed Database Management Systems (DDBMS)' where data can be shared, read, written, and maintained in a consensus and decentralized manner. Furthermore, the integrity of the data is maintained without any third-party involvement (Meunier, 2018). Blockchain technology assists the organization in checking the authenticity of the candidate via a distributed third-party ledger or candidate's access (token or key). It can also facilitate the recruiters in verifying the educational skills of the candidate through a distributed ledger. Such pre-verification helps in reducing 'time, effort and price' and hence speeding the job matching process (Yi et al., 2020). Filtering out and hiring the worthiest job applicants from the pool of online postings of job applications became quite hefty and required a lot of recruiters. To mediate such challenges, the use of AI has become very prominent. AI-enabled tools create job descriptions with proper keywords that are bias-free and gender-neutral. AI-based CV screening tool ATS (Applicant Tracking System) can identify and analyze the keywords for placing the right candidate in the specific job opening (Brishti & Javed, 2020). The chatbots are part of natural language processing systems (NLPS) that act as virtual conversational agents imitating human interactivity. While in the developing phase, health chatbots are viable enough for improving doctor-patient and clinic-patient communications, enabling health services via remote testing, medicinal monitoring, and teleconsultations (Nadarzynski et al., 2019).The application of AI is revolutionizing and improving various aspects of candidate selection and hiring. The AI's ability to quickly process and interpret the information increases the impartiality and efficiency of the recruitment procedure. Alignment of AI with the hiring process enables organizations to enhance their talent acquisition strategy, improve hiring decisions, and increase organizational performance (The Perfect Blend: Balancing AI and Human Touch in Modern Recruitment, n.d.)

3.4 Hypotheses Development

3.4.1 Culture Fit

Culture can be defined as the set of meanings and behavioral norms shared by group members. Cultural fit includes two aspects: cognition and behavior. The cognitive aspect explains the mental descriptions, values, and beliefs drawn by individuals to frame a perception of everyday occurrences. The behavioral aspect is related to the expectations and beliefs that are confined to the actions of an individual (Mobasseri et al., 2019).

The process of recruitment needs to be concerned with going for the right candidate who not only fits best with the vacancy but also fits best for the culture of the organization. AI-enabled chatbots leverage recruiters at the first stage of recruitment, such as replying and asking frequently asked questions related to employee benefits or the culture of the organization, enabling the human recruiters in the later recruitment stages. Moreover, natural language processing and machine learning frame a psychological profile that predicts if a person will fit into the organizational culture (Javed & Brishti, 2020). AI has rejuvenated culture fit assessment in hiring. It supplements traditional methods of interviews with skill tests and simulations. Utilizing the data from top performers, AI predicts job success traits. Tools such as Pymetrics use this data to gauge candidate fit. AI also evaluates digital records to assess culture alignment. This data-driven approach improves the probability of finding culturally fitting candidates (Hunkenschroer & Luetge, 2022).

Through the lens of existing literature, the following hypothesis has been put forth:

H_1 *There exists a positive relationship between culture fit and Hiring satisfaction.*

3.4.2 AI-Enabled Job Fit

Job fit or person-job fit refers to the alignment between the individual's personal knowledge, skills, and abilities and the demands of the job. It can also be defined as the aspiration of a candidate that the job can supply (Deniz et al., 2015).

Although there is less information available on the response of candidates towards the organizations that make use of AI recruitment for catering towards characteristics and universalizing the probable behaviors regarding job fit and performance, yet, facial recognition can help in determining a candidate's sexual orientation with noteworthy accuracy. During the recruitment process, if the organization can gather supplementary information related to demography, health, and sexual orientation, this data can be utilized to classify the job candidates further (Van Esch et al., 2019). The study by Delecraz et al. (2022) elucidates the role of AI, specifically machine and deep learning algorithms in enhancing the job-matching procedure between job seekers and jobs. AI improves both the qualitative and quantitative facets through the automation of tasks, facilitating insights from aspects like Natural Language Processing (NLP) and behavioral sciences. ML algorithms outshine at exploring the skills and motivation of a candidate, leading to better predictions of job fit and performance, signifying a crucial shift in talent acquisition practices.On the basis of our current understanding of the literature, we conclude that

H_2 *There exists a positive relationship between AI-enabled job fit and Hiring Satisfaction.*

3.4.3 AI-Enabled Applicant's Reaction

'Attraction and Fit Theory' and related research have found that there is a strong correlation between candidates' likelihood of joining the job and their perception of both the attractiveness of the organization as well as the job application process itself. The organization that employs innovative and leading-edge AI technology during the recruitment process signifies that it is open to change and is aligned with the dynamic environment. This postulation is consistent with the findings that organizations employing AI in their hiring process make the organization attractive and appealing to new talent (Van Esch et al., 2021). The reactions of applicants toward the AI-enabled hiring process can have important implications for the organization. The reactions of applicants can distort the hiring process, including how the organization is being perceived by the applicant, his attraction towards the organization, and the intention to accept the job offer (Gonzalez et al., 2022).

Although technology-based interviews can stimulate flexibility and efficiency but can also have some negative reactions. For example, interviews via videoconferencing are perceived as less fair and less opportunity-driven in comparison to face-to-face interviews. On the other hand, candidates also recognize that automated interviews are quite consistent (Langer et al., 2019).

Based on the existing literature, we hypothesize that

H_3 *There exists a positive relationship between AI-enabled applicant's reactions and Hiring Satisfaction.*

3.4.4 AI-Enabled Induction Process

Induction or orientation is the planned introduction of new employees to their jobs, co-workers, and the culture of the organization to familiarize them with the workplace and its environment. The induction process significantly impacts the hiring satisfaction of the employees as the quality of this process allows employees to make decisions on their employment, thereby impacting the retention rates of the organization (Bharthvajan & Kavitha, 2019). With the advancement in science and technology, AI is being implemented in various domains of human resource management from workload reduction to matching best-suited profiles. Yet, the maximum assistance of AI has been seen within the induction process. The team HR is required to introduce the new recruitees to the organization and its policies, this can be done with the help of bots, leveraging the AI usage (Garg et al., 2021). AI can create personalized induction or onboarding protocols for every position allotted to the new entrants as a well-structured induction system will yield better retention rates (Wassan, 2021). The study by Kleinings highlights that automated onboarding has been employed by businesses by around 68%, through the use of chatbots, smart checklists, and remote support applications. Also, AI can be used in evaluating the completeness of employee onboarding documents and helping in interpreting feedback forms (How to Automate Your Employee Onboarding Process with AI, n.d.).

On the basis of the aforementioned, we posit the following hypothesis:

H_4 *There Exists a Positive Relationship Between AI-Enabled Induction Process and Hiring Satisfaction.*

3.4.5 AI-Enabled Recruitment Process

Automation in the recruitment process simplifies human efforts to allocate their time and efforts for realizing the dynamic needs of employees for retaining them and henceforth reducing the training and developmental costs (Raveendra et al., 2020). Fairness in the talent acquisition process will lead to positive work behavior and attitude, implying a sense of job satisfaction. Organizational policies must facilitate fairness during the hiring process as a fair hiring policy correlates to the satisfaction of the employee (Park, 2021).

The application of AI leverages simplification of the recruiter's day-to-day tasks and stimulates efficiency by replacing time-consuming approaches. In the AI-enabled recruitment process, for job advertisements, ML, and language analysis are used to identify keywords and structure job descriptions to attract qualified candidates. For the process of job search, AI identifies the job seeker's profile for suggesting particular job openings. For the communication and information phase, AI is capable of answering FAQs from prospective candidates and suggesting prominent job roles. During the application stage, AI phrases resumes for highlighting relevant skills and AI assistants or chatbots can help in framing applications. For candidate evaluation and selection, AI analyzes the application components and predicts the job fit. Various tests and assessments can also be evaluated by AI (Laurim et al., 2021).Based on the literature, the following hypothesis has been formulated:

H_5 *There exists a positive relationship between the AI-enabled recruitment process and Hiring Satisfaction.*

Building upon the background information and hypotheses formation, Fig. 3.2 illustrates the conceptual model, where factors such as culture fit, job fit, applicant's reaction, induction process, and AI-enabled recruitment process are the independent variables, while hiring satisfaction represents the dependent variable. This model is subsequently analyzed using empirical data.

3.5 Research Method

The study employs a cross-sectional survey design to explore the application of hiring satisfaction based on AI-enabled factors within the healthcare sector. Drawing from a comprehensive literature review, a survey instrument has been developed, including parameters identified through prior research and specific considerations from the current study. With a sample size of 106 respondents, the pilot survey was conducted

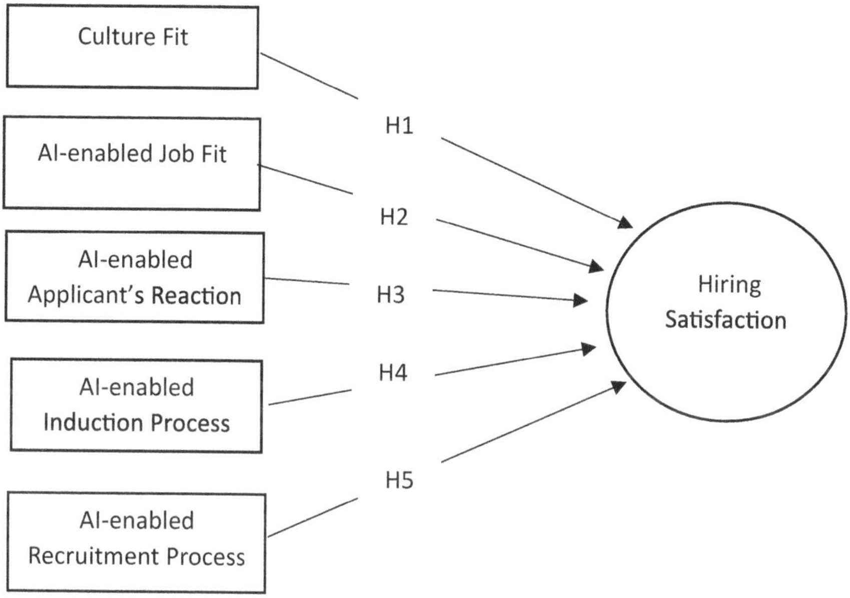

Fig. 3.2 Source: Author's creation

from various levels in the hospitals. A pilot survey is a concise feasible study applied to analyze various facets of methods lined up for a bigger, more definite, or confirmatory exploration. The aim of a pilot study is not to figure out a particular research question but to restrict researchers from leading a large-scale analysis without proper knowledge of the proposed research method (Lowe, 2019). The survey instrument for culture fit included questions such as 'Does the organization value the perspectives of employees from different backgrounds?'. Further, for the AI-enabled induction process, there are questions such as 'How satisfied are you with the technology-enabled information received during the hiring process?' Moreover, the participants were asked to rate the organization's efficiency in optimizing the recruitment process, including tracking time to fill and time to hire and following AI-enabled recruitment instructions.

3.5.1 *Sampling Method*

The convenience sampling method has been adopted. Convenience sampling is a non-random sampling method wherein individuals who fit the sample criteria are identified in any possible way (Emerson, 2015). Two healthcare organizations in India have been approached and asked to participate in the survey. The organizations agreed and gave their consent, after which the sample was obtained from employees working in different departments.

3.6 Data Analysis

3.6.1 Regarding the demographics mentioned in Table 3.1, out of 106 respondents, 54.7% are male, 38.6% are female, and 1% belong to the transgender community. Regarding the age groups, 23.5% are aged 20–30, 38.6% are aged 31–40, 17.9% are aged 41–50, and 14.1% are aged 51 or older. In terms of educational levels, 33% hold bachelor's degrees, 41.5% hold master's degrees, and 19.8% hold doctorates. Regarding years of experience, 28.4% have 0–5 years, 36.7% have 6–10 years, 21.6% have 11–15 years, and 9.4% have more than 15 years of experience.

3.6.2 Before the model's analysis, the constructs' reliability and validity have been evaluated. This includes conducting Cronbach's alpha and composite reliability for the constructs. For both the coefficients, values above 7.0 indicate that the construct has high reliability. The limit value of 0.7 and more implies that is acceptable, while the value above 0.8 and 0.9 indicates that is very satisfying (Rahayu et al. 2022). The results, as shown in Table 3.2, indicate that all variable values exceed 0.7. This implies that the five independent variables and one dependent variable utilized in the study depict reliability.

Table 3.1 Demographic details of respondents

Sample characteristics	Frequency	Percentage
Gender		
Male	58	54.7
Female	41	38.6
Transgender	01	0.9
Age		
20–30	25	23.5
31–40	41	38.6
41–50	19	17.9
51 +	15	14.1
Level of education		
Bachelor's	35	33
Master's	44	41.5
Doctorate	21	19.8
Experience level		
0–5	28	28.4
6–10	39	36.7
11–15	23	21.6
15 +	10	9.4

Number of respondents = 106

Table 3.2 Reliability, validity, and cronbach's alpha

Construct	Cronbach's alpha	Composite reliability	Composite validity	Average variance extracted (AVE)
Culture fit	0.708	0.825	0.829	0.810
AI-enabled job fit	0.977	0.967	0.969	0.846
AI-enabled applicant's reaction	0.883	0.730	0.859	0.858
AI-enabled induction process	0.710	0.747	0.758	0.929
AI-enabled recruitment process	0.839	0.802	0.936	0.687
Hiring satisfaction	0.730	0.950	0.715	0.616

3.6.3 The structural model estimation has been performed through the Smart PLS 3 software for evaluating the model in which culture fit, job fit, applicant's reaction, induction process, and AI-enabled recruitment process are described as the independent variables, while AI-enabled hiring has been described as the dependent variable. PLS-SEM is an approach within structural equation modelling (SEM), aiming at maximizing the explained variance of dependent variables in a path model. It is commonly used by scholars and practitioners for the identification of success factors and outcomes of competitive advantage. PLS-SEM facilitates the researcher in estimating complex models having many variables and indicator constructs (Ringle, 2015).

The findings indicate that if the other factors remain constant and there is one unit increase in the applicant's reaction, there is an increase in hiring satisfaction by 0.252. In the same manner, if there is a unit increase in the AI-enabled recruitment process, then hiring satisfaction would increase by 0.432. However, if culture fit increases by one unit, then the hiring satisfaction would increase by 0.282 indicating that the hypothesis framed is supported. For the AI-enabled job fit and AI-enabled induction process, an increase in these variables by one unit decreases the hiring satisfaction by −0.073 and −0.127, respectively.

For testing hypotheses, the acceptable p-value threshold is 0.05, which is used for statistical significance, implying that if the p-value is less than 0.05, the observed effect is unlikely to be due to random chance alone (Thiese et al., 2016).Table 3.3 provides data that is sufficient for conducting path analysis to test the hypothesis from H1 to H5. It shows the P-value, path coefficient, and T-statistic value for each hypothesis. The direct effects of culture fit, AI-enabled job fit, and AI-enabled recruitment process toward hiring satisfaction were found to be significant ($p < 0.05$), hence supporting H1, H2, and H5, respectively. However, the hypotheses related to the AI-enabled applicant's reaction and AI-enabled induction process were not supported ($p > 0.05$), indicating that these two variables would be rejected.

Additionally, a bootstrap test was conducted with a sample size of 106 respondents to determine the t-values of the model parameters for testing the research hypotheses. The validated hypotheses are those having p-values less than 0.05, while those with p-values greater than 0.05 are considered valid. Henceforth, it can be concluded that H1, H3, and H4 are the supported hypotheses while H2 and H5 are not supported. Therefore, it can be inferred that culture fit, job fit, and AI-enabled recruitment process positively influence hiring satisfaction in the healthcare industry.

3.6.5. R-square measures the proportion of variance in the dependent variable that is explained by the independent variables (Chicco et al., 2021). The outcome of the R-square value is 0.54, indicating that the 54% variability in hiring satisfaction is being accounted for by the variability in the independent variables (Table 3.4). This indicates that the strength of the relationship between the model and hiring satisfaction is 54%, which is considerably moderate and is on a positive scale. The scatter points on the regression line would be closer with the increase of variance.

Table 3.3 Direct effects and hypotheses testing

Hypothesis	Relationship effect	Original sample(O)	Sample mean (M)	T statistics	P values	Acceptance
HI	Culture Fit → Hiring satisfaction	0.282	0.082	0.842	0.027	Yes
H2	AI-enabled job fit → Hiring satisfaction	0.252	0.276	1.218	0.045	Yes
H3	AI-enabled applicant's reaction → Hiring satisfaction	−0.127	−0.075	0.526	0.599	No
H4	AI-enabled induction process → Hiring satisfaction	−0.073	−0.095	0.320	0.749	No
H5	AI-enabled recruitment process → Hiring satisfaction	0.433	0.220	1.286	0.009	Yes

Table 3.4 Model results

Dependent variable	R-square	R-square adjusted
Hiring satisfaction	0.541	0.449

3.7 Discussion

Our analysis of hiring satisfaction in the healthcare sector has revealed intriguing results. In the gender distribution, the majority of respondents are male (54.7%), followed closely by females (38.6%), and a smaller portion identified as transgender (0.9%), implying a balanced representation of genders within the sample frame. The largest segment of participants falls within the 31–40 age group (38.6%), signifying that mid-career professionals are prominently represented in the study. The educational levels of the respondents indicated that the sample is well-educated, implying the requirement of high skill levels required within the healthcare sector. The experience levels represented a blend of early-career and mid-career professionals in the study.

The relationship of the antecedents of hitting satisfaction, i.e., culture fit, AI-enabled job fit, AI-enabled applicant's reaction, AI-enabled induction process, and AI-enabled recruitment process of the employees working in the Healthcare industry of India has been analyzed. The findings revealed that culture fit (H1), AI-enabled job fit (H2), and AI-enabled recruitment process (H5) significantly influence hiring satisfaction, as depicted by their statistically significant p-values ($p < 0.05$). On the contrary, hypotheses concerning the applicant's reaction (H3) and induction process (H4) are not supported, as their p-values exceed the pre-determined significance levels. It has been analyzed that the AI-enabled recruitment process is the strongest predictor of hiring satisfaction followed by culture fit. The dependent variable, namely, hiring satisfaction reveals a considerable R-square value of 0.541, implying that the proposed model explains approximately 54.1% of the variance in hiring satisfaction. The adjusted R-square value (0.449) exhibits that the model adequately accounts for the complexity of factors impacting hiring satisfaction while reducing the risk of overfitting.

Through leveraging the insights from the data, healthcare organizations can redefine and delve deeper into AI-oriented recruitment strategies, attract top talent, and eventually enhance organizational performance.

3.7.1 Further Research and Implications

Future research can further enrich the understanding of hiring satisfaction based on AI-enabled factors among the employees of the healthcare industry by examining the personality traits, culture of the particular organization, and impact of leadership. Also, longitudinal studies can prove to be insightful for tracking satisfaction over time. Moreover, qualitative methods can help in exploring nuanced insights. Based on the findings, it is indicated that hiring satisfaction improves retention of the workforce, thereby leading to improved productivity. Since the AI-enabled recruitment process and culture fit are the two strongest variables showcasing a direct impact on hiring satisfaction, therefore, the HR manager would more likely prioritize these two attributes to enhance hiring satisfaction and related AI-enabled hiring practices among the workforce. Furthermore, the HR managers should consider the AI-enabled recruitment process and the culture fit in increasing the hiring satisfaction among the employees of the healthcare industry in India, to ensure better retention of the workforce, leading to a lower attrition rate and maximizing the productivity of the workforce. The theoretical implication of the study lies in understanding the role of AI-enabled factors in talent acquisition within the healthcare sector. By investigating such factors, the study contributes to the adoption of AI in the hiring process. The findings highlight the insights the importance of optimizing recruitment strategies, eventually enhancing organizational performance and employee well-being. It also emphasizes the importance of integrating AI in hiring practices tailored to the healthcare sector's requirements.

3.8 Conclusion

The healthcare sector stands to gain from leveraging AI-driven HR practices, particularly in the aspects of hiring satisfaction and talent acquisition and in the aspects of hiring satisfaction and talent acquisition. The healthcare sector usually engages technical excellence in hiring strategies. However, in the current dynamic environment, AI-led talent acquisition should be adapted to meet the shifting expectations and needs of top talent.

By channeling AI technologies, healthcare organizations can increase their employer branding endeavor. This requires exhibiting certifications and creating a positive work environment to enhance the reputation of the organization and appeal to skilled professionals. Furthermore, AI can simplify the talent acquisition process, ensuring that job seekers are identified and engaged productively.

AI-enabled talent acquisition can help healthcare firms not only attract competent candidates but also nurture and retain them within the organization. By providing personalized career development opportunities and encouraging a supportive work environment, healthcare organizations can differentiate themselves as the 'employers of choice'.

Conclusively, with the adoption of AI-enabled hiring practices, healthcare organizations can gain a competitive advantage, attracting innovative individuals who could propel organizational growth and success. This would help the organizations build a strong foundation for sustainable success in the dynamic and cut-throat competitive industry.

References

Abdelraouf, M., Emad, A., & Kadry, M. (2024). The impact of artificial intelligence (A.I) on recruitment and selection of human resources management (HRM). https://doi.org/10.21203/rs.3.rs-3956032/v1

Aggerholm, H. K., & Andersen, S. E. (2018). Social media recruitment 3.0. *Journal of Communication Management, 22*(2), 122–137. https://doi.org/10.1108/jcom-11-2017-0131

Banerjee, P., & Gupta, R. (2019). Talent attraction through online recruitment websites: Application of web 2.0 technologies. *Australasian Journal of Information Systems, 23*. https://doi.org/10.3127/ajis.v23i0.1762

Bharthvajan, R., & Kavitha, S. F. (2019). Impact of employee satisfaction towards induction training program. *International Journal of Innovative Technology and Exploring Engineering, 8*(11), 3598–3605. https://doi.org/10.35940/ijitee.k2479.0981119

Black, J. S., & van Esch, P. (2020). AI-enabled recruiting: What is it and how should a manager use it? *Business Horizons, 63*(2), 215–226. https://doi.org/10.1016/j.bushor.2019.12.001

Brishti, J. K., & Javed, A. (2020). The viability of AI-based recruitment process: A systematic literature review.

Burrell, D. N., & Mcandrew, I. (2023). Exploring the ethical dynamics of the use of artificial intelligence (AI) in hiring in healthcare organizations. *Land Forces Academy Review, 28*(4), 309–321. https://doi.org/10.2478/raft-2023-0037

Chandra, M., Kumar, K., Thakur, P., Chattopadhyaya, S., Alam, F., & Kumar, S. (2022). Digital technologies, healthcare and Covid-19: Insights from developing and emerging nations. *Health and Technology, 12*(2), 547–568. https://doi.org/10.1007/s12553-022-00650-1

Chen, M., & Decary, M. (2019). Artificial intelligence in healthcare: An essential guide for health leaders. *Healthcare Management Forum, 33*(1), 10–18. https://doi.org/10.1177/0840470419873123

Chen, Z. (2022). Collaboration among recruiters and artificial intelligence: Removing human prejudices in employment. *Cognition, Technology & Work, 25*(1), 135–149. https://doi.org/10.1007/s10111-022-00716-0

Chicco, D., Warrens, M. J., & Jurman, G. (2021). The coefficient of determination R-squared is more informative than SMAPE, MAE, MAPE, MSE and RMSE in regression analysis evaluation. *PeerJ Computer Science, 7*, e623. https://doi.org/10.7717/peerj-cs.623

Deniz, N., Noyan, A., & Ertosun, Z. G. (2015). Linking person-job fit to job stress: The mediating effect of perceived person-organization fit. *Procedia Social and Behavioral Sciences, 207*, 369–376. https://doi.org/10.1016/j.sbspro.2015.10.107

Delecraz, S., Eltarr, L., Becuwe, M., Bouxin, H., Boutin, N., & Oullier, O. (2022). Responsible artificial intelligence in human resources technology: An innovative inclusive and fair by design matching algorithm for job recruitment purposes. *Journal of Responsible Technology, 11*, 100041. https://doi.org/10.1016/j.jrt.2022.100041

Emerson, R. W. (2015). Convenience sampling, random sampling, and snowball sampling: How does sampling affect the validity of research? *Journal of Visual Impairment & Blindness, 109*(2), 164–168. https://doi.org/10.1177/0145482x1510900215

Garg, A., Gaur, S., & Sharma, P. (2021). A review paper: Role of artificial intelligence in recruitment process. *Anwesh, 6*(1), 33.

Gonzalez, M. F., Liu, W., Shirase, L., Tomczak, D. L., Lobbe, C. E., Justenhoven, R., & Martin, N. R. (2022). Allying with AI? Reactions toward human-based, AI/ML-based, and augmented hiring processes. *Computers in Human Behavior, 130*, 107179. https://doi.org/10.1016/j.chb.2022.107179

How to Automate Your Employee Onboarding Process with AI. (n.d.). How to Automate your employee onboarding process with AI. Retreived from https://levity.ai/blog/automate-your-employee-onboarding-process-with-ai

Hummelsberger, P., Koch, T. K., Rauh, S., Dorn, J., Lermer, E., Raue, M., Hudecek, M. F. C., Schicho, A., Colak, E., Ghassemi, M., & Gaube, S. (2023). Insights on the current state and

future outlook of AI in health care: Expert interview study. *JMIR AI*, *2*, e47353. https://doi.org/10.2196/47353
Hunkenschroer, A. L., & Luetge, C. (2022). Ethics of AI-Enabled recruiting and selection: A Review and research agenda. *Journal of Business Ethics*, *178*(4), 977–1007. https://doi.org/10.1007/s10551-022-05049-6
Iot and AI in Healthcare: A systematic literature review. (2018). *Issues in Information Systems*. https://doi.org/10.48009/3_iis_2018_33-41
John, J. (2019). Effective recruitment strategy to improve your hiring process. *ZENITH International Journal of Multidisciplinary Research*, *9*(3), 352–357.
Langer, M., König, C. J., Sanchez, D. R. P., & Samadi, S. (2019). Highly automated interviews: Applicant reactions and the organizational context. *Journal of Managerial Psychology*, *35*(4), 301–314. https://doi.org/10.1108/jmp-09-2018-0402
Laurim, V., Arpaci, S., Prommegger, B., & Krcmar, H. (2021). *Computer, whom should i hire?—acceptance criteria for artificial intelligence in the recruitment process*. http://hdl.handle.net/10125/71288
Li, L., Lassiter, T., Oh, J., & Lee, M. K. (2021). Algorithmic hiring in practice. *Proceedings of the 2021 AAAI/ACM Conference on AI, Ethics, and Society*. https://doi.org/10.1145/3461702.3462531
Lowe, N. K. (2019). What is a pilot study? *Journal of Obstetric, Gynecologic & Neonatal Nursing*, *48*(2), 117–118. https://doi.org/10.1016/j.jogn.2019.01.005
Mandal, A., & Shrivastava, D. R. (2023). The Hiring Process and Its Significance. *IRE Journals*, *7*(1).
Meunier, S. (2018). Blockchain 101. *Transforming Climate Finance and Green Investment with Blockchains, 23–34*. https://doi.org/10.1016/b978-0-12-814447-3.00003-3
Mobasseri, S., Goldberg, A., & Srivastava, S. B. (2019). What is cultural fit. *The Oxford Handbook of Cognitive Sociology*, 305.
Nadarzynski, T., Miles, O., Cowie, A., & Ridge, D. (2019). Acceptability of artificial intelligence (AI)-led chatbot services in healthcare: A mixed-methods study. *Digital Health*, *5*, 205520761987180. https://doi.org/10.1177/2055207619871808
O'Neil, J. (2023). How recruitment and staffing in the healthcare Industry has been impacted by the covid-19 pandemic. In *BSU honors program theses and projects*. Item 624. https://vc.bridgew.edu/honors_proj/624
Nikolaou, I. (2021). What is the role of technology in recruitment and selection? *The Spanish Journal of Psychology, 24*. https://doi.org/10.1017/sjp.2021.6
Park, H. Y. (2021). The association between fair hiring policy and employee job satisfaction: Theoretical approach in the literature analysis. *East Asian Journal of Business Economics (EAJBE)*, *9*(2), 43–54.
Rahayu, A., Sulastri, S., Wibowo, L., & Gaffar, V. (2022). Effect of technology readiness towards acceptance technology in using information system. *Proceedings of the 6th global conference on business, management, and entrepreneurship* (GCBME 2021). https://doi.org/10.2991/aebmr.k.220701.101
Raveendra, P. V., Satish, Y. M., & Singh, P. (2020). Changing landscape of recruitment industry: A study on the impact of artificial intelligence on eliminating hiring bias from recruitment and selection process. *Journal of Computational and Theoretical Nanoscience*, *17*(9), 4404–4407. https://doi.org/10.1166/jctn.2020.9086
Ringle, C. M. (2015). Partial least squares structural equation modelling (PLS-SEM) using SmartPLS 3. In *Computational data analysis and numerical methods VII WCDANM*. Portugal
Shahzad, M. F., Xu, S., Naveed, W., Nusrat, S., & Zahid, I. (2023). Investigating the impact of artificial intelligence on human resource functions in the health sector of China: A mediated moderation model. *Heliyon*, *9*(11), e21818. https://doi.org/10.1016/j.heliyon.2023.e21818
Wassan, S. (2021). How artificial intelligence transforms the experience of employees. *Turkish Journal of Computer and Mathematics Education (TURCOMAT)*, *12*(10), 7116–7135. https://doi.org/10.17762/turcomat.v12i10.5603

The Perfect Blend: Balancing AI and Human Touch in Modern Recruitment. (n.d.). *The perfect blend: Balancing AI and human touch in modern recruitment*. https://www.remoti.io/blog/balancing-ai-and-human-touch-in-recruitment

Thiese, M. S., Ronna, B., & Ott, U. (2016). P value interpretations and considerations. *Journal of Thoracic Disease*, *8*(9), E928–E931. https://doi.org/10.21037/jtd.2016.08.16

Van Esch, P., Black, J. S., & Arli, D. (2020). Job candidates' reactions to AI-enabled job application processes. *AI and Ethics*, *1*(2), 119–130. https://doi.org/10.1007/s43681-020-00025-0

Van Esch, P., Black, J. S., & Ferolie, J. (2019). Marketing AI recruitment: The next phase in job application and selection. *Computers in Human Behavior*, *90*, 215–222. https://doi.org/10.1016/j.chb.2018.09.009

Wang, W., Chen, L., Xiong, M., & Wang, Y. (2021). Accelerating AI adoption with responsible AI signals and employee engagement mechanisms in health care. *Information Systems Frontiers*, *25*(6), 2239–2256. https://doi.org/10.1007/s10796-021-10154-4

Yi, C. S. S., Yung, E., Fong, C., & Tripathi, S. (2020). Benefits and use of blockchain technology to human resources management: A critical review. *International Journal of Human Resource Studies*, *10*(2), 131. https://doi.org/10.5296/ijhrs.v10i2.16932

Chapter 4
Artificial Intelligence-Driven Technologies for Environmental Sustainability in the Healthcare Industry

Narayanage Jayantha Dewasiri, Mananage Shanika Hansini Rathnasiri, and Karunarathnage Sajith Senaka Nuwansiri Karunarathna

Abstract This chapter broadly focuses on the state of artificial intelligence (AI) in environmental sustainability in the healthcare sector. It also explores different AI technologies for analysing the efficient use of resources, waste minimisation, and improving the overall organisational operation by explaining the risks of ethical issues and potential regulation. A systematic literature review (SLR) was performed according to the Preferred Reporting Items for Systematic Reviews and the Population-Intervention-Comparison-Outcome (PICO) model to search for papers on AI-driven technologies in healthcare sustainability. Accordingly, forty-one (41) studies were considered as the study's sample. The findings accentuated the collaboration of AI with blockchain, IoT, and other advancements to study the effect of the resulting solutions on energy consumption, waste treatment, and ecological tracking. Several examples are used in the study: energy consumption forecasting using AI, biomedical waste management through deep learning, and policy analysis based on natural language processing. The findings further revealed that AI-integrated blockchain improved transparency and accountability by creating a tracking system in the supply chain and carbon credit system. The chapter also outlines some of the risks inherent in the deployment of AI in healthcare, including data privacy issues, algorithm bias, and energy concerns about the use of AI in healthcare. The study's results imply that AI can significantly impact healthcare facilities' environmental impact and upgrade efficiency and sustainability work. We encourage government authorities to call for creating more responsive regulations and incentives to integrate AI technologies in healthcare. The chapter also emphasises the importance of a multidisciplinary approach with researchers, technologists, and clinicians to optimise the roles of AI for sustainability.

N. J. Dewasiri (✉) · M. S. H. Rathnasiri · K. S. S. N. Karunarathna
Sabaragamuwa University of Sri Lanka, Belihuloya, Sri Lanka
e-mail: jayantha@mgt.sab.ac.lk

N. J. Dewasiri
Adjunct Faculty, University Center for Research and Development (UCRD), Chandigarh University, Chandigarh, Punjab, India

R. Singh et al. (eds.), *Transforming Healthcare Sector Through Artificial Intelligence and Environmental Sustainability*, Approaches to Global Sustainability, Markets, and Governance, https://doi.org/10.1007/978-981-97-9555-0_4

Keywords Artificial intelligence (AI) · Environmental sustainability · Blockchain · Healthcare · Internet of Things (IoT) · Systematic literature review (SLR)

JEL Codes Q56 · I10 · I15 · L86 · C63

4.1 Introduction

Although the role of healthcare is to enhance people's health standards, the industry needs to work on protecting the environment (Macpherson et al., 2024). Energy and water are considerable consumers in healthcare facilities such as hospitals, clinics and polyclinics, producing large quantities of waste, including biomedical hazards (Reddy et al., 2024). The sector also contributes to massive greenhouse gas (GHG) emissions because it utilises power-intensive equipment and mechanisms. These disadvantages are again magnified by the demands of the food production industry in terms of being always available, together with adherence to high hygiene and safety standards that, in most cases, translate to high consumption and waste. Thus, as population growth and ageing increase worldwide and the need for healthcare services rises, increasing the burden on the environment will become a definitive question for sustainable development (MacNeill et al., 2021).

In this regard, it is impossible to overemphasise the importance of such innovative concepts as healthcare sustainability. The conventional approaches of mitigating the harms of the environment by applying efficiency have to do with energy as well as the improved management of waste, which, despite being critical, need to be revised to support the increasing needs and expectations. Therefore, there is an increased need to apply higher technologies to enhance conservation and reduce wastage and impacts to healthcare activities and facilities without compromising healthcare quality. Advancements in this area are beneficial as they result in environmental conservation while at the same time improving functionality and economical operations, making the health system sustainable.

Consequently, AI's potential to resolve these environmental problems is unique and valuable. Machine learning, deep learning, and natural language processing present some of the AI technologies for analysing large datasets, output predictions, and improving processes. Furthermore, sustainability can be applied to healthcare in many aspects to improve its outcomes with the help of AI. For example, using AI in predictive models can enhance the energy management of health facilities, and deep learning algorithms can improve the management and disposal of medical waste. Natural language processing (NLP) can dissect policies and opinions on healthcare sustainability to assist critical officials. Enhancing healthcare operations by combining AI with promising technologies like blockchain and IoT to increase transparency and traceability is also possible.

In the future advancement of the healthcare setting, the integration of AI will be a mandate to meet healthcare delivery targets while minimising environmental effects. Hence, adopting AI can move healthcare towards environmentalism. AI-driven innovations can also reduce hospitals' environmental impact and improve the sustainable management of resources within the sector. Therefore, this study explores how AI-driven technologies can be utilised for environmental sustainability in the healthcare industry.

4.2 Methodology

This chapter adopts the PRISMA, PICO, and SLR approaches to perform an extensive research synthesis to identify, assess, and integrate the existing studies focusing on AI-driven technologies for environmental sustainability in the healthcare sector. The first procedure was manually extracting academic databases such as Google Scholar using relevant keywords. Thus, the combination of phrases like 'artificial intelligence,' 'environmental sustainability,' 'healthcare industry,' 'machine learning,' 'deep learning,' 'natural language processing,' 'blockchain,' and 'IoT' was employed to make the search more comprehensive. The search was limited to articles in peer-reviewed English language journals from January 2010 through July 2024 because the most recent literature was presumed to be most relevant.

The PRISMA framework was used for the sample selection in the current study as used by Karunarathna et al. (2024), Dewasiri et al. (2024), Rathnasiri et al. (2024), Jayawardena and Dewasiri (2023), Mishra et al. (2024), and Dewasiri et al. (2022). Full-text articles in English were assessed for eligibility against predefined inclusion criteria. Accordingly, sample studies should have focused on AI-driven technologies to outline ways of minimising emissions and helping environmental sustainability within the healthcare industry. The articles that did not focus on the criteria identified above were removed; hence, only the most relevant research was included.

Accordingly, some of the valuable information considered for each study involved the objective and methods of the study, information about the AI techs used, the healthcare contexts they applied to, the environmental impact report, and any issues or restrictions encountered. PICO model was applied where the population entailed the healthcare facilities and systems; intervention involved the implementation of AI-based technologies; control included conventional methods or the pre-implementation scenario; and the outcomes explored the measures of environmental sustainability, including energy use, waste minimisation, and carbon footprint.

The collected data were analysed to categorise them based on themes, trends and research gaps highlighted in the literature search. This synthesis offered a broad glimpse into incorporating AI technologies to increase the healthcare system's environmental sustainability. The SLR approach guaranteed a very systematic and bias-free assessment of the literature. The credibility and dependency of the studies incorporated were evaluated based on characteristics such as study type, sample and outcome verity. The papers were classified according to the AI technology employed (e.g., machine learning, deep learning, natural language processing) and the corresponding application to healthcare sustainability.

The data obtained during the systematic review was considered to understand which AI-driven technologies are most effective and their relevance to healthcare. This account also revealed the effectiveness or otherwise of these technologies, their upsides, and their contributions or threats to environmental conservation. New trends and decision-making technologies were discussed, outlining the future concepts to work on and implement.

Ethical considerations were integral to the review, addressing data privacy, security, and algorithmic bias in using AI technologies. The potential environmental impact of AI itself, such as energy consumption, was also examined to ensure a balanced evaluation of its overall sustainability benefits. Following this rigorous methodology, we selected forty-one (41) sample studies for the current study. This chapter aims to provide a thorough and objective review of the current state of AI-driven technologies for environmental sustainability in the healthcare industry. The findings will inform researchers, policymakers, and healthcare practitioners about the effective integration of AI to achieve sustainable healthcare practices.

4.3 AI Technologies and Their Environmental Applications in Healthcare

Applying AI technologies can improve environmental sustainability in the health sector (Kulkov et al., 2024). There are several distinctive benefits of using, for example, machine learning (Garlík &Přívětivý, 2024), deep learning (Abdel-Jaber et al., 2022), and natural language processing (Zhu et al., 2020) to address various environmental issues healthcare facilities encounter. All these AI technologies can then be used exclusively in the healthcare sector, enhancing energy efficiency, resources, waste management, and policy analysis.

4.3.1 Machine Learning (ML)

Machine learning is a sub-field of AI that seeks to get systems or models to find information independently without being programmed (Štuikys & Burbaitė, 2024). Machine learning can be applied to forecast energy use in hospitals and healthcare centres in a medical setting. Data gathered from such energy consumption patterns in the past can be used to build machine learning models that can predict the energy that will be used in the future, discover trends and make recommendations on how energy usage can be done efficiently. For instance, machine learning algorithms can estimate when the energy demand is most likely high and manage the heating, ventilation, and air conditioning (HVAC) systems to avoid energy wastage. Moreover, one can analyse patient flow in hospitals and clinics, control the number of workers, and use resources in real time with the help of machine learning (El-Bouri et al., 2021).

This can result in the effective utilisation of medical equipment, cutting down on unnecessary energy consumption and thus reducing the effects of the health facility on the environment using machine learning.

4.3.2 Deep Learning (DL)

Deep learning, another subfield of AI, utilises neural networks that consist of at least two layers to be trained so that it may analyse complicated information (Kumar et al., 2023). In healthcare sustainability, deep learning can help detect biomedical waste and the arrangements to take care of it. Such images of waste containers can be subjected to deep learning algorithms for the analysis and segregation of wastes based on specific types. This can help contain the spread of infectious diseases, encourage reuse and recycling of materials and minimise hazardous waste that pollutes the environment. Also, deep learning can improve the ecological data in the healthcare environment regarding identifying and recognising images and speeches (Li et al., 2023a). For example, intelligent cameras and sensors can track the attributes of air and water in and around healthcare centres, which will help manage the environment. Similarly, speech recognition systems can capture oral reports from the healthcare staff on the state of the environment, thus enabling quick action.

4.3.3 Natural Language Processing (NLP)

NLP may be the study of the interaction between computers and natural language to provide instructions or commands for computers to understand and translate human language (Mohamed et al., 2024). In this area of healthcare, NLP is indispensable in understanding the guidelines and rules of sustainability (Roy et al., 2021). To elaborate, a large amount of textual content obtained from policy documents, research papers, and regulatory guidelines can be analysed by NLP algorithms to determine the topics most relevant to sustainability practices today. Healthcare organisations can adhere to the current sustainable practices or incorporate them into the market through various plans or strategies. In addition, NLP can also be applied to gather opinions and sentiments of the public over sustainability issues in healthcare (Zunic et al., 2020). NLP tools can identify the feelings of the public and their awareness of the existing environmental change campaigns through Twitter posts, customer reviews, and surveys. The data gathered can be helpful for legislation and chiefs of medical and hospitable facilities to create and implement comprehensible sustainable initiatives for the community.

Hence, ML, DL, and NLP are valuable instruments for increasing environmental sustainability in healthcare. The investigated technologies help healthcare facilities minimise energy use, better manage resources, deal with biomedical waste, and meet the core requirements of sustainable activities and public demand. The growing applications of AI in healthcare sustainability assure the net addition to the healthcare industry's environmental impact and enhanced organisation of its ventures.

4.4 AI-Driven Blockchain Technologies for Environmental Sustainability in Healthcare

AI-driven blockchain is one of the new technologies that can revolutionise improving health and the environment in healthcare services provision (Ramachandran, 2024). Implementing AI with blockchain enhances many environmental uses through increased procedural openness, audibility, and operational proficiency. This section explains how these integrated technologies may be helpful in energy management, environmental maintenance, waste disposal, and carbon credits in healthcare institutions.

4.4.1 Overview of Blockchain and AI Integration

Blockchain technology, a decentralised digital system of recording data, ensures that records cannot be altered (Dewasiri et al., 2023). Integrating blockchain with AI improves data processing, analysis and decision-making functions. In healthcare delivery, this synergy can satisfactorily solve environmental issues by offering an open and traceable system of handling resources, waste, and emissions. AI allows the analysis of immense amounts of data accumulated through blockchain technology, provides forecasts, and improves organisational performance factors while guaranteeing data credibility and protection.

4.4.2 Enhancing Transparency and Traceability in Healthcare Supply Chains

Transparency and traceability of the supply chain are some of the essential domains where AI-driven blockchain technology plays a vital role in healthcare (Ramachandran, 2024). Supply chain management is necessary for the healthcare industry because it embraces many players ranging from manufacturers of products to suppliers, healthcare providers, and end-users. This is a great difficulty in the globalisation of business, knowing that these supply chains should be sustainable and ethical. With the help of the blockchain, every transfer and action with the materials can be documented deliberately. Through data analysis, AI can aid in eliminating the relevant sources of waste, improving the supply chain, and guaranteeing that all acquired and manufactured products are sustainable. This will also help lower the carbon footprint linked to developing and transporting medical items and gadgets.

4.4.3 Energy Management

Blockchain platforms powered by AI are essential in enhancing energy efficiency in health facilities (Li et al., 2023b). Due to the constant use of artificial lighting, heating and cooling of enclosed spaces, and uses by various medical equipment, hospitals and clinics consume a significant amount of energy. Optimising energy consumption with AI and increasing the reliability of recommendations, AI algorithms can obtain data regarding energy consumption and identify periods of maximum load with the help of big data predictive models. However, this integrated approach enables varying demand control of energy such that the energy utilised in every process and at different times is optimised. For instance, AI can determine when some hospital zones will be underutilised or vacant, regulating the conditions of heating, ventilation, and air conditioning, thus cutting expenses considerably.

4.4.4 Waste Management

Biomedical waste is a top issue that healthcare facilities face regarding environmental management (Odonkor & Mahami, 2020). Therefore, it is essential to ascertain that garbage in this aspect is managed effectively and in the least harm to the environment and humanity. AI through blockchain is a perfect solution for managing biomedical waste from its creation until disposal (Bamakan et al., 2022). Blockchain also sees that the waste management process is documented to eliminate situations where some people dump waste where they are not supposed to or against the set environmental laws. Thus, AI can analyse the patterns by which waste is produced and how they can be more efficiently managed, such as by recycling or changing the volume of garbage pumped out. In this regard, the integrated manner guarantees the environmentally sensitive disposal of hazardous wastes while reducing the effects of wastes produced within the healthcare facilities.

4.4.5 Carbon Credits and Offsetting

Carbon credit tracking and management can also be another sector significantly transformed by AI-enhanced blockchain technology (Mascarenhas et al., 2023). Hospitals and medical institutions can reduce their carbon emission through carbon offsetting schemes. Blockchain creates a transparent and highly secure account of carbon credits that enhances the validity of all transactions. It can be extended even further with the help of AI, which can analyse carbon emissions data and forecast future offsets to fulfil the company's sustainability goals. Integrating these technologies helps efficiently administrate carbon offsetting projects so that supply solutions benefit the healthcare industry in its efforts to minimise its carbon emissions.

Therefore, the proposed AI and blockchain integration solutions provide a solid foundation for improving environmental responsibility in the healthcare sector. These technologies solve many of the industry's critical environmental issues, such as transparency and traceability of the supply chain, rational use of energy, proper waste disposal, and the management of carbon credits. Healthcare organisations that integrate AI and blockchain technologies successfully increase the speed of operations, minimise adverse environmental effects, and make all processes sustainable.

4.5 Other AI-Driven Technologies in Healthcare

Several other AI techniques have significant potential to improve environmental sustainability in the healthcare sector. These technologies provide new opportunities for instant control and observation, sharing, and measurement of resources and evaluating the effects on the ecosystem, thus enhancing sustainability strategies in healthcare facilities.

4.5.1 Edge Computing

Edge computing holds that data is processed closer to where it is generated at the sensors or devices (Yu et al., 2017). This eliminates the relay of data and constant requests to the source system and thus enhances real-time processing and decision-making. In the healthcare sector, edge computing can be used to check the energy status, working conditions, and other measurements of environmental surroundings in real time at healthcare establishments (Abirami & Chitra, 2020). This is possible because localised assessments of such data allow individual healthcare facilities to quickly recognise potential issues or practices that waste energy resources and apply corrective measures. For instance, edge computing can be used to better control the settings of the heating, ventilation, and air conditioning systems through occupancy and environmental data, which would lead to saving on energy and hence make it more sustainable.

4.5.2 Swarm Intelligence

Swarm intelligence also mimics the social behaviour of insects like ants and bees (Ribino et al., 2022). It is defined as a collective way of problem-solving by several squads of people or artificial systems. To monitor the environment, swarm solutions in healthcare can be implemented through robotic devices and other swarm commanding accessories, such as AI-driven drones. Equally noteworthy, quality sensors mounted on drones help measure air quality near healthcare facilities, identify pollutants, and estimate environmental effects. Besides, they can be applied in localising waste sources and streams to control medical waste by identifying concentration areas and proper disposal. These AI-driven drones operate in a swarm, which means they are capable of collective decision-making, integration of the acquired data and reaction to new circumstances, which gives this approach in the sphere of environmental management in healthcare facilities very high efficiency.

4.5.3 Generative Adversarial Networks (GANs)

Generative adversarial networks (GANs) are neural structures formed by two sub-neural networks, a generator and discriminator, optimised to produce high-quality synthetic data (Wu et al., 2022). Regarding the use of environmentally sustainable concepts in healthcare, GAN can synthesise environmental realities such as the dispersion of pollutants or the effects of the disposal of wastes. Some of these uses include the cradle-to-grave depiction of a patient as a way of helpful training that simulates environmental dangers to healthcare givers so that they can come up with measures to eradicate them. Furthermore, satellite images and all the other environmental data can also be improved by GANs to provide better details for understanding the environment and decision-making processes. Thus, due to the enhanced quality of received environmental data, GANs can provide more accurate evaluations and targeted actions in the framework of healthcare sustainability.

4.5.4 Reinforcement Learning

Reinforcement learning can be described as training algorithms on rewards and penalties to improve business decisions (Metzger et al., 2020). Thus, this approach may enhance the natural resources in healthcare facilities. For example, reinforcement learning algorithms can be applied to designing self-governing systems related to water management and to control and optimise the usage of water and water resource management. Such systems can develop ways of dealing with the environment and can become better in terms of allocating resources for environmental sustainability. Another potential area for the use of reinforcement learning may be related to cleaning services where self-driven robots or systems that learn to recognise the presence of various contaminants in healthcare environments and remove them with the help of state-of-the-art technologies, which would contribute to the general decrease in the adverse effects on the environment and make the promoting of sustainability measures feasible.

4.5.5 *Digital Twins*

Digital twins are the digital representations of physical objects and functions used for analysis and modelling (Tao et al., 2022). In the healthcare system, it is possible to build digital twins for facilities, instruments, and logistics chains to control environmental conditions. Based on various scenarios, digital twins enable the analysis of the impact of sustainability activities, such as energy efficiency actions or a waste management scheme. This helps healthcare providers manage the resources better, minimise emissions and implement proper sustainable practices. For instance, in the case of a hospital, the digital twin can mimic its energy consumption characteristics, and it is recommended that the heating, ventilation, and air conditioning or lighting be altered to reduce energy usage while maintaining patients' comfort.

Hence, edge computing, swarm intelligence, GANs, reinforcement learning, and digital twins are innovative AI-based solutions that may help the healthcare sector increase environmental sustainability. These technologies can help healthcare providers track data in real-time, efficiently utilise resources, and systematically evaluate environmental effects. These advances help decrease the ecological impact of an organisation's operations and performance in healthcare but also relate to the global aim of constructing a sustainable and responsive healthcare system.

4.6 Case Studies

AI technologies help create or improve different machines used in the environmental sustainability sectors and can also be used practically. AI, especially AI-driven blockchain technologies, show a profound transactional influence on intelligent grids and renewable energies, sustainable healthcare practices, and water and waste management, as presented in this section in a detailed case-by-case manner. For instance, AI-driven blockchain applications, such as Medrec, were introduced to manage electronic health records with high data security (Ekblaw, 2017).

4.6.1 *Smart Grids and Renewable Energy in Healthcare*

Regarding healthcare structures, it is significant to note the opportunities for integrating AI with smart grids and related renewable energy systems, which present a concept for the efficient distribution of energies and timely maintenance amongst healthcare facilities. Conventional energy consumption patterns can be analysed, and peak usage times can be predicted using AI algorithms; thus, energy resources can be distributed efficiently. Kahwash et al. (2020) proposed an intelligent grid-driven renewable energy system for a hospital in the United Kingdom. For example, AI can control how the HVAC systems and lights operate and adapt to people's presence and the rest of the environment's conditions, thus saving energy. Also, exceptional predictive maintenance driven by AI can detect possible equipment failures before they happen, preventing frequent energy production breaks. AI, through blockchain, enhances this system by making records of energy generation, supply, and consumption in hospitals more transparent and secure. This, in turn, helps efficiently monitor the levels of renewable energy and incorporate solar, wind, and other renewable power structures, adding to the sustainability standards of healthcare services and hence minimising carbon emissions within the facilities.

4.6.2 Sustainable Healthcare Practices (SHP)

SHP reveals how AI mainly targets the precise management of healthcare assets, enhancing the sustainable utilisation of healthcare services (Haleem et al., 2022). Using the information regarding the patients' traffic in healthcare organisations, the usage of the resources, and the efficiency of the whole process, AI can determine the distribution of medical equipment, employees, and others in the best possible manner. This is because resources are used proficiently to mitigate wastage, hence offering quality services to the patients. For this reason, AI-powered blockchain technology can be utilised to have proper records of resource consumption and waste produced in healthcare institutions. The sunlight character allows healthcare providers to enact and assess sustainable programs like decreasing singular-use plastics, bolstering recycling, and reducing energy use. The integration of AI and blockchain guarantees responsibility and fosters the constant enhancement of sustainable conduct in the sphere of healthcare.

4.6.3 Water Management in Healthcare

Water is vital in any facility, especially healthcare since it is used in cleaning and sterilisation, patient handling, and various company activities. AI technologies can identify leakage incidences and integrate water utilisation in health facilities, hence conserving water (Nova, 2023). Through integrating sensors and AI algorithms, water consumption can be constantly checked to ensure that remedial measures can be immediately implemented in case of wastage. The professional analysis also indicates that when applying predictive analytics to flooding and drought affecting healthcare facilities, possible measures can be taken in advance to provide for the necessary water supply. This is facilitated by integrating artificial intelligence in blockchain technology to manage hospital water data. This guarantees the efficiency of the water bill and the observation of water consumption to work towards water conservation within the country.

4.6.4 *Waste Management in Healthcare*

Medical waste can be produced as a byproduct of healthcare activities, contributing to the deterioration of the environment if not well disposed of. AI technologies can make helpful recommendations about identifying materials that can be recycled and improve the methods of sorting medical waste (Kumar et al., 2021). The AI-generated predictive models can predict waste generation trends by healthcare facilities, and ways and mechanisms can be developed to manage the waste effectively. For instance, AI can set out the most appropriate time in which points to collect ashes are most likely needed to ensure the most efficient work. This technology is based on AI in a blockchain structure and guarantees the efficiency of tracking wastes from production to disposal. This system effectively ensures that there is no case of people dumping hazardous wastes or mishandling them, hence making healthcare organisations adopt ethical means and the right ways of handling waste.

Thus, the described usable examples reveal the positive impact of AI-driven technologies in enhancing environmental sustainability within the healthcare field. These ideas revolve around AI, where the overall efficiency improvement related to energy distribution and resource management exists through novel ideas to solve more significant environmental problems within the industry, such as improving water and waste management solutions through AI and blockchain technologies. Thus, by implementing these technologies, healthcare facilities can decrease their negative impact on the environment, increase efficiency in functioning, and become environmentally friendly.

4.7 AI-Driven Environmental Monitoring and Data

Healthcare is experiencing advanced technological amendments that feature AI in analysing and optimising environmental observations and information. This section focuses on using remote sensing, satellite imagery, IoT, and sensor networks and utilising AI-blockchain for secure decentralised intelligent health records.

4.7.1 *Remote Sensing and Satellite Imagery*

Geographical information systems using AI help interpret large-scale environmental data and data gathered from satellites (Himeur et al., 2022). Similarly, in healthcare, AI algorithms analyse the data collected from satellites and give information on the conditions of the healthcare region's environment. For instance, AI may be applied to identify land-use changes, quantify deforestation, and determine the effects of urbanisation on ecosystems from satellite images. This data is helpful for the assessment of such concerns as healthcare facilities' carbon footprint, water consumption and land use. Thus, using such precise and up-to-date information, healthcare facilities, with the help of AI satellite imagery, can optimise their activities and reduce the negative impacts on the environment by employing improved sustainability solutions.

4.7.2 IoT and Sensor Networks

Since IoT devices, including sensor networks, are incorporated into every aspect of a healthcare facility, integrating AI technologies improves real-time data acquisition and evaluation maximally. The interconnected sensors can detect different environmental parameters such as air quality, temperature, humidity, level of pollution, and other necessary information that flows to AI algorithms in constant streams with tendencies and fluctuations (Garrido-Momparler & Peris, 2022). For example, it is possible to use data from air quality sensors to find out which pollutants are present and analyse potential threats to patients' and staff's health. This real-time monitoring ability also allows healthcare facilities to act quickly when dealing with environmental threats, thereby preventing the exposure of patients and facility workers to hazardous health stimuli. Also, AI-advanced analytics helps to fine-tune the efficiency of the HVAC systems, lighting, and other energy-consuming systems to decrease the impact of healthcare organisations on the environment.

4.7.3 AI-Driven Blockchain for Secure and Decentralised Data Management in Healthcare

Integrating AI and blockchain technology presents a reliable and effective solution for properly and securely decentralising healthcare information. Based on environmental data availability, blockchain offers a trustworthy, decentralised, and secure repository to record data. AI takes this system a notch higher by analysing voluminous data from the environment through IoT and sensors, which gives insights into health facilities. Holistically, this integration guarantees that records of air quality, water usage, and energy consumption, amongst other aspects of sustainability, are well stored and can be retrieved anytime for relevant actions to be considered. In addition, the AI-enabled blockchain can also help automate compliance with environmental standards and facilitate reporting requirements, promoting multi-stakeholder engagement in the healthcare organisation. It not only enhances the level of privacy and decentralisation but also encourages the creation of radical initiatives to create sustainable healthcare programs.

It is worth concluding that contemporary technological advances, such as implementing the AI system in environmental monitoring and data gathering, influence the healthcare industry's ability to be sustainable. With the help of the Internet of Things, sensor networks, remote sensing, satellite imagery, and AI based on a blockchain, healthcare facilities will achieve the aims of high accuracy and efficiency in assessing and managing environmental impact. These technologies equip healthcare providers to initiate preventive strategies and increase organisational resilience, besides helping the globe in ecological conservation.

4.8 Challenges and Ethical Considerations

The use of intelligent technologies in environmental sustainability in healthcare delivery has its benefits, but it comes with various issues of concern. As helpful as these technologies are, they carry certain controversies in areas such as privacy, unfair reasoning by artificial intelligence algorithms, power consumption, and laws. Mitigating these areas of risk is critical to ensuring that AI works well and is ethically sound.

4.8.1 Data Privacy and Security

A general issue that often arises when applying AI to enhance environmental sustainability in healthcare (HC) is data privacy and security (Bublitz et al., 2019). HC data should, therefore, be integrated with environmental data to produce a wealth of information concerning patients' health status and the general operation of a health facility. Thus, the protection of this data is something that we can always maintain. Advanced artificial intelligence in healthcare services requires sound encryption and data protection mechanisms to ensure unauthorised individuals or organisations cannot access patient records. Also, since data is being dealt with primarily in the medical sphere, the standards for data privacy, including the USA's Health Insurance Portability and Accountability Act (HIPAA) rules, must be honoured. Appropriate steps in handling data are imperative to ensure and regain the public's confidence in compliance with compliance concerning artificial intelligence's use in monitoring the environment and conserving resources.

4.8.2 Algorithmic Bias and Fairness

AI's high susceptibility to algorithmic bias is a prominent ethical concern when applying AI to healthcare sustainability (Kerasidou, 2021). However, AI models can take prejudices from the datasets through which they learn and thus result in bias, often prejudicial or otherwise. In the case of environmental sustainability, this would lead to a lending bias or predictions and advice damaging to specific groups of people and places. To bring fairness and accuracy, a proper method has to be followed to collect data and then train and evaluate these models. It becomes the challenge and responsibility of healthcare providers to try to use diverse datasets and ensure that techniques to reduce bias in AI are incorporated in performing those algorithms. There is also a need to be open to the processes that the AI systems and models employ in coming up with their decisions; this would help stakeholders understand and have confidence in the decision-making of the said models.

4.8.3 Energy Consumption of AI

On the one hand, the application of AI technologies generates unprecedented advantages for environmental sustainability; on the other, deploying those technologies implies an enormous expenditure in energy, especially regarding data analysis and model training moments. The sustainability of AI applications is one area of concern, as they impact the environment and demand power usage. Maintaining energy efficiency in healthcare settings where AI resources are utilised is crucial, as it mandates the reduction of carbon impact. Some practices like cognitive computing can ask questions and convey information through an inference engine instead of using energy-consuming pathways such as Internet lines. Also, more appropriate choices of energy-efficient hardware and the utilisation of renewable energy sources when powering AI applications help reduce the adverse effects of AI in the healthcare industry on the environment.

4.8.4 Regulatory and Policy Challenges

Using AI and blockchain technologies in healthcare for environmental conservation has regulation-bound conditions (Mennella et al., 2024). An area that may limit the broad use of such technologies may be legal and policy-related challenges. There is a need to regulate AI and Blockchain since they have their peculiarities; however, the regulation should be done correctly using the technology and its merits. While compliant with current laws, healthcare providers can also help spearhead the making of favourable policies that allow sustainable practices in the future. This includes issues as to who owns data, who is liable for it, and the ethicality of AI. Thus, integrating healthcare stakeholders, including physicians, policymakers, and developers of AI technologies, is needed to foster a favourable environment for the operation of eco-friendly AI solutions for healthcare and respect the rights of patients and the environment.

Therefore, to increase the efficiency of AI technology use, it is necessary to focus on the issues and ethical implications of its application in healthcare sustainability. Data privacy and protection, avoiding algorithmic bias, managing power consumption, and complying with existing legislation are the critical factors to follow while implementing AI with high efficiency and morality. Thus, the regular management of the mentioned challenges will help healthcare providers take advantage of AI to increase the efficiency of healthcare activities and achieve large-scale environmental objectives.

4.9 Future Directions

Self-driving cars and robots gradually develop new AI and climate change integration opportunities. These opportunities are being searched for individually and jointly with healthcare. They present the chance to build on previous developments to increase the sustainability of healthcare delivery and operations, lower the impact on the environment, and provide better care to patients.

4.9.1 Emerging AI Technologies

Quantum computing is considered one of the newest revolutionary ideas in AI and adds value to environmental modelling in healthcare. In contrast to classical computers, quantum computers can solve many problems with significant amounts of data at incredibly high speeds to create more elaborate simulations. Applying a quantum computed approach in the provision of healthcare, the new healthcare environment improvement models may influence pollution levels, changes in climate, and consumption levels, amongst others. For instance, quantum algorithms could be used to determine the best approaches towards energy efficiency in a hospital by modelling different options. Furthermore, quantum computing applies to improving the methods used in outbreak predictions concerning diseases associated with the environment and, in turn, helps create better and faster strategies in public health systems.

4.9.2 Cross-Disciplinary Collaboration

The adoption of AI in combination with other advanced technologies, including blockchain and the IoT, is considered to open new opportunities for improving the sustainability of healthcare. The synergistic operation of these technologies requires the integration of multiple disciplines. For instance, the integration of AI and IoT can result in intelligent health sectors whereby devices and sensors are connected and constantly check on the usage rates of resources. There is, however, the added application of blockchain security and data transparency, which would allow feasible and reliable management of environmental resources as data is collected. Cross-disciplinary initiatives have been assessed as vital for identifying interrelated issues and proposing linked and interrelated strategies for improving the sustainability of healthcare. These collaborations can result in use cases like a blockchain-based IA platform for monitoring the usage of renewable energy or an IoT-based system to monitor the environment in healthcare facilities in real-time.

4.9.3 *Policy and Regulation*

By suggesting the need to embrace AI for sustainability in healthcare, there is a need to have policies and regulations that support innovation and the use of advanced technology in healthcare provision while meeting the set ethical standards and gaining the public's trust. Governments are critical in creating conditions for applying artificial intelligence systems. This involves entailing rules governing data protection and security, AI governance and transparency, and encouraging sustainability by providing grants, subsidies or tax credits. For instance, healthcare policies could require adopting AI systems for energy control in ample health facilities or include grants for developing AI solutions to lessen the effects of healthcare activities on the climate. In this regard, policymakers must set enhanced AI rules, regulations, and initiatives to improve runs for a sustainable healthcare environment.

Therefore, further development of AI in environmental sustainability in healthcare is quite promising, as new technologies and joint efforts are expected to bring considerable shifts. Ecological modelling, resource optimisation via quantum computing, and combining AI and blockchain/IoT technology result in more innovative and secure health facilities. Multidisciplinary work and good policies would be needed to make the best out of such innovations so that adopting artificial intelligence can contribute to the positive transformation of the globe's healthcare. When healthcare providers and policymakers accept these future directions, they will set the pace to improve the health delivery system's efficiency, sustainability, and effectiveness.

4.10 Conclusion

This chapter has analysed how AI can transform the HC industry's sustainability by addressing environmental issues. Some points to note are the analytical solutions with machine learning, deep learning, and natural language processing as AI technologies in energy use efficiency, biomedical waste management, and environmental policy analysis. The combination of AI and blockchain provides end-to-end solutions for increasing trustworthiness, accountability, and transparency in the healthcare industry, focusing on domains like energy consumption, waste disposal, and carbon credits. Furthermore, other advanced AI technologies, such as edge computing, swarm intelligence, and digital twins, provide novel strategies for time-sensitive tracking, resource management, and evaluating the ecosystem's effects in healthcare organisations.

AI holds great potential for influencing the improvement of environmental sustainability in healthcare. It has proven that integrating AI and similar technologies can bring about a drastic change prevalent in the healthcare industry by lessening the ecological impact, raising verdant proficiency, and promoting sustainability. Implementing a system with artificial intelligence allows healthcare providers to acquire valuable information from large databases and make decisions that may bring good compromise between the influence and the need to save the environment and guarantee the optimum level of patients' treatment. In addition, when combined with other emerging technologies like IoT and blockchain, superiority enhanced by AI results from multiplicative interactions that enhance the benefits of sustainable practices. These advances help make the healthcare industry environment-friendly and help to have better health returns through healthier surroundings, be it patients, clients, or healthcare workers.

Thus, the present chapter offers a call to arms for researchers, policymakers, and healthcare practitioners to fully support and promote the presence of AI for environment enhancement. Specifically, researchers are prompted to expand the current state of AI-based solutions and develop new ideas to foster HC sustainability. The authorities should establish appropriate legal conditions and favourable stimuli for AI technologies' implementation while maintaining non-negotiable ethical and data protection guidelines. Healthcare practitioners should always consider embracing techniques driven by artificial intelligence to try to minimise their impact on the environment constantly. The result of these stakeholders' cooperative effort is enhanced utilisation of AI with meaning and purpose to transform the healthcare system to one that is sustainable and resilient.

Therefore, AI-related technologies provide a robust set of tools to counteract the problems in environmental management in the healthcare industry. AI applied to strategic change in management, and cross-disciplinary involvement has the potential to propel sustainable improvement of healthcare systems. This proves advantageous for the environment, the quality of care, and the operational functioning of healthcare facilities. It is a continuous process to have a sustainable applied healthcare industry, and AI will have a critical function.

References

Abdel-Jaber, H., Devassy, D., Al Salam, A., Hidaytallah, L., & EL-Amir, M. (2022). A review of deep learning algorithms and their applications in healthcare. *Algorithms, 15*(2), 71.

Abirami, S., & Chitra, P. (2020). Energy-efficient edge-based real-time healthcare support system. *Advances in Computers, 117*(1), 339–368.

Bamakan, S. M. H., Malekinejad, P., & Ziaeian, M. (2022). Towards blockchain-based hospital waste management systems; applications and future trends. *Journal of Cleaner Production, 349*, 131440.

Bublitz, F. M., Oetomo, A., Sahu, K. S., Kuang, A., Fadrique, L. X., Velmovitsky, P. E., Nobrega, R. M., & Morita, P. P. (2019). Disruptive technologies for environment and health research: An overview of artificial intelligence, blockchain, and internet of things. *International Journal of Environmental Research and Public Health, 16*(20), 1–24.

Dewasiri, N. J., Baker, H. K., Banda, Y. K. W., & Rathnasiri, M. S. H. (2022). The dividend decision model: A possible solution for the dividend puzzle. In S. Rana, Sakshi, J. Singh (Eds.),

Exploring the latest trends in management literature. Review of Management Literature, vol. 1 (pp. 249–267). Emerald Publishing Limited, Leeds. https://doi.org/10.1108/S2754-586520220000001013
Dewasiri, N. J., Karunarathne, K. S. S. N., Menon, S., Jayarathne, P. G. S. A., & Rathnasiri, M. S. H. (2023). Fusion of artificial intelligence and blockchain in the banking industry: Current application, adoption, and future challenges. In A. Saini, & V. Garg (Eds.), *Transformation for sustainable business and management practices: Exploring the spectrum of industry 5.0* (pp. 293–307). Emerald Publishing Limited. https://doi.org/10.1108/978-1-80262-277-520231021
Dewasiri, N. J., Dharmarathna, D. G., & Choudhary, M. (2024). Leveraging artificial intelligence for enhanced risk management in banking: A systematic literature review. In R. Singh, S. Khan, A. Kumar, & V. Kumar (Eds.), *Artificial intelligence enabled management: an emerging economy perspective* (pp. 197–213). De Gruyter. https://doi.org/10.1515/9783111172408
Ekblaw, A. C. (2017). *MedRec: Blockchain for medical data access, permission management, and trend analysis*. Doctoral dissertation, Massachusetts Institute of Technology.
El-Bouri, R., Taylor, T., Youssef, A., Zhu, T., & Clifton, D. A. (2021). Machine learning in patient flow: A review. *Progress in Biomedical Engineering, 3*(2), 022002.
Garlík, B., & Přívětivý, J. (2024). Artificial intelligence algorithms for prediction and diagnosis of air pollution affecting human health. *Journal of Physics: Conference Series, 2701*(1), 012072. IOP Publishing.
Garrido-Momparler, V., & Peris, M. (2022). Intelligent sensors in environmental/water quality monitoring using IoT and cloud services. *Trends in Environmental Analytical Chemistry, 35*, e00173.
Haleem, A., Javaid, M., Singh, R. P., & Suman, R. (2022). Medical 4.0 technologies for healthcare: Features, capabilities, and applications. *Internet of Things and Cyber-Physical Systems, 2*, 12–30.
Himeur, Y., Rimal, B., Tiwary, A., & Amira, A. (2022). Using artificial intelligence and data fusion for environmental monitoring: A review and future perspectives. *Information Fusion, 86*, 44–75.
Jayawardena, N. S., & Dewasiri, N. J. (2023). Food acquisition and consumption issues of South Asian countries: A systematic literature review and future research agenda. *FIIB Business Review*. https://doi.org/10.1177/23197145231194113
Kahwash, F., Barakat, B., Taha, A., Abbasi, Q. H., & Imran, M. A. (2020). Optimising electrical power supply sustainability using a grid-connected hybrid renewable energy system—an NHS hospital case study. *Energies, 14*(21), 7084. https://doi.org/10.3390/en14217084
Karunarathna, K. S. S. N., Dewasiri, N. J., Singh, R., & Rathnasiri, M. S. H. (2024). What does artificial intelligence–powered ChatGPT bring to academia? A review. In R. Singh, S. Khan, A. Kumar, & V. Kumar (Eds.), *Artificial intelligence enabled management: An emerging economy perspective* (pp. 3–13). De Gruyter. https://doi.org/10.1515/9783111172408
Kerasidou, A. (2021). Ethics of artificial intelligence in global health: Explainability, algorithmic bias and trust. *Journal of Oral Biology and Craniofacial Research, 11*(4), 612–614.
Kulkov, I., Kulkova, J., Rohrbeck, R., Menvielle, L., Kaartemo, V., & Makkonen, H. (2024). Artificial intelligence-driven sustainable development: Examining organisational, technical, and processing approaches to achieving global goals. *Sustainable Development, 32*(3), 2253–2267.
Kumar, V., Azamathulla, H. M., Sharma, K. V., Mehta, D. J., & Maharaj, K. T. (2023). The state of the art in deep learning applications, challenges, and prospects: A comprehensive review of flood forecasting and management. *Sustainability, 15*(13), 10543.
Kumar, N. M., Mohammed, M. A., Abdulkareem, K. H., Damasevicius, R., Mostafa, S. A., Maashi, M. S., & Chopra, S. S. (2021). Artificial intelligence-based solution for sorting COVID-related medical waste streams and supporting data-driven decisions for competent circular economy practice. *Process Safety and Environmental Protection, 152*, 482–494.
Li, M., Jiang, Y., Zhang, Y., & Zhu, H. (2023a). Medical image analysis using deep learning algorithms. *Frontiers in Public Health, 11*, 1273253.
Li, J., Herdem, M. S., Nathwani, J., & Wen, J. Z. (2023b). Methods and applications for artificial intelligence, big data, internet of things, and blockchain in smart energy management. *Energy and AI, 11*, 100208.

MacNeill, A. J., McGain, F., & Sherman, J. D. (2021). Planetary health care: A framework for sustainable health systems. *The Lancet Planetary Health, 5*(2), e66–e68.
Macpherson, C. C., Smith, E., & Rieder, T. N. (2024). Does health promotion harm the environment? In E*nvironmental bioethics* (pp. 13–30). Routledge.
Mascarenhas, M., Ribeiro, T., Afonso, J., Mendes, F., Cardoso, P., Martins, M., Macedo, G., et al. (2023). Smart endoscopy is greener endoscopy: Leveraging artificial intelligence and blockchain technologies to drive sustainability in digestive health care. *Diagnostics, 13*(24), 3625.
Mennella, C., Maniscalco, U., De Pietro, G., & Esposito, M. (2024). Ethical and regulatory challenges of AI technologies in healthcare: A narrative review. *Heliyon, 10*(4), e26297.
Metzger, A., Kley, T., & Palm, A. (2020). Triggering proactive business process adaptations via online reinforcement learning. In *International conference on business process management* (pp. 273–290). Springer International Publishing.
Mishra, U., Rathnasiri, M. S. H., & Dewasiri, N. J. (2024). Net-zero economy and energy sustainability: A Nepalese perspective. In R. Singh, S. Khan, A. Kumar, S. Luthra, H. Chokshi (Eds.), *Net zero economy, corporate social responsibility and sustainable value creation.* CSR, Sustainability, Ethics & Governance (pp. 145–161). Springer. https://doi.org/10.1007/978-3-031-55779-8_9
Mohamed, Y. A., Khanan, A., Bashir, M., Mohamed, A. H. H., Adiel, M. A., & Elsadig, M. A. (2024). The impact of artificial intelligence on language translation: A review. *IEEE Access, 12*, 25553–25579.
Nova, K. (2023). AI-enabled water management systems: An analysis of system components and interdependencies for water conservation. *Eigenpub Review of Science and Technology, 7*(1), 105–124.
Odonkor, S. T., & Mahami, T. (2020). Healthcare waste management in Ghanaian hospitals: Associated public health and environmental challenges. *Waste Management & Research, 38*(8), 831–839.
Ramachandran, M. (2024). AI and blockchain framework for healthcare applications. *Facta Universitatis Series: Electronics and Energetics, 37*(1), 169–193.
Rathnasiri, M. S. H., Dewasiri, N. J., & Kumar, A. A. (2024). Policy framework and implementation strategies for Sri Lanka's transition to a net-zero economy. In R. Singh, S. Khan, A. Kumar, S. Luthra, H. Chokshi (Eds.), *Net Zero economy, corporate social responsibility and sustainable value creation.* CSR, Sustainability, Ethics & Governance (pp. 43–60), Springer. https://doi.org/10.1007/978-3-031-55779-8_3
Reddy, C. O., Doolam, B., Reddy, N. R., Surabhi, D., Kambhampati, J. M., Nagamanju, S., & Rajagopal, K. (2024). Exploring the potential of biomedical waste management: Transforming challenges into opportunities for sustainable solutions. In *From waste to wealth* (pp. 1179–1206). Springer Nature.
Ribino, P., Ciampi, M., Islam, S., & Papastergiou, S. (2022). Swarm intelligence model for securing healthcare ecosystem. *Procedia Computer Science, 210*, 149–156.
Roy, K., Debdas, S., Kundu, S., Chouhan, S., Mohanty, S., & Biswas, B. (2021). Application of natural language processing in healthcare. *Computational Intelligence and Healthcare Informatics*, 393–407.
Štuikys, V., & Burbaitė, R. (2024). Introduction to artificial neural networks and machine learning. In *Evolution of STEM-driven computer science education: The perspective of big concepts* (pp. 311–346). Springer Nature Switzerland.
Tao, F., Xiao, B., Qi, Q., Cheng, J., & Ji, P. (2022). Digital twin modelling. *Journal of Manufacturing Systems, 64*, 372–389.
Wu, X., Xu, D., Yuan, S., & Zhang, L. (2022). Fair data generation and machine learning through generative adversarial networks. *Generative adversarial learning: Architectures and applications* (pp. 31–55). Springer International Publishing.
Yu, W., Liang, F., He, X., Hatcher, W. G., Lu, C., Lin, J., & Yang, X. (2017). A survey on the edge computing for the Internet of Things. *IEEE Access, 6*, 6900–6919.

Zhu, R., Tu, X., & Huang, J. (2020). Using deep learning-based natural language processing techniques for clinical decision-making with EHRs. *Deep Learning Techniques for Biomedical and Health Informatics*, 257–295.

Zunic, A., Corcoran, P., & Spasic, I. (2020). Sentiment analysis in health and well-being: Systematic review. *JMIR Medical Informatics, 8*(1), e16023.

Chapter 5
The Influence of Artificial Intelligence on the Healthcare Industry in Relation to Sustainability

Mohamed Irfan Shaikh, Jacinta Dsilva, and Fahad Ziaul Ain Usmani

Abstract The rapid advancement of artificial intelligence (AI) technologies is significantly transforming the healthcare industry, particularly in the realm of sustainability. This study investigates the profound impacts of AI on healthcare operations, emphasising sustainable development. It explores how AI-driven innovations such as robotics, deep learning, and predictive analytics are revolutionising clinical management systems, patient engagement, chronic care models, and clinical trials, thereby enhancing efficiency and outcomes. Despite AI's potential to reduce operational costs and improve service delivery, its integration poses challenges, particularly concerning environmental sustainability and long-term viability. The healthcare sector's transition towards AI encompasses the adoption of digital solutions aimed at achieving the World Health Organization's (WHO) goals, such as universal health coverage and reduced global health expenditure. The research underscores the need for sustainable AI solutions that align with the principles of the circular economy and social well-being. It also highlights successful case studies demonstrating AI's role in analysing large data volumes, improving decision-making processes, and developing self-learning algorithms. In the short term, the study aims to disseminate findings to the scientific community and identify opportunities for further research and partnerships. In the long term, it envisions AI refining healthcare management and contributing to a sustainable future where social well-being is prioritised. This study emphasises the importance of ethical and legal considerations, data security, and the need for a balanced approach to AI adoption in healthcare to ensure sustainability and equity. The findings provide a comprehensive understanding of AI's potential and challenges in the healthcare sector, offering insights for future research and policy development.

Keywords Artificial intelligence · AI · Healthcare · Sustainability · Ethical considerations · Sustainable development · Healthcare management

---

M. I. Shaikh · J. Dsilva (✉) · F. Z. A. Usmani
SEE Research Centre, SEE Institute, The Sustainable City, Dubai, United Arab Emirates
e-mail: jacinta@seeinstitute.ae

R. Singh et al. (eds.), *Transforming Healthcare Sector Through Artificial Intelligence and Environmental Sustainability*, Approaches to Global Sustainability, Markets, and Governance, https://doi.org/10.1007/978-981-97-9555-0_5

5.1 Introduction

The growing demand for sustainable systems has been the main challenge for many nations worldwide. Despite considerable progress in the field of sustainability, many growing nations still struggle to make use of a resource-efficient economic system to resolve the problem of sustainable development. Human society is currently grappling with an environmental crisis that is causing a massive transformation in the healthcare system. With increased attention on sustainable growth in healthcare over the last two decades, the question is: what does the paradigm shift in the healthcare sector indicate in terms of allowing sustainable growth with a focus on artificial intelligence (AI)? (Di Vaio et al., 2020; Palomares et al., 2021).

The healthcare industry is undergoing significant transformation with the advent of a new generation of analytics, like those in use throughout various industrial sectors, that rely on AI and machine learning to drive additional insights from healthcare data. Predictive technologies can be utilised to anticipate events, determine needs, and drive value in a variety of other industries. Recent improvements in predictive technologies in healthcare, such as efforts in several health systems to predict which patients are at risk for cardiac arrest, sepsis, or other conditions, provide hope that a more innovative, more proactive healthcare model is possible. This article provides an analysis of the effect of AI solutions centred on healthcare in view of a sustainable scale of standards (Patil & Shankar, 2023; Shaheen, 2021).

5.1.1 Background and Significance

The process of how technology has penetrated healthcare, with an emphasis on the surgical department, has attracted the attention of modern scientists. The integration of robotics into medical care has transformed various fields, including surgery, from a perspective of sustainability. The design of the robotic type of surgery meets all three critical roles of sustainability. However, the role of AI in the sustainability of surgery has received minimal focus from scholars (Secinaro et al., 2021). Understanding the current impact of AI in medicine and healthcare can be a flagship for the fruitful and sustainable development of AI's role in healthcare in terms of profit, environment and society. Therefore, there is an urgency for scholars to have an express study on the impact of AI in healthcare in the light of sustainability. Due to the above-mentioned encouraging reasons, the researchers are motivated to carry out the current study (Amann et al., 2020).

Significance of the study: Recently, the healthcare sector has been described as the 'gold mine' for AI. In several countries, significant public health organisations and healthcare partners have presented their AI strategies and a range of AI interventions in the different operational sectors of healthcare. This implies that artificial intelligence (AI) is making a significant impact on the healthcare sector today. To investigate the impact, academic literature with experimental research was consulted;

this led to the study of the link between AI and healthcare (Alowais et al., 2023). The healthcare provider community and conglomerates have tried and developed better clinical management systems with the assistance of AI robotics and deep learning, which include better patient engagement with those AI services, enhanced chronic care data models, efficient clinical trial models, and participant staple for the evident proofs, more comprehensive drug outcome models, and continued registrations that save (Ali et al., 2023).

The term "sustainability" is an umbrella that is mainly described in companies and economic development. Healthcare sustainability is undoubtedly a "necessary endpoint for healthcare and community well-being," which is getting increasing interest among healthcare organisations with an emphasis on a strategic plan to ascertain long-term viability (Sterling et al., 2020). The healthcare sector around the globe is constantly transitioning, and part of the world has had to cope with societal challenges due to the increasing stress of diseases on a regular basis. The healthcare sector is similarly evolving into technological advances, accompanied by rising expenses, resulting in a group paying more than half the world's total healthcare expenditure. There is more possibility of rising costs associated with technological advances in the healthcare sector. Research has demonstrated a significant reduction in costs for surgeries using robotics, despite the function as a CSF, that the surgery market is changing due to changes in consumer demand in terms of costs, risk, and convenience and that new CSFs will develop for this service in the longer term. Regulation and consumer behaviour, however, have potential short-term impacts on demand. Consequently, a carefully controlled introduction of robotics to cater to specific consumer segments may enable sufficient demand to be generated to compensate for costs in one or two years (Macpherson et al., 2024; Palozzi et al., 2020).

5.1.2 Research Aim and Objectives

This paper aims to present the main findings of an interdisciplinary research project driven by the specific context in the healthcare sector, where companies demand innovative solutions that address both business needs, as well as the increasing expectations of the population and social stakeholders about health-related services. Furthermore, it is relevant to enable the positioning of current and future health sector managers regarding sustainable solutions that optimise processes and result in qualitative and quantitative improvements using information and artificial intelligence—both of which are expensively developed by the health industry itself. The integration of sustainable digital solutions in the health sector also directly contributes to the World Health Organization's goals of reducing two-fold the increase in global health spending by 2019 and ensuring that an additional 1 billion people benefit from universal health coverage by 2023 (Alowais et al., 2023). The study identifies and disseminates successful cases of research projects in the healthcare sector that have used sophisticated tools to analyse large volumes of data used by different companies in the different links in the healthcare chain to gain insight, improve decision-making processes, and develop self-learning algorithms (Alami et al., 2020; Secinaro et al., 2021). From a long-term perspective, the study aims to present a critical view of how artificial intelligence can be used to refine the management in the healthcare sector and how healthcare contributes to the emergence of a sector of the future where we will prioritise social well-being based on concepts of sustainability and circular economy (Ali et al., 2023).

5.2 Artificial Intelligence in Healthcare

Artificial intelligence (AI) refers to how computers are programmed to perform tasks that usually require human intelligence. These tasks include decision-making, speech recognition, and language translations. When this is applied within a healthcare context, it is referred to as 'AI in Healthcare' or 'AI Health' (Singh & Hamid, 2022). The types of artificial intelligence most cited in the context of healthcare begin with the definition of four types of artificial intelligence. These are known as reactive machines (meaning that the AI system is programmed to respond to specific inputs, but it has no memory, which entails no ability to use past experiences to inform future decisions); those with limited memory (these AI systems can use past experiences to inform current decisions); those with the ability to detect human emotion; and self-aware artificial intelligence (Bohr & Memarzadeh, 2020).

In healthcare, this can be seen in a variety of applications, such as in robot-assisted surgeries and the detection of fraud in insurance claims. One of the most popular uses of AI in health is in the early detection of disease, especially in medical imaging. In addition, AI is often used in virtual health assistance as well as in automated external defibrillators (Castiglioni et al., 2021). A vast amount of research has been conducted delineating the impact of AI on the healthcare sector, focusing predominantly on providing a complete profile of how AI is used, the status of its use, and how effective it is (Zhou et al., 2021). However, most existing research predominantly investigates high-income countries.

5.2.1 Definition and Types of AI in Healthcare

Artificial intelligence for healthcare (AI4H) refers to the use of machine learning methods to identify relationships in observational clinical data and effectively perform the tasks of improving clinical care (Ruggerio, 2021). AI4H is not a single technology; instead, it is a result of combining tools from computer science, statistics, evidence-based medicine, information retrieval, informatics, and other disciplines to create a new class of algorithms (Dzhengiz & Niesten, 2020). Machine learning, a collection of data-driven tools and techniques, lies at the heart of this innovation. Rapid progress in this area has prompted a wave of research on the subject and has raised the possibility of using AI algorithms in various clinical and non-clinical scenarios (Van Wynsberghe, 2021). The application of AI in healthcare has long been neglected; however, its potential to assist both medical professionals and individual patients in clinical diagnosis, healthcare management, and public health is now well established (Ullah & Garcia-Zapirain, 2024).

Artificial intelligence for the healthcare (AI4H) domain has a pivotal role in enhancing resource utilisation, reducing medical assistance and treatment costs, improving patient prognosis, and enhancing the well-being of the surrounding environment. By making the healthcare sector even more efficient and equitable, AI can assist with maintaining sustainable services from a social point of view (Oala et al., 2021). AI4H is a mixture of various informatics subfields such as imaging, genomic, and clinical informatics and computational approaches like machine learning and knowledge discovery paradigms specifically tailored to this application domain.

5.2.2 Applications of AI in Healthcare

The use of AI technologies is seen as inevitable for the utilisation of large data sets in modern healthcare systems in which sustainable costs and an efficient but precise provision of services combined with high customer satisfaction are the focus. In healthcare, these technologies can be categorised into two major groups: diagnostic and psychological. The American Food and Drug Administration (FDA) has articles about the classification of computer-aided detection (CADe) as well as computer-aided diagnosis (CADx) (Zhu et al., 2022).

First, the CADe systems mark or highlight the positions in an image in which there could be a specific type of lesion or tumour. Second, the CADx systems provide a so-called diagnosis by just analysing the findings from CADe, making use of machine learning (ML) and its particular subset deep learning (DL) algorithms like convolutional neural networks (CNN) (Sujith et al., 2022). It can be observed that deep learning is mainly used when actual decision-making, suggestions, or conclusions are being made. For example, DeepMind, which was acquired by Google in 2014, created an ophthalmology diagnostic. Moreover, IBM Watson Health uses AI to interpret data with the purpose of building a database of medical evidence (Castiglioni et al., 2021). A general commercialised application is in a chatbot on presenting symptoms or in post-surgery recovery. This psychological AI, which touches the patient directly, is more sensitive than the diagnostic AI. This will then shape the development of the framework of a sustainable healthcare AI.

5.2.3 Sustainability in Healthcare

The term sustainability has always been used by economists, environmentalists, and political analysts in the context of corporate strategies or governments' future.

When the idea is translated to the healthcare sector, the understanding becomes about sustainable financing or any vertical perspective that is about resourcing. However, the fact is that the actual definition of sustainability is the capacity of something to be maintained or to sustain itself (Luna-Nemecio et al., 2020). It' is about how to generate more access and availability with the same resource or less. Furthermore, when it comes to healthcare, it' is about delivery and operations.

In healthcare, the World Health Organization (WHO) constitution declares health as 'A state of complete physical, mental, and social well-being and not merely the absence of disease or infirmity'. This also answers the question of coverage, which is also the determinant of availability effect sustainability (Billings et al., 2021). The availability of specific interventions and broader programs for the appropriate level of care should not only be present but should be consistently and constantly available when needed. The quality of the care against effectiveness and safety that is delivered must be evaluated by a mix of both outcome, process and performance indicators. Forward on this ideology and beyond, the practice of customising healthcare to all its stakeholders is evaluated. Preventive measures, mental health, community medicine, patient rights, and the enabling of patients and caregivers have been designed for the individual (Tanwar et al., 2020).

It' is not only about human resources, budgets, equipment, or facilities. Traditionally, economic development, industrial growth, and infrastructure developments have been predicated on a future whose gross national product or inflation rate is determined and forecasted based on past experiences. In the context of health, how we have delivered and operated health systems is predicated upon the availability of resources and their anticipated requirement based on the demographic transition (Haldane et al., 2021). The models of health services and the level of care, the pattern of urbanisation, and the distribution of services are all based on these projections.

5.3 Concept and Importance of Sustainability in Healthcare

The term sustainability in healthcare is essential, especially with the current changes in population and healthcare systems. Although some countries use the term more frequently, different aspects and scopes of sustainability in healthcare have been presented. The WHO proposed the framework of 'Health in all Policies', which confirms that health, environment, and socio-economic are sustainable development goals and should not be looked at alone (Greer et al., 2022). In 2019, the European Union announced a 'One Health' research cluster that covers health-related settings for humans, animals and the environment and is convincing as a green paper for the future of 'Food 2030', probably focusing on the sustainability of food (diet) and agriculture next (Ståhl & Koivusalo, 2021). All these efforts confirm the need for a transdisciplinary approach to sustainability in healthcare. Hence, for the assessment of IA's impact on healthcare, the Smart Process follows the pathway of 'What is the intersection concept between healthcare and sustainability?' This will help set the basis for the importance of adopting sustainability principles when developing AI tools in healthcare settings.

5.3.1 Challenges and Opportunities

Developing and managing sustainable healthcare involves several challenges. First, operational challenges suggest that investment in healthcare facilities, as well as transportation of personnel and patients in some instances, can lead to increased pollution. This problem arises because both the demand and supply of individual healthcare will increase the need for transportation (Fuller et al., 2022). Second, clinical challenges reveal that progress in curing certain diseases has led to an increase in life expectancy. This, in turn, creates a population that is not only growing in number but also includes older people (Eckelman et al., 2020). The expansion of this population boom requires an increase in health and social facilities and services, such as long-term care. Third, community health challenges indicate how it is predicted or expected that direct health impacts of global climate change will include an increase in fatalities and injuries in severe weather events (Mujtaba & Shahzad, 2021).

This issue is fundamental since it is consistent with the concern that the current trend of increased emissions and environmental degradation could make humanity sick in the future. Furthermore, delays in the mitigation measures necessary to provide healthcare activities may potentially require more resources in the future. Managing this expansion using the current healthcare system and practices can be seen as unsustainable. For this reason, sustainability should also always be associated with the control and management of territory as related to the particular social environment of healthcare sectors. However, in terms of opportunities, sustainability not only reveals or explores the concern for social good (e.g. care) but also reveals that manners of demonstrating responsible decision-making resonate within its social institution (Alowais et al., 2023). It concerns the way health care is managed, not just the departmental vision of care. Because of that, practice in healthcare is essential to monitor and manage the way resources are used. This practice is fast becoming the role of hospital managers and possibly care services alongside clinical governance (Bohr & Memarzadeh, 2020). These challenges and opportunities for sustainability present several critical questions for academics and practitioners within specific sectors. Considering these challenges and opportunities, the central objective of this paper is to discuss the roles of technology in healthcare delivery, with a focus on the use of artificial intelligence (AI) in the diagnosis of bone age.

5.4 Intersection of AI and Sustainability in Healthcare

There is a fundamental intersection of 'Artificial Intelligence' and 'Sustainability' in the healthcare sector to achieve universal healthcare, including social, operational, and fiscal sustainability. The purpose of this article is to investigate research studies that explain the potential benefits, challenges, and suitability of artificial intelligence with respect to the United Nations' Sustainable Development Goals (SDGs) and the need for sustainability in healthcare. We conduct a critical, evidence-informed, thematic analysis of peer-reviewed articles on AI and healthcare issues. We conclude that AI has the ability not only to propel advanced technologies into healthcare functions but also to achieve enhanced delivery models for providers and patients. However, AI faces ethical, moral, and legal challenges related to data security, privacy and data sharing, transparency, and operational issues. To this end, we call for more interdisciplinary research to undertake AI-induced dynamic performance, the economic sustainability of AI adoption in practice, the sensitivity of market valuation towards AI-induced benefits and building sustainable ecosystems.

Even though the artificial intelligence-sustainability paradigm in healthcare is extensively debated and advancing well with mutual technological development and enhancement in various processes and systems, the development also shares abutments in terms of ethical emphasis in defining the change or transformation (Alami et al., 2020). The aim of AI in healthcare serves as the cornerstone for equity, accessible healthcare delivery, affordability, and continuity of care as mandated in the healthcare's commitment towards sustainability, with modern ICT infrastructures enabling communication between disparate machines and systems and ensuring the data integrity, privacy, and confidentiality of patients. This realisation of modern healthcare delivery has seen digital networks merging seamlessly with healthcare infrastructure in innovative hospital and city initiatives in line with existing standards, such as ISO-27000, to follow approved and certified knowledge management controls (Karimian et al., 2022).

5.4.1 Enhancing Healthcare Delivery and Efficiency

Given the immense amount of electronic health data that can be directly accessed, AI applications can be developed to deliver necessary, enhanced, and augmented healthcare. These applications, in a wide range of scenarios, can improve the speed while maintaining or improving the quality and safety of healthcare delivery (Shelmerdine et al., 2022). Similarly, in scenarios ranging from radiology to cardiology, AI can deliver high-quality diagnostics. Early research demonstrated that AI was more adept at diagnosing wrist fractures than experienced radiologists (Tobler et al., 2021). To prioritise the attention of the radiologists providing the diagnosis, the AI system was designed as a decision-support tool. Overall, this methodology indicates that AI can be leveraged to enhance speedy accessibility to healthcare in settings where there are too few specialists to keep pace with the demand. GPs were supported in the UK by AI chat box applications during the COVID-19 pandemic, with patient triage enabling infected patients to be isolated and tested earlier.

One of the most visible uses of AI in healthcare is chatbots that play a role across the 'health continuum', from telemedicine to support such needs as drug prescription concerns or therapy management. In providing 24/7 real-time support for consumers of health products and services, chatbots can augment human service delivery systems while reducing the non-essential demand for 'human' health workers. Chatbots are a readily accessible tool that is already improving access to immediate health support in mature and emerging economies (Xu et al., 2021). AI-based optimisation approaches have also been applied to a variety of frontline emergency medical treatment tasks to improve healthcare delivery in the context where resources are limited. Similarly, several models to improve emergency room performance, based on AI algorithms, are currently being researched and refined, while the use of smartphones and AI to predict the blood alcohol concentration of methanol-poisoned emergency patients to aid medical staff and conserve vital resources has produced promising results (Aggarwal et al., 2023). International research is also exploring the potential rate of acceptance of AI in the emergency room medical decision-making process.

5.4.2 Reducing Environmental Impact

The new technologies, especially artificial intelligence, are being used by hospitals to conduct waste audits to determine what is or is not being recycled. The results of an audit can mean the difference between a hospital wasting a $700 a month garbage haulier fee when as much as 17% of their 'garbage' is discovered to be recyclable or redirecting the payment to buy items needed most—perhaps fresh fruit for a paediatric ward that has banned all sweet things from the premises (Hsu et al., 2021). This will lead hospitals to have a lower environmental impact and can be regarded as sustainable themselves. Based on the model, over 100,000 lives per year could be improved if companies aligned their interest to actively work in the direction of the United Nations' Sustainable Development Goals (SDGs) using AI to identify the material issues that they are well positioned to address.

It cannot be denied that the entire population across the globe is increasingly expanding, and the demands for healthcare facilities have exploded in recent years, resulting in boosted energy consumption and a large amount of hazardous waste, leading to a heavy carbon footprint (Holman, 2020). Recent trends integrate more sustainable practices and operations in healthcare management. In such a context, AI should become the most suitable technology for healthcare operations in general to decrease the total carbon footprint. The present paper continues earlier analyses dealing with AI in the healthcare systems. The goal is to examine to what extent the analytical choices impact the overall system layout and on the reduction in carbon emissions (Filip et al., 2022).

An article mentioned four real-world examples of the applications of artificial intelligence in the healthcare sector within the framework of SD to show the usefulness of these technologies in the future. One such study involves interpretations of brain trauma and trauma in imaging. The importance of AI in pioneering early warning systems, which can be used to identify—as possibly deceive—early warning signs using deep learning techniques for brain injuries and assist clinicians in making the appropriate decisions with an increase in the clinical outcome for patients (Yuan et al., 2020). In addition, the early warning detection of trauma in imaging is another study that involves artificial intelligence (Kang et al., 2020; Wells et al., 2024).

The text insists on the issue of sustainability, which is seen as a significant tool to represent an ethical and social stance. One could argue that since the interest and focus of the article were on sustainability rather than artificial intelligence, the technologies were mentioned only fleetingly in the additional information. Regards the three primary biases present in artificial intelligence (Galaz et al., 2021). Each bias contributes to the detrimental effects of AI and causes a loss of resources as well as negative externalities that can be deemed unsustainable. Not only are biases perceived as threats to regulatory financial integrity, but they also pose a risk to both the privacy of the data and the reputation of the company or government using AI (Roberts et al., 2024). Providing AI with large amounts of information, in addition to failing to unify all potentially disjointed sources, affects the accuracy and consistency of the information and possibly deteriorates the reliability of the data, making it more complex to spot untruths or confirm reliability. (Nishant et al., 2020).

5.5 Ethical and Legal Considerations

Subsequently, to the analysis of some of the primary social considerations that might critically impact the potential positive outputs of AI in healthcare, also in affiliated relation to the accomplishment of the most renowned sustainability dimensions, it is essential to focus on another two aspects that are considered crucial to fully understand the comprehensiveness and interconnectedness of sustainability in combination with healthcare and AI (Gerke et al., 2020). Indeed, ethical concerns and legal issues deserve a more profound reflection for a comprehensive perspective concerning the implementation of AI in the medical environment.

Addressing the topic from an ethical point of view, it can be easily argued that, as the employment of digital health technologies in healthcare is gaining momentum, questions of an ethical nature are bound to be raised related to the potential ways AI can infringe upon privacy rights, compromise patient or healthcare data protection, exacerbate rather than mitigate the digital divide, encroach upon patient autonomy, and a long list of more fine-grained issues that might serve to shape regulatory provisions (Amann et al., 2020).

It has also been indicated that the big data potentially available can be utilised by AI to increase knowledge in the medical community, paving the way for personalised medicine and generating socio-economic effects such as the development of personalised health systems, reducing inequalities and enhancing solidarity. Nonetheless, the golden era of personalised health should immediately comply with some harsh ethical, social, and legal prerequisites, matching privacy, data protection, and net-neutrality standards to fairness within the framework provided by bioethics, not least considering how to secure some degree of equity as regards access to such benefits (Johnson et al., 2021; Subramanian et al., 2020).

From a needlepoint view on data-sharing, indeed, once an agreement among the involved parties achieves this ethical and legal point, the AI-driven analysis will be allowed to benefit from enhanced medical data banks, as well as AI-based translational research, generation, and checking of potentially valuable hypotheses reducing experimental to bioethical animal testing.

In conclusion, AI in healthcare, within a long-term crucial ethical perspective—mainly focusing on the implementation of co-responsibility among different stakeholders—might be framed by four axes as proposed by the European Group on Ethics in Science and New Technologies mandate on personalised medicine: quality, access, sustainability of health systems (as well as economic sustainability, another proper key element), and solidarity in healthcare. The current large number of AI applications, resulting likewise in an equally significant personal and socio-economic problems, requires a shared terminology, a proper ethical compression of those risks, and an international legal framework, as well as some substantial and sometimes sore changes in the healthcare systems.

5.6 Future Trends and Directions

Clearly, the impact of increasing artificial intelligence applications in a domain as critical as the health sector, which aims to develop in the light of sustainability, requires much greater attention. In the future, it is anticipated that advances in hardware, sensors, battery life, and the expansion of quantum computing can allow for more effective, predictive, personalised treatment and social interventions for health (Hartmann et al., 2022). The trend towards a digital and decentralised identity by combining blockchain and other technologies such as multi-sig, Shamir secrets, and smart contracts has already begun. The Security Operations Center (SOC), in addition to being trained in hacking defence systems and tactics, will have to be prepared for the vast amount of new data that is already being produced and is expected to be created in the future by cyber threats (Pramanik et al., 2020). Additionally, the increasing importance of educational bodies in shaping the aptitude and attitudes of the future workforce in the public sector towards the SDGs and the innovative use of AI should not be underestimated. Regulations concerning AI's application in the health sector are also needed. Finally, and perhaps most importantly, the future challenge to hybrid business models that provide healthcare resources based on sustainable development goals will need to be addressed, which encourages responsible research and innovation (RRI).

In terms of future business models for the use of AI in healthcare, the personalisation of healthcare according to the needs of individual patients and the primary care market with the emergence of e-health companies and individual AI doctors seem to be the most distinguished business models, so far (Fiani et al., 2020). In terms of the impact level of applied AI, the key investors in merged or acquired companies appear to favour procedures and diagnosis the most. The use of robotics in the health sector is currently limited, and it is used only in a few diagnostic and procedural business models. Managing liability and legal issues and securing a data rights management system or blockchain seems to be the dominant approach to ensure privacy and data security. Finally, the refinement of current medical practices and the search for effective treatments or cures for rare diseases appear to be the critical trend underlying all future developments (Wang & Wang, 2021). We have already discussed how applying AI in the healthcare sector is linked to and guided by sustainable development as defined by the Sustainable Development Goals. The chapter outlines six essential principles, expressing our initial insights into how AI applications influence organisations and society and what tendencies are already shaping the landscape. As several international organisations are currently seeking out companies in which these trends are present, the outcomes are relevant for practitioners and academics alike. In addition, the identified principles can be used in corporate strategy to align an organisation's future business processes with global trends (Tavakoli et al., 2020).

5.7 Conclusion and Recommendations

The main motive for doing this research was to understand the impact of artificial intelligence on the healthcare sector in the light of sustainability. From the discussions and reviews of artificial intelligence, artificial intelligence in healthcare, and sustainability in healthcare, it is found that the use of AI in the healthcare sector is just entering its development stage, and it does have some potential negative consequences on the environment in the long run. The present state of AI adoption in healthcare has no adverse impacts related to sustainability because the focus is on data mining, analytics, and the detection of certain diseases from available medical records. However, future adoption in enhancing the accuracy and treatment of chronic diseases would need healthcare robots, which would have negative consequences on the environment. However, technology is growing, and the future impacts can become positive if potential advances are used to also adhere to the goals of sustainability. Studies on the integration of AI with E-Health, explicitly focusing on remote monitoring, early detection, and lifestyle management with an aim for rehabilitation and personalised treatment, would be a good area for further work. More case studies are needed to scan the broader spectrum of AI applications in healthcare.

While the significance of AI in the healthcare industry is irrefutable, there must be a critical examination of sustainability concerns linked to this transcendent trend. The UN has emphasised the need for more AI applications, and this beckons the global world to conduct more research, deliberations, and considerations on AI and the health industry, focusing on the objectives and approaches employed. Such a research agenda could enable the collaboration between innovation industries and the sustainable development industries for win–win corporations. The footprint of ICT and its various related applications remains undetermined, and this correspondingly portents research on the negative and positive consequences of each AI application, delivering a guide that discusses critical passengers that threaten sustainability in a few affected cases by AI machine applications. Such studies are triggered by the sustainable development goal of good health and well-being, which forms part of the 2030 agenda, specifying the conscious inclusion approach for the vulnerable in terms of gender, age, mental and physical challenges, racial, biological and immunologically different people in the AI policy, trend applications and consumer clubbing experiences to meet every sustainable development goal for the people in a sustainable way. These research directions are expected to take care of the health and safety of all people of the United Nations, including children, staff, mail, and disabled people in all works of life.

References

Aggarwal, A., Tam, C. C., Wu, D., Li, X., & Qiao, S. (2023). Artificial intelligence–based chatbots for promoting health behavioural changes: Systematic review. *Journal of Medical Internet Research, 25*, e40789 (jmir.org)

Alami, H., Rivard, L., Lehoux, P., Hoffman, S. J., Cadeddu, S. B. M., Savoldelli, M., & Fortin, J. P. (2020). Artificial intelligence in health care: Laying the foundation for responsible, sustainable, and inclusive innovation in low- and middle-income countries. *Globalisation and Health, 16*, 1–6 (springer.com)

Ali, O., Abdelbaki, W., Shrestha, A., Elbasi, E., Alryalat, M. A. A., & Dwivedi, Y. K. (2023). A systematic literature review of artificial intelligence in the healthcare sector: Benefits, challenges, methodologies, and functionalities. *Journal of Innovation & Knowledge, 8*(1), 100333 (sciencedirect.com)

Alowais, S. A., Alghamdi, S. S., Alsuhebany, N., Alqahtani, T., Alshaya, A. I., Almohareb, S. N., Albekairy, A. M., et al. (2023). Revolutionising healthcare: The role of artificial intelligence in clinical practice. *BMC Medical Education, 23*(1), 689 (springer.com)

Amann, J., Blasimme, A., Vayena, E., Frey, D., Madai, V. I., & Precise4Q Consortium. (2020). Explainability for artificial intelligence in healthcare: a multidisciplinary perspective. *BMC Medical Informatics and Decision Making, 20*, 1–9 (springer.com)

Billings, J., Ching, B. C. F., Gkofa, V., Greene, T., & Bloomfield, M. (2021). Experiences of frontline healthcare workers and their views about support during covid-19 and previous pandemics: a systematic review and qualitative meta-synthesis. *BMC Health Services Research, 21*, 1–17 (springer.com)

Bohr, A. & Memarzadeh, K. (2020). The rise of artificial intelligence in healthcare applications. *Artificial Intelligence in Healthcare* (nih.gov)

Castiglioni, I., Rundo, L., Codari, M., Di Leo, G., Salvatore, C., Interlenghi, M., Sardanelli, F., et al. (2021). AI applications to medical images: From machine learning to deep learning. *Physica Medica, 83*, 9–24 (physicamedica.com)

Di Vaio, A., Palladino, R., Hassan, R., & Escobar, O. (2020). Artificial intelligence and business models in the sustainable development goals perspective: A systematic literature review. *Journal of Business Research, 121*, 283–314 (HTML)

Dzhengiz, T. & Niesten, E. (2020). Competences for environmental sustainability: A systematic review on the impact of absorptive capacity and capabilities. *Journal of Business Ethics* (springer.com)

Eckelman, M. J., Huang, K., Lagasse, R., Senay, E., Dubrow, R., & Sherman, J. D. (2020). Health care pollution and public health damage in the United States: An update study examines health care pollution and public health damage in the United States. *Health Affairs, 39*(12), 2071–2079 (healthaffairs.org)

Fiani, B., Quadri, S. A., Farooqui, M., Cathel, A., Berman, B., Noel, J., & Siddiqi, J. (2020). Impact of robot-assisted spine surgery on health care quality and neurosurgical economics: A systemic review. *Neurosurgical Review, 43*, 17–25. (HTML)

Filip, R., Gheorghita Puscaselu, R., Anchidin-Norocel, L., Dimian, M., & Savage, W. K. (2022). Global challenges to public health care systems during the Covid-19 pandemic: a review of pandemic measures and problems. *Journal of Personalized Medicine, 12*(8), 1295 (mdpi.com)

Fuller, R., Landrigan, P. J., Balakrishnan, K., Bathan, G., Bose-O'Reilly, S., Brauer, M., Yan, C., et al. (2022). Pollution and health: A progress update. *The Lancet Planetary Health, 6*(6), e535–e547 (thelancet.com)

Galaz, V., Centeno, M. A., Callahan, P. W., Causevic, A., Patterson, T., Brass, I., Levy, K., et al. (2021). Artificial intelligence, systemic risks, and sustainability. *Technology in Society, 67*, 101741 (sciencedirect.com)

Gerke, S., Minssen, T., & Cohen, G. (2020). Ethical and legal challenges of artificial intelligence-driven healthcare. *Artificial Intelligence in Healthcare.* (nih.gov)

Greer, S. L., Falkenbach, M., Siciliani, L., McKee, M., Wismar, M., & Figueras, J. (2022). From health in all policies to health for all policies. *The Lancet Public Health, 7*(8), e718–e720 (thelancet.com)

Haldane, V., De Foo, C., Abdalla, S. M., Jung, A. S., Tan, M., Wu, S., Legido-Quigley, H., et al. (2021). Health systems resilience in managing the covid-19 pandemic: Lessons from 28 countries. *Nature Medicine, 27*(6), 964–980 (nature.com)

Hartmann, M., Hashmi, U. S., & Imran, A. (2022). Edge computing in smart health care systems: Review, challenges, and research directions. *Transactions on Emerging Telecommunications Technologies, 33*(3), e3710 (wiley.com)

Holman, H. R. (2020). The relation of the chronic disease epidemic to the health care crisis. *ACR Open Rheumatology* (wiley.com)

Hsu, S., Banskota, S., McCormick, W., Capacci, J., Bustamante, C., Moretti, K., Martin, K. D., et al. (2021). Utilization of a waste audit at a community hospital emergency department to quantify waste production and estimate environmental impact. *The Journal of Climate Change and Health, 4*, 100041 (sciencedirect.com)

Johnson, K. B., Wei, W. Q., Weeraratne, D., Frisse, M. E., Misulis, K., Rhee, K., Snowdon, J. L., et al. (2021). Precision medicine, AI, and the future of personalized health care. *Clinical and Translational Science, 14*(1), 86–93 (wiley.com)

Kang, D. Y., Cho, K. J., Kwon, O., Kwon, J. M., Jeon, K. H., Park, H., Oh, B. H., et al. (2020). Artificial intelligence algorithm to predict the need for critical care in prehospital emergency medical services. *Scandinavian Journal of Trauma, Resuscitation and Emergency Medicine, 28*, 1–8 (springer.com)

Karimian, G., Petelos, E., & Evers, S. M. A. A. (2022). The ethical issues of the application of artificial intelligence in healthcare: A systematic scoping review. *AI and Ethics* (springer.com)

Luna-Nemecio, J., Tobón, S., & Juárez-Hernández, L. G. (2020). Sustainability-based on socioformation and complex thought or sustainable social development. *Resources, Environment and Sustainability, 2*, 100007 (sciencedirect.com)

Macpherson, C. C., Smith, E., & Rieder, T. N. (2024). Does health promotion harm the environment? *Environmental Bioethics* (researchgate.net)

Mujtaba, G., & Shahzad, S. J. H. (2021). Air pollutants, economic growth and public health: Implications for sustainable development in OECD countries. *Environmental Science and Pollution Research, 28*, 12686–12698 (springer.com)

Nishant, R., Kennedy, M., & Corbett, J. (2020). Artificial intelligence for sustainability: Challenges, opportunities, and a research agenda. *International Journal of Information Management, 53*, 102104 (HTML)

Oala, L., Murchison, A. G., Balachandran, P., Choudhary, S., Fehr, J., Leite, A. W., Goldschmidt, P. G., Johner, C., Elora D. M. Schörverth, Nakasi, R., Meyer, M., Cabitza, F., Baird, P., Prabhu, C., Weicken, E., Liu, X., Wenzel, M., Vogler, S., Akogo, D., & Alsalamah, S. (2021). Machine learning for health: Algorithm auditing & quality control. *Journal of Medical Systems, 45*(12). https://doi.org/10.1007/s10916-021-01783-y

Palomares, I., Martínez-Cámara, E., Montes, R., García-Moral, P., Chiachio, M., Chiachio, J., Herrera, F., et al. (2021). A panoramic view and swot analysis of artificial intelligence for achieving the sustainable development goals by 2030: Progress and prospects. *Applied Intelligence, 51*, 6497–6527 (springer.com)

Palozzi, G., Schettini, I., & Chirico, A. (2020). Enhancing the sustainable goal of access to healthcare: Findings from a literature review on telemedicine employment in rural areas. *Sustainability* (mdpi.com)

Patil, S., & Shankar, H. (2023). Transforming healthcare: harnessing the power of AI in the modern era. *International Journal of Multidisciplinary Sciences and Arts, 2*(1), 60–70 (itscience.org)

Pramanik, P. K. D., Solanki, A., Debnath, A., Nayyar, A., El-Sappagh, S., & Kwak, K. S. (2020). Advancing modern healthcare with nanotechnology, nanobiosensors, and internet of nano things: Taxonomies, applications, architecture, and challenges. *IEEE Access, 8*, 65230–65266 (ieee.org)

Roberts, H., Zhang, J., Bariach, B., Cowls, J., Gilburt, B., Juneja, P., Floridi, L., et al. (2024). Artificial intelligence in support of the circular economy: Ethical considerations and a path forward. *AI & Society, 39*(3), 1451–1464 (springer.com)

Ruggerio, C. A. (2021). Sustainability and sustainable development: A review of principles and definitions. *Science of the Total Environment* (HTML)

Secinaro, S., Calandra, D., Secinaro, A., Muthurangu, V., & Biancone, P. (2021). The role of artificial intelligence in healthcare: A structured literature review. *BMC Medical Informatics and Decision Making, 21*, 1–23 (springer.com)

Shaheen, M. Y. (2021). Applications of artificial intelligence (AI) in healthcare: A review. *ScienceOpen* Preprints (scienceopen.com)

Shelmerdine, S. C., White, R. D., Liu, H., Arthurs, O. J., & Sebire, N. J. (2022). Artificial intelligence for radiological paediatric fracture assessment: A systematic review. *Insights into Imaging, 13*(1), 94 (springer.com)

Singh, J., & Hamid, M. A. (2022). Cognitive computing in mental healthcare: A review of methods and technologies for detection of mental disorders. *Cognitive Computation* (HTML)

Ståhl, T., & Koivusalo, M. (2021). Health in all policies: Concept, purpose, and implementation. *Handbook of Global Health* (academia.edu)

Sterling, E. J., Pascua, P., Sigouin, A., Gazit, N., Mandle, L., Betley, E., McCarter, J., et al. (2020). Creating a space for place and multidimensional well-being: lessons learned from localizing the SDGs. *Sustainability Science, 15*, 1129–1147. (springer.com)

Subramanian, M., Wojtusciszyn, A., Favre, L., Boughorbel, S., Shan, J., Letaief, K. B., Chouchane, L., et al. (2020). Precision medicine in the era of artificial intelligence: implications in chronic disease management. *Journal of Translational Medicine, 18*, 1–12 (springer.com)

Sujith, A. V. L. N., Sajja, G. S., Mahalakshmi, V., Nuhmani, S., & Prasanalakshmi, B. (2022). Systematic review of smart health monitoring using deep learning and artificial intelligence. *Neuroscience Informatics, 2*(3), 100028 (sciencedirect.com)

Tanwar, S., Parekh, K., & Evans, R. (2020). Blockchain-based electronic healthcare record system for healthcare 4.0 applications. *Journal of Information Security and Applications, 50*, 102407 (academia.edu)

Tavakoli, M., Carriere, J., & Torabi, A. (2020). Robotics, smart wearable technologies, and autonomous intelligent systems for healthcare during the covid-19 pandemic: An analysis of the state of the art and .... *Advanced Intelligent Systems.* (wiley.com)

Tobler, P., Cyriac, J., Kovacs, B. K., Hofmann, V., Sexauer, R., Paciolla, F., Hirschmann, A., et al. (2021). AI-based detection and classification of distal radius fractures using low-effort data labeling: Evaluation of applicability and effect of training set size. *European Radiology, 31*, 6816–6824 (springer.com)

Ullah, U., & Garcia-Zapirain, B. (2024). Quantum machine learning revolution in healthcare: A systematic review of emerging perspectives and applications. *IEEE Access* (ieee.org)

Van Wynsberghe, A. (2021). Sustainable AI: AI for sustainability and the sustainability of AI. *AI and Ethics* (springer.com)

Wang, X. V., & Wang, L. (2021). A literature survey of the robotic technologies during the COVID-19 pandemic. *Journal of Manufacturing Systems.* (sciencedirect.com)

Wells, A. J., Viaroli, E., & Hutchinson, P. J. A. (2024). The management of traumatic brain injury. *Surgery (Oxford).* (HTML)

Xu, L., Sanders, L., Li, K., & Chow, J. C. L. (2021). Chatbot for health care and oncology applications using artificial intelligence and machine learning: systematic review. *JMIR Cancer* (jmir.org)

Yuan, K. C., Tsai, L. W., Lee, K. H., Cheng, Y. W., Hsu, S. C., Lo, Y. S., & Chen, R. J. (2020). The development of an artificial intelligence algorithm for early sepsis diagnosis in the intensive care unit. *International Journal of Medical Informatics, 141*, 104176 (sciencedirect.com)

Zhou, S. K., Greenspan, H., Davatzikos, C., Duncan, J. S., Van Ginneken, B., Madabhushi, A., Summers, R. M., et al. (2021). A review of deep learning in medical imaging: Imaging traits, technology trends, case studies with progress highlights, and future promises. *Proceedings of the IEEE, 109*(5), 820–838 (nih.gov)

Zhu, S., Gilbert, M., Chetty, I., & Siddiqui, F. (2022). The 2021 landscape of FDA-approved artificial intelligence/machine learning-enabled medical devices: An analysis of the characteristics and intended use. *International Journal of Medical Informatics, 165*, 104828 (henryford.com)

Part II
Patient Engagement and Data Security

Chapter 6
Transforming Healthcare Through Smart Health Systems: Harnessing Technology for Enhanced Patient Care

Remya Jayachandran, Bhargava Dantapur, Aneeta S. Antony, and Rohini Nagapadma

Abstract Smart healthcare enhances medical service provision by integrating advanced technologies and intelligent systems within the healthcare sector. This chapter provides an overview of the concept of smart healthcare and its potential benefits by incorporating artificial intelligence, machine learning, and internet of things in early detection and diagnosis of various diseases. The advent of smart healthcare has transformed the landscape of healthcare delivery, allowing for enhanced patient care, tailored medical treatments, and streamlined healthcare administration. Leveraging cutting-edge AI algorithms, healthcare professionals can now achieve unparalleled precision in disease diagnosis through the comprehensive analysis of medical images and clinical data. This not only serves to minimize diagnostic inaccuracies but also significantly enhances treatment outcomes, ensuring patients receive the most effective care tailored to their specific needs. In addition to these sophisticated advancements, the realm of smart healthcare encompasses an array of wearable devices and mobile applications. In this chapter, we discuss smart healthcare systems used in various applications namely the prediction of Alzheimer's disease at various stages, the prediction of the behavior of patients with mental illness, analysis of the health condition of premature babies in smart NICU, applications of AI and ML in hearing aids and in discerning speech of people with speech difficulty. For each application, the conventional methods in diagnosing the disease and the existing methods using advanced technology are discussed. The challenges and the future directions are discussed at the end of the chapter.

Keywords Smart healthcare · Smart NICU · Hearing aids · Mental-illness diagnosis · Alzheimer's disease · Dysarthric speech · AI algorithms

---

Present Address:
R. Jayachandran (✉) · B. Dantapur · A. S. Antony · R. Nagapadma
Department of Electronics and Communication Engineering, The National Institute of Engineering Mysore, Mysore, Karnataka, India
e-mail: remyajayachandran@nie.ac.in

R. Singh et al. (eds.), *Transforming Healthcare Sector Through Artificial Intelligence and Environmental Sustainability*, Approaches to Global Sustainability, Markets, and Governance, https://doi.org/10.1007/978-981-97-9555-0_6

6.1 Introduction: Background and Driving Forces

Conventional disease diagnosis methods involve a blend of physical examination, medical history review, and diverse diagnostic test outcomes. Physical examination encompasses a comprehensive patient inspection, checking for indicators of illness such as fever, swelling, or irregularities in the skin, organs, or bodily movements. Medical history documentation involves manually recording the patient's symptoms, past medical history, family history, and lifestyle habits that may influence the disease. Diagnostic tests encompass blood tests, imaging studies, and biopsies, providing doctors with additional information about the patient's condition to aid in making a diagnosis. These traditional methods, having been utilized for centuries, continue to be a crucial component of contemporary disease diagnosis. The traditional approach to disease diagnosis presents both advantages and disadvantages.

Advantages include the following:

i. Experience and expertise: Traditional diagnostic methods heavily rely on human expertise, allowing experienced healthcare professionals to use their knowledge and judgment to diagnose diseases accurately.
ii. Comprehensive assessment: Conventional diagnostic approaches typically entail a meticulous physical examinations coupled with an in-depth review of the patient's medical background, symptoms, and individual observations. This approach can help detect subtle signs and symptoms that may be missed by automated or technology-driven diagnostic methods.
iii. Flexibility: Conventional diagnostic methods are customizable and individualized to accommodate the unique situations and medical histories of each patient. Healthcare providers have the freedom to choose suitable diagnostic instruments and evaluations that are tailored to their assessment of the patient's specific condition.

However, there are a few disadvantages in traditional methods like

i. Subjectivity: Conventional diagnostic methods are customizable and individualized to accommodate the unique situations and medical histories of each patient. Healthcare providers have the freedom to choose suitable diagnostic instruments and evaluations that are tailored to their assessment of the patient's specific condition.
ii. Limited access: Traditional diagnostic methods often require access to specialized medical equipment and expertise, which may not be easily available in all areas. This can limit access to accurate diagnoses, particularly in remote or resource- poor regions.
iii. Time-consuming: Traditional diagnostic methods can take significant time to complete, especially when multiple tests and consultations are required. This delay in diagnosis may lead to delayed treatment and potential worsening of the patient's condition.

iv. Cost: Traditional diagnostic methods can be expensive, involving several tests, consultations, and follow-ups. This cost burden can be a barrier for patients, especially those without adequate healthcare coverage.

Modern advancements, such as technological innovations and AI-powered systems, can complement traditional methods to improve accuracy and efficiency of the diagnosis methods. Smart health systems, fueled by cutting-edge technology, hold the promise of revolutionizing healthcare and significantly elevating the quality of treatment and enhancing patient care (Barragán-Montero, et al., 2021). These systems leverage various technological innovations to improve efficiency, accuracy, and accessibility in healthcare delivery. Artificial intelligence is currently driving significant changes across various industries, with the healthcare sector emerging as the next major focus. The healthcare realm is currently witnessing a significant uptick in the integration of machine learning and deep learning techniques, representing a transformative shift in medical practice. Deep learning, a subset of machine learning and a cornerstone of AI, operates through intricate layered architectures to autonomously analyze vast datasets using advanced algorithms (Lin et al., 2018). In contrast, machine learning involves tailored computer programming that statistically processes inputs to generate outputs with minimal human intervention. Despite the advancements in machine learning models, they still require human oversight for guidance and control. Conversely, deep learning models exhibit a remarkable capacity for independent accuracy assessment through complex neural networks. While machine learning algorithms necessitate human intervention, often from individuals with engineering acumen and specialized domain knowledge, to preprocess data for pattern recognition, deep learning algorithms demand less preprocessing. They can autonomously discern patterns directly from raw data, reducing the need for extensive human intervention. However, recent years have witnessed a notable surge in the adoption of deep learning methodologies, particularly in tasks involving intricate image processing. This shift signifies a progressive move towards leveraging advanced technologies to enhance diagnostic accuracy and patient care standards (Jiang et al., 2017).

Different ways of utilizing the modern technology to achieve these goals are:

i. **Electronic Health Records (EHRs)**: EHRs replace traditional paper-based medical records with digital records that can be easily accessed and shared among healthcare providers. This enables seamless communication and coordination between different healthcare facilities, with high data security about a patient is made available to the required people.
ii. **Telemedicine and Remote Patient Monitoring**: Facilitating remote access to healthcare services, often through video consultations with healthcare professionals, telemedicine proves invaluable for individuals residing in remote areas or experiencing limited mobility.
iii. **AI and ML**: Utilizing AI technologies, healthcare data undergoes extensive analysis to offer insightful guidance for decision-making. Machine learning

algorithms discern patterns and forecast outcomes, aiding in diagnosis, treatment strategies, and personalized medicine.

iv. **Internet of Things (IoT) in Healthcare**: Wearable fitness trackers, smart medical devices, and interconnected hospital equipment constitute IoT devices, facilitating live monitoring and data acquisition (Gopalan et al., 2021). This data aids in early health issue detection, patient progress tracking, and treatment plan optimization.

v. **Data Security and Privacy**: Smart health systems gather and store vast amounts of sensitive healthcare data, underscoring the importance of robust security and privacy measures (Nagendran et al., 2020). This entails deploying stringent encryption, access controls, and protocols to safeguard patient information against unauthorized access or breaches.

Overall, smart health systems stand poised to revolutionize healthcare, delivering exceptionally efficient patient care. By harnessing EHR, telemedicine, AI and machine learning, IoT, and prioritizing data security, the healthcare industry can work towards delivering more personalized and effective healthcare services to people worldwide (Akundi et al., 2020).

This chapter focuses on harnessing technology for enhanced patient care using smart health care systems in various applications such as neonatal intensive care units to predict the health condition of premature baby, Alzheimer's disease prediction using MRI images, performance improvement of hearing aids using AI and ML for speech processing, predicting the mood variations of patients with mental illness, etc. In the healthcare industry, a variety of AI algorithms are utilized, including decision trees, support vector machines, and neural networks, for tasks ranging from disease diagnosis to patient analysis and drug discovery. The selection of an algorithm depends on factors such as the complexity of the problem and the available data type. Among the commonly used algorithms, support vector machines excel in protein classification, image segmentation, and text categorization. Artificial neural networks, notably convolutional neural networks (CNN) and recurrent neural networks (RNN), find extensive application in computer vision (e.g., medical image analysis), natural language processing (e.g., speech recognition), and early drug discovery (e.g., predicting medical benefits of substances). Logistic regression, employing regression to forecast the state of a categorical dependent variable, is instrumental in classifications and event probability prediction (Kelly et al., 2019, Datta et al., 2019). In addition to these prominent AI algorithms, other notable methodologies in healthcare include ensemble learning techniques such as random forests and gradient boosting machines, which combine multiple models to enhance predictive accuracy. Furthermore, deep learning architectures like long short-term memory (LSTM) networks are gaining traction for tasks involving sequential data analysis, such as time-series forecasting and medical record processing. These diverse AI approaches collectively contribute to advancing diagnostic precision, treatment efficacy, and healthcare decision-making in today's rapidly evolving medical landscape (Castiglioni et al., 2021, Rizk et al., 2023).

Discriminated analysis leverages electronic health records to forecast the onset of mental disorders in their early stages. Random forest algorithms are applied extensively in the analysis of ECG, EEG, and MRI images, as well as for disease prediction tasks. Linear regression models excel in predicting diseases based on identified risk factors. Naïve Bayes, known for its efficiency in classification, finds utility in medical data classification and disease prognosis. Additionally, K-nearest neighbor algorithms are employed for both classification and regression tasks in data mining, particularly in the classification of heart diseases. These diverse machine learning techniques play pivotal roles in advancing healthcare analytics and decision-making processes. These are some of the most frequently used AI algorithms in the healthcare sector for the development of smart healthcare system. Incorporating IoT and AI into diagnostic devices enables doctors to remotely analyze patients' data in real-time. The algorithms are designed to predict early disease detection, while IoT facilitates effective real-time patient monitoring. This capability supports medical teams in delivering timely interventions to save patients.

6.2 Application of AI and ML in neonatal intensive care unit (NICU) for Monitoring Premature Babies

AI and ML have emerged as transformative forces shaping the way we interact with and leverage computing systems. AI performs these complex tasks using algorithms and predicts the various possibilities. These tasks encompass a wide spectrum, spanning from visual perception and speech recognition to decision-making and language translation. ML is dedicated to creating computer systems capable of learning and adapting autonomously, without the need for explicit programming instructions. This learning process is facilitated by algorithms and statistical models that analyze data patterns, allowing machines to draw inferences to improvise the performance of the diagnosis. Neonatal intensive care units (NICUs) stand at the forefront of healthcare, especially when it comes to caring for fragile newborns facing health challenges. The incorporation of AI and ML technologies in NICUs has propelled neonatal care into a new era (Adegboro et al., 2022, Malak et al., 2019). A smart NICU is a technologically advanced environment where AI and ML seamlessly integrate into the fabric of neonatal care. It surpasses traditional monitoring systems, incorporating data analytics, predictive modeling, and automation to enhance monitoring, diagnostics, and treatment protocols (Henry et al., 2022, Gulzar Ahmad et al., 2022).

The foundation of a smart NICU lies in its components:

- Sensors and IoT Devices: These are the eyes and ears of the system, continuously monitoring vital signs such as heart rate, respiratory rate, temperature and oxygen saturation (Zhang et al., 2023, Abbas & Leonhardt, 2014, Cay et al., 2021, Kumar et al., 2023, Chandrasekaran et al., 2021).
- Data Processing Unit: A centralized system processes real-time data from sensors, enabling quick and accurate decision-making.

- AI Algorithms: Intelligent algorithms play a pivotal role in predictive analytics, risk assessment, and decision support, ensuring timely interventions. Table 6.1 depicts the algorithms used for detection of various diseases in premature babies.
- Automation Systems: Robotic systems are integrated for routine tasks, allowing healthcare professionals to focus on critical aspects of patient care.
- Body and Frame: Interfacing the sensors, actuators, boards onto a specific designed body for functioning and support.

6.2.1 Neonatal Resuscitation

Neonatal resuscitation is a critical aspect of newborn care, acknowledging the profound physiologic changes that accompany the birth process, often revealing conditions not apparent during intrauterine life. The neonatal period encompasses the first 28 days of life, and during this critical phase, some infants may require assistance to initiate or sustain vital physiological functions. The need for neonatal resuscitation arises when a newborn exhibits signs of respiratory distress, such as inadequate breathing, poor muscle tone or a weak heartbeat. This may result from various factors, including prematurity, complications during delivery, infections, or congenital anomalies. The imperative for skilled individuals with neonatal resuscitation expertise to be present at every birth is emphasized, underlining the potential challenges that may arise during this transitional phase. Roughly 10% of neonates require respiratory assistance at birth, with less than 1% needing extensive resuscitation. The Apgar score serves as a crucial tool for evaluating a neonate's cardiorespiratory and neurologic status at birth. Table 6.2 depicts the Apgar score of the premature baby. Its primary purpose is to offer a quick evaluation of the newborn's overall health status during the critical moments of transition from intrauterine to extrauterine life. The Apgar score, which assigns 0 to 2 points for each of the five measures—appearance, pulse, grimace, activity, and respiration—is influenced by factors like physiologic maturity, birth weight, maternal perinatal therapy, and fetal cardiorespiratory and neurologic conditions. Normally, Apgar scores are assigned at 1 and 5 minutes after birth. However, if the scores remain at 5 or below at the 5-minute mark, ongoing assessments should persist every 5 minutes until the score exceeds 5. This continuous monitoring allows healthcare providers to gauge the neonate's response to interventions and identify any evolving health concerns. Importantly, a low Apgar score, falling within the 0 to 3 range, is considered a clinical finding rather than a definitive diagnosis. It indicates that the newborn requires further evaluation and potential intervention. Diverse factors contribute to low Apgar scores, including severe, chronic problems with a poor prognosis, alongside acute issues that may be promptly resolved with a favorable prognosis. It assigns scores based on five measures, providing a quick snapshot of the newborn's health.

- **Continuous Monitoring and Early Detection**: In the realm of neonatal care, a smart NICU equipped with artificial intelligence brings a transformative impact

Table 6.1 Comprehensive overview of neonatal intensive care unit (NICU) algorithms: applications and impact on infant health

Algorithm	Area of application	Purpose/functionality	Benefits/outcomes
Apnea detection algorithm	Apnea detection in premature infants	Detects and monitors apnea episodes in newborns, triggering timely interventions such as stimulation or respiratory support	Early intervention in apnea events helps prevent complications such as hypoxia and bradycardia, improving overall respiratory management
Respiratory distress syndrome (RDS)	Predicting RDS in premature infants	Assesses risk factors and predicts the likelihood of Developing RDS, allowing healthcare providers to implement proactive measures and tailored respiratory support	Enables early identificationof infantsat risk for RDS facilitating timely interventions such a surfactant administration and mechanical ventilation, ultimately improving respiratory outcomes
Jaundice risk assessment algorithm	Assessing risk of jaundice in newborns	Evaluates factors contributing to jaundice, aiding in monitoring and managing bilirubin levels	Facilitates timely interventions, such as phototherapy or exchange transfusions, reducing the risk of severe hyperbilirubinemia and associated neurotoxicity
Weight gain prediction algorithm	Predicting weight gain patterns	Analyzes growth parameters to predict weight gain, assisting in the development of personalized feeding plans and monitoring growth progress	Supports individualized nutritional strategies, optimizing weight gain and promoting overall growth and development in premature infants
Neuro-developmental outcome prediction algorithm	Predicting long-term neuro-developmental outcomes	Integrates various data points to predict neuro-developmental outcomes, aiding in early identification and intervention for infants at risk for developmental challenges	Facilitatesearly access to intervention services, enhancing theoverall neuro-developmental trajectory and improvinglong-term outcomes for high-risk infants

Table 6.2 Apgar score describing neonate's cardiorespiratory and neurologic condition at birth

Criteria	Mnemonic	Apgar score		
		0	1	2
Color	Appearance	All blue, pale	Pink body, blue extremities	All pink
Heart rate	Pulse	Absent	<100 beats/minute	>100 beats/minute
Reflex response to nasal catheter/tactile stimulation	Grimace	None	Grimace	Sneeze, cough
Muscle tone	Activity	Limp	Some flexion of extremities	Active
Respiration	Respiration	Absent	Irregular, slow	Good, crying

* A total score of 7–10 at 5 min is considered normal; 4–6, intermediate; and 0–3 low

through continuous monitoring and early detection capabilities. This innovative approach harnesses the power of AI algorithms to analyze data streaming from various sensors, providing unprecedented advantages in terms of early anomaly detection. The real-time monitoring capabilities of a smart NICU allow for a meticulous examination of vital signs, oxygen levels, and other crucial parameters. AI algorithms sift through this influx of data, swiftly identifying deviations from the norm. The integration of AI in continuous monitoring not only enhances the speed of detection but also minimizes the chances of oversight that may occur with human monitoring alone.

- **Predictive Analytics for High-Risk Conditions**: ML models, a branch of AI, are proficient in recognizing patterns and making predictions. In smart NICU settings, these models function as robust tools for anticipating high-risk conditions. Trained using extensive datasets covering a wide spectrum of neonatal cases, ML models can identify subtle correlations and patterns that might elude human detection. The predictive analytics capacity of ML proves invaluable in foreseeing complications or high-risk situations. These models can anticipate issues like infections, respiratory problems, or neurological conditions based on early signals within the data. This predictive ability equips healthcare professionals with the foresight to proactively implement preventive measures, start interventions early, or adjust treatment strategies to match personalized risk evaluations. Such a proactive predictive method not only enhances neonatal care effectiveness but also aids in optimizing healthcare system resources.
- **Personalized Treatment Plans**: In a smart NICU, AI and ML have a groundbreaking impact on personalized treatment planning. Unlike traditional healthcare methods, AI algorithms can analyze extensive patient data to tailor interventions based on each newborn's unique characteristics. For instance, if a newborn has a genetic predisposition to a certain condition, the AI system can recommend customized interventions or monitoring protocols specific to that risk. This level of personalization not only enhances treatment effectiveness but also reduces the

likelihood of adverse reactions or unnecessary interventions. It signifies a shift towards precision medicine in neonatal care, where each treatment decision is finely tuned to the individual characteristics and needs of the newborn.

6.2.2 Benefits of Smart NICU Implementation

The implementation of smart NICUs yields a multitude of benefits that span improved patient outcomes, efficient resource utilization, and data-driven decision-making.

- Improved Patient Outcomes: The primary aim of any healthcare system is to improve patient outcomes, and a smart NICU accomplishes this by harnessing state-of-the-art technologies. Early detection, a fundamental aspect of smart NICU capabilities, significantly influences outcomes for neonates. Studies have consistently demonstrated a tangible reduction in mortality rates and long-term health issues among neonates receiving care in smart NICU environments (Aviles-Espinosa et al., 2019). The proactive approach enabled by AI-driven continuous monitoring allows healthcare providers to intervene promptly, mitigating the impact of complications before they escalate. This not only enhances survival rates but also contributes to superior overall health outcomes for neonates receiving critical care.
- Efficient Resource Utilization: The implementation of AI-driven automation in a smart NICU optimizes resource utilization, addressing a critical challenge in healthcare settings. By automating routine tasks and streamlining workflows, healthcare professionals can redirect their time and attention to critical aspects of patient care. In a traditional NICU, healthcare providers may spend considerable time on administrative tasks, data entry, or manual monitoring. The introduction of automation through AI allows these tasks to be handled efficiently, freeing up valuable human resources for more complex and specialized care. This efficient utilization of resources ensures that the limited personnel and equipment available in healthcare settings are maximized to benefit patients in the NICU. This not only improves the efficiency of care delivery but also contributes to cost-effectiveness in healthcare management.
- Data-Driven Decision Making: Major part of smart NICU lies in its ability to generate, collect, and analyze extensive data in real-time. This abundance of information is evaluated by AI algorithms, furnishing healthcare providers with actionable insights that guide data-informed decision-making. Real-time insights empower healthcare professionals to make timely and well-informed decisions. Continuous monitoring and analysis of patient data enable the early detection of potential complications, response patterns, and trends. Consequently, this elevates the overall standard of care offered in the NICU.

The ability to analyze patterns and correlations in real-time data equips healthcare providers with the knowledge needed to adapt treatment plans, respond to emerging complications, and optimize care strategies. The iterative feedback loop created by data-driven decision-making ensures a dynamic and responsive approach to neonatal care.

6.2.3 Challenges in Smart NICU

The integration of smart NICUs with AI and ML technologies brings forth a new era in neonatal care, but it is not without its challenges. As we delve into the complexities of implementing smart NICUs, several considerations and challenges need to be addressed to ensure the smooth integration and effectiveness of these innovative technologies. Data security and privacy concerns, integration with existing systems, ethical considerations, cost and resource allocation, continuous monitoring challenges, regulatory compliance etc. are some of the challenges in implementing smart NICU.

The applications of AI and ML in smart NICUs are diverse and impactful. Continuous monitoring and early detection capabilities, powered by AI algorithms, provide unprecedented advantages in terms of early anomaly detection. Predictive analytics for high- risk conditions, facilitated by ML models, empower healthcare professionals to anticipate complications and initiate preventive measures. The global adoption and standardization of smart NICU technologies will ensure consistent quality of care across different healthcare systems, benefiting newborns worldwide. As technology continues to advance, the future holds the promise of even more sophisticated and intelligent healthcare systems, ensuring the well-being of our most vulnerable patients.

6.3 Applications of AI and ML in the Prediction of disease from medical Images

Deep learning, a form of AI, is utilized in medical image analysis to examine CT, MRI, and X-ray images (Castiglioni et al., 2021, Barragán-Montero et al., 2021). These models are trained to detect patterns and features within the images, enabling them to identify potential findings that may elude human radiologists. The application of AI and ML in medical imaging encompasses a broad range of uses with the potential to significantly impact healthcare.

i. Disease Detection and Diagnosis: AI and ML technologies are instrumental in examining various imaging modalities like X-rays, CT scans, MRI scans, and mammograms. These technologies aid in identifying anomalies and facilitating the diagnosis of illnesses from cancer, cardiovascular diseases to neurological disorders.

ii. Image Segmentation and Quantification: AI and ML algorithms can segment medical images to identify and quantify specific structures or lesions, for example, in tumor volume measurement or neuroimaging for brain structure analysis (Kim et al., 2019)
iii. Personalized Treatment Planning: These advancements can support healthcare professionals in formulating individualized treatment strategies through the analysis of imaging data and the prediction of patient responses to specific therapeutic approaches.
iv. Radiology Workflow Optimization: AI and ML technologies have the potential to automate mundane tasks within radiology, such as prioritizing studies, annotating images, and managing cases, thus enhancing operational efficiency and alleviating the workload of radiologist.
v. Predictive Analytics: Through the examination of extensive datasets comprising medical images and patient records, AI and ML methodologies offer insights into the progression of diseases, treatment efficacy, and potential complications, enabling proactive intervention strategies.
vi. Drug Development and Research: The fusion of AI and ML technologies with medical imaging enables progress in drug discovery and research by offering insights into disease mechanisms and treatment effects, ultimately expediting the creation of innovative therapeutic interventions.
vii. Drug Development and Research: The amalgamation of AI and ML with medical imaging can aid in the process of drug discovery and development by enhancing comprehension of disease mechanisms and treatment impacts.
viii. Point-of-Care Imaging: AI and ML technologies can be integrated into point-of-care devices to provide real-time analysis of medical images, particularly beneficial in resource-limited settings.

The integration of AI and ML technologies into medical imaging offers a significant opportunity to improve patient outcomes, enhance diagnostic accuracy, and increase healthcare efficiency. However, before clinical implementation, it is essential to ensure their ethical use and robust validation processes. Alzheimer's disease (AD) presents a significant global health challenge impacting individuals, families, and healthcare systems worldwide. With no current cure for AD, early detection and prediction are vital for enabling timely interventions and improving patient outcomes. Recent research highlights a growing interest in utilizing convolutional neural networks (CNNs) for early prediction and diagnosis of AD. Several studies have pointed out the limitations of traditional diagnostic techniques and recognized CNNs as a promising tool in AD diagnosis. For instance, one study focused on using CNNs to predict AD from mild cognitive impairment (MCI) using MRI images, showcasing the superiority of CNNs over traditional machine algorithms (Babu et al., 2023). Another study employed existing CNN architectures and transfer learning techniques to improve the accuracy of AD prediction (Patil et al., 2022). Additionally, another study explored the use of CNNs for detecting Alzheimer's disease in MRI images, achieving notable accuracy in distinguishing between AD patients and healthy control subjects (Lin et al., 2018; Ebrahimi et al., 2021). AI and ML methodologies have demonstrated potential in predicting and diagnosing AD through MRI scans. Here are some notable applications of AI and ML in AD prediction from MRI:

- Image Segmentation: AI algorithms can segment brain MRI scans into different regions, such as the hippocampus and cortex, which are commonly affected by AD. This enables quantitative analysis of these regions and identification of potential abnormalities or signs of AD progression.
- Feature Extraction: ML techniques can automatically extract relevant features from MRI images that may be indicative of AD pathology. These features can include measures of brain volume, thickness, shape, or texture. By capturing subtle patterns and changes in brain structures, these features can help differentiate between healthy individuals and those at risk or in the early stages of AD.
- Disease Classification Models: Machine learning (ML) models can undergo training using MRI datasets to categorize individuals into two groups: healthy controls or those affected by Alzheimer's disease (AD) or predisposed to it. These models acquire patterns from the imaging data and can subsequently predict disease status when presented with new MRI scans. Standard ML algorithms employed for classification encompass support vector machines (SVM), random forests, and deep learning architectures like convolution neural networks (CNNs).
- Longitudinal Analysis: AI algorithms can analyze temporal changes and trajectories of AD pathology by comparing MRI scans taken at different time points. By tracking the progression of brain structural changes over time, ML models can provide insights into disease progression and predict future cognitive decline in individuals.
- Multi-modal Approaches: AI techniques can integrate MRI data with other modalities such as genetics, cerebrospinal fluid biomarkers, or neuropsychological assessments. By combining information from multiple sources, ML models can improve the accuracy of AD prediction and provide a more comprehensive assessment of disease risk.
- Transfer Learning: Transfer learning involves the practice of fine-tuning pre-trained models, typically trained on extensive datasets, using smaller datasets specifically tailored for Alzheimer's disease (AD) MRI analysis. This approach helps overcome limitations in the availability of labelled data, enabling more robust and accurate predictions, as shown in Fig. 6.1.

It is imperative to stress that although AI and ML methods show potential in predicting AD from MRI images, they are not intended to take the role of clinical diagnosis. Although these technologies support healthcare practitioners in risk assessment and early identification, comprehensive clinical examinations are still necessary for a conclusive diagnosis. A significant development is the Convolution Neural Network (CNN) model called "NeuroPredict", which uses MRI scans to predict AD and has an 80% accuracy rate, according to Babu et al. (2023). Figure 6.1 shows the flowchart for NeuroPredict. In order to better understand neurodegenerative disease studies and interventions, this study highlights the value of exploratory data analysis

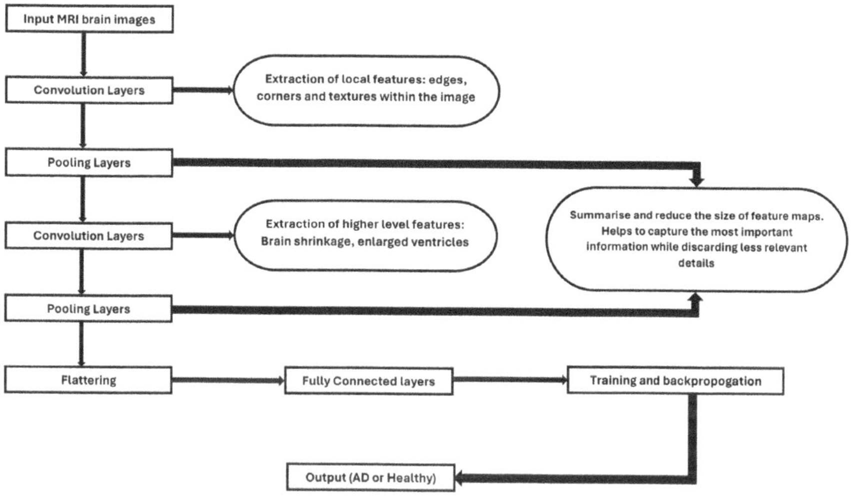

Fig. 6.1 Typical CNN architecture of predicting AD based on MRI images

(EDA) in identifying patterns and linkages within the AD patient dataset. NeuroPredict addresses the critical need for early prediction and individualized care in AD by integrating CNN models, EDA findings, and cutting-edge hardware and online applications. The outcomes that the NeuroPredict system produced when the MRI picture was uploaded are displayed in Fig. 6.2.

In conclusion, AI and ML techniques offer valuable tools for AD prediction from MRI. By leveraging image segmentation, feature extraction, classification models, longitudinal analysis, multi-modal approaches, and transfer learning, these technologies have the potential to enhance AD diagnosis, facilitate early intervention, and improve patient outcomes.

6.4 Applications of AI and ML in hearing Aids

AI and ML are seeing growing use in hearing aid applications aimed at enhancing device performance, functionality, and user experience (Iliadou et al., 2022; Venkatesh 2019; Nayak et al., 2024). Following are the ways in which AI and ML are revolutionizing the field of hearing aids, offering innovative solutions and advancements:

i. Speech Enhancement: AI algorithms can enhance the speech signal by reducing background noise and improving speech clarity. ML techniques are used to train models that can effectively separate the audio signal from the background noise, allowing the device to selectively amplify speech and suppress unwanted sounds. This enables users to understand speech better, particularly in noisy environments.

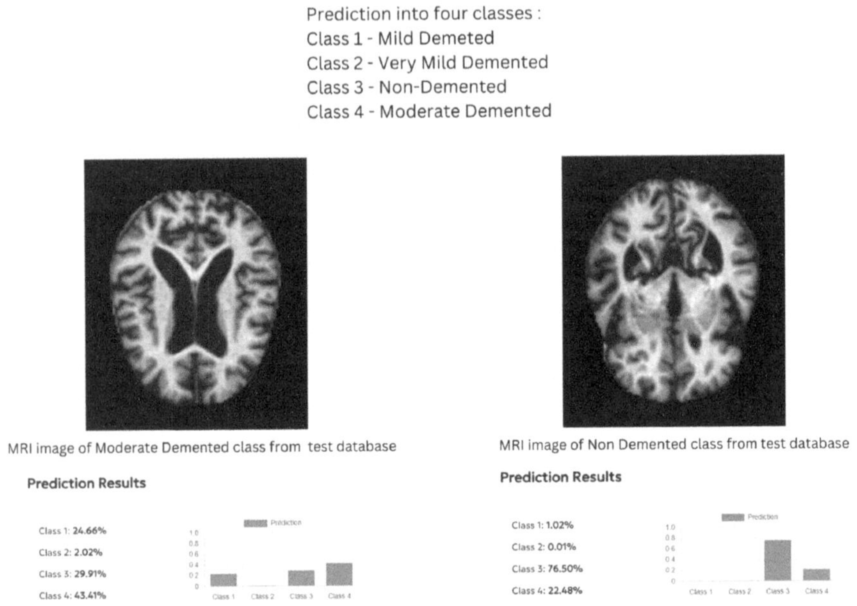

Fig. 6.2 Prediction method in the neuro predict model (Babu et al., 2023)

ii. Adaptive Sound Processing: AI can enable hearing aids to automatically adjust their settings based on the wearer's listening environment. ML algorithms can continuously analyze the acoustic environment to identify different listening scenarios (e.g., quiet, noisy, and music) and optimize the hearing aid's settings accordingly. This adaptive sound processing improves the overall listening experience and ensures optimal sound quality (Hamacher et al., 2006).

iii. Personalization and User Preferences: AI and ML techniques can be utilized to make personalized hearing profiles for individual users. By analyzing the user's listening habits, preferences, and feedback, algorithms can automatically adjust the hearing aid's settings to match their specific needs. This personalization helps users receive a tailored listening experience and can contribute to increased user satisfaction.

iv. Feedback and Whistle Management: Hearing aids can sometimes produce feedback or whistling sounds, which can be uncomfortable and disruptive for the wearer. AI algorithms can detect and identify these feedback signals and suppress them in real time, reducing the occurrence of annoying feedback and enhancing user comfort.

v. Data Logging and Analysis: Hearing aids equipped with AI capabilities can log and analyze data about the user's listening behavior, including the environments they encounter and the settings they prefer. Audiologists and hearing healthcare professionals can utilize this information to improve the user's hearing aid experience by tuning the device during follow-up visits, thereby enhancing patient care and satisfaction.

vi. Remote Support and Firmware Updates: AI-powered hearing aids can enable remote support and firmware updates. AI empowers users to connect with professionals remotely, facilitating adjustments to their hearing aids' settings, receiving guidance, or performing software updates, all conveniently from their own homes. This feature simplifies the process of fine-tuning and troubleshooting hearing aids and enhances user convenience. By leveraging these technologies, hearing aids can become more effective, user-friendly, and customizable. Ultimately, these advancements improved the quality of life for individuals grappling with hearing loss.

6.5 Applications of AI and ML in the Diagnosis of mental Illness

Mental health is one of the major concerns of human beings. Due to today's fast-moving and competitive world, people often suffer from stress, anxiety and demotivation, which lead to serious mental health issues. In some cases, it is observed that the person may behave indifferently and show violent tendencies. Once the mental health issue is identified, then the approach of society towards them changes, which creates a significant impact on the patients as well as their family members. To assist the caregivers and make their lives easy, we require a health monitoring system for both the patient and the caregiver. The violent behavior of the patients can be for many reasons, and it varies with respect to each patient. Dopamine hormones produced by the human brain help to understand a person's emotions. Dopamine level is directly related to the mental stress of a person. Along with other health parameters, monitoring dopamine levels can predict the more accurate mental stage of the patient, as in Fig. 6.3. By understanding the patient's medical history and daily routines, it will be easy to predict the mood swings of the patient accurately using AI and ML (Graham et al., 2019; Tutun et al., 2023; Vaishnavi et al., 2022; Fareesha et al. 2023; Su et al., 2020). This can make the caregiver take the necessary steps to calm down the patient. Real-time monitoring of patients with mental illness is proposed in Fareesha et al. (2023). Incorporating AI and ML algorithms in data analysis can make the predictions accurate. Vaishnavi et al. (2022) provide a comprehensive overview of machine learning algorithms utilized for the detection and diagnosis of depression and also present a structured examination of ML-based algorithms for detecting depression, categorizing them into classification, deep learning, and ensemble approaches. Notably, machine learning techniques can support the diagnosis of various mental health conditions, including PTSD, schizophrenia, depression, ASD, and bipolar disorders. Recent investigations emphasize support vector machine (SVM) as a frequently used classifier, valued for its accuracy in depression detection, mainly due to its adaptability to unstructured and high-dimensional data. This critical evaluation contributes significantly to our comprehension of the current research landscape. Overall, this chapter presented a valuable synthesis of

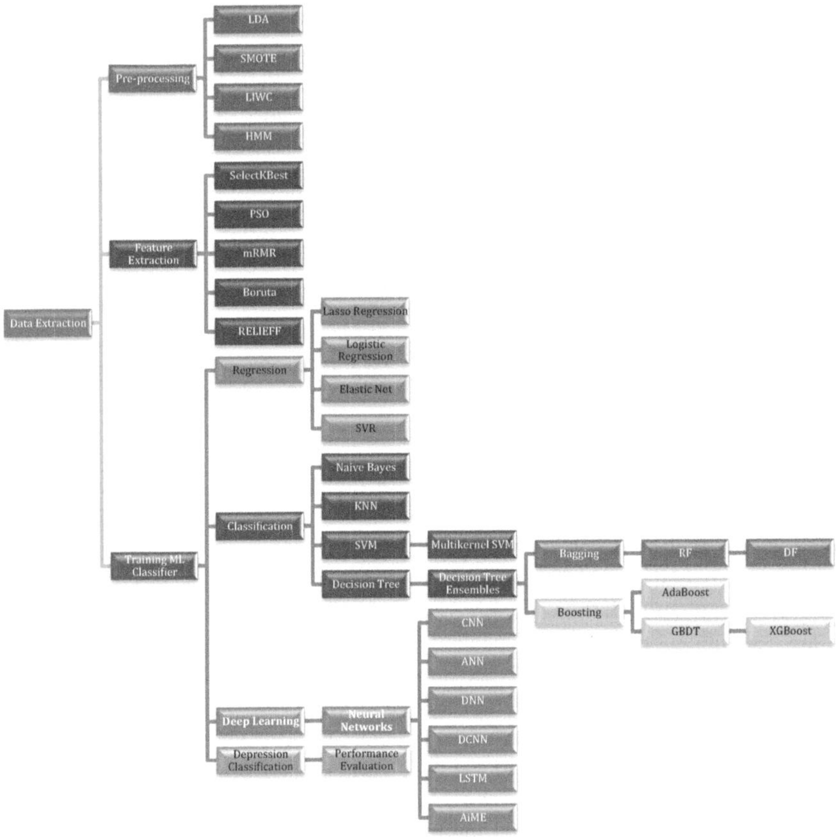

Fig. 6.3 AI and ML models used in depression diagnosis

current research, set the stage for future investigations, and provided a roadmap for developing effective depression detection systems.

6.6 Applications of AI and ML in Discerning Speech of People with Speech Difficulty

In a fast-paced world where people wish to listen, comprehend, and act quickly, it is indeed a challenge for people with speech disabilities like dysarthria, stuttering, and apraxia to communicate with normal people. Comprehending the speech of people with speech difficulties is a challenge indeed. Those who are close kin to such a person may be able to decipher what is being said for such a person to effortlessly communicate with the fast- paced rank and file their rises the need to develop speech recognition systems for the several speech disabilities. But that being the scenario, if the same speech is passed through any speech recognition engine, that would definitely not be able to catch the words the person has spoken. Recognizing speech of people with speech difficulties is indeed a challenge using most of the conventional techniques (Vadwala et al., 2017). Deep learning techniques have been well deployed in the last decade to tailor-make dysarthric speech recognizers. Acoustic, lexical, and language models are designed using deep neural networks, DNN. Convolutional neural networks (CNN) and recurrent neural networks (RNN) have played a pivotal role in advancing speech recognition models across various vocabularies, ranging from small to large. Researchers have explored a variety of acoustic features to train these models, including linear predictive coding (LPC), linear prediction cepstral coefficients (LPCC), discrete wavelet transform (DWT), Mel-frequency cepstrum coefficients (MFCC), perceptual linear prediction (PLP), and linear discriminant analysis (LDA).

Based on the type of speech difficulty, the most optimum feature has to be chosen. At times, a combination of features would be required to train a model. The complexity of the training process abounds from the fact that the variability involved is enormous, especially for a disability like apraxia or dysarthria. Often, MFCC features and wavelet-based features are used to train models for dysarthric speech recognition. Exploring the difficulties encountered by individuals with dysarthria, often resulting from damage to bilateral superior motor neurons, speakers with spastic dysarthria frequently experience erratic speech patterns and uncontrolled loudness. This can lead to increased strain on articulation muscles, reduced muscular strength, and diminished soft palate elevation. Additionally, less common types, such as atonic and ataxic dysarthrias, present their unique challenges. Symptoms of atonic dysarthria may include slow speech, irregular loudness, sudden pronunciation, inappropriate pauses, and an overabundance of nasal speech sounds. The impairment is caused by injury to the cerebellum, which affects the regulation of articulatory muscles and makes directing and scaling motions difficult. This makes it challenging to control tongue movement and alternate directions. The type of articulation problems differs from patient to patient (Qian & Xiao, 2023). This highlights the role of proper feature selection and the intervention of a deep learning model where one need not worry about customizing the model too much. Either a generic model can be built, or a user-specific one can be made using a deep neural network model. The application of the Kullback–Leibler divergence-based hidden Markov model (KL-HMM) for recognizing dysarthric speech exhibits promising results, showcasing its potential to effectively discern patterns within dysarthric speech (Rampello et al., 2016). Automatic speech recognition accuracy can be improved for patients suffering from aphasia, apraxia, and dysarthria speech using electroencephalography (EEG) data in unison with a deep learning approach to improve automatic speech recognition. The technique improves decoding efficiency by 50% for isolated tasks and has tentative results for continuous activities (Kim et al, 2019). The goal of this study is to create real-time strong speech prostheses for stroke patients recovering from these disorders, thus raising the bar for them to communicate with the people around them. AI plays a significant role in the assessment of speech disabilities as well. Ataxia is a disorder where one loses muscle control in the arms and legs. A patient with the same can eventually develop dysarthria. A groundbreaking hierarchical machine learning model (HMLM) has been created to enhance the identification and evaluation of dysarthria through a structured speech disturbance test (Krishna et al., 2021; Tartarisco et al., 2021). The model demonstrates a remarkable accuracy of around 90% in its initial stage, where it carefully selects an optimal subset of traditional features. Progressing to the second stage, it employs a deep learning feature extraction technique to assess the severity of speech disturbances, achieving a commendable maximum accuracy of approximately 80%. Through empirical evidence, the study highlights the superior performance of the HMLM over deep learning approaches, positioning it as an up-and-coming tool for the early diagnosis of ataxia. Such advancements offer significant reassurance in achieving precise diagnoses and facilitating subsequent therapeutic interventions for affected individuals. Thus, AI-ML improves the quality of life for people with speech disabilities.

6.7 Conclusion

In this chapter, the applications of AI and ML in various health sectors like neonatal care, Alzheimer's disease prediction, improving hearing aids, predictions of mental health conditions, and applications in speech recognition is detailed. The integration and evolution of AI and ML within healthcare represent a monumental shift, significantly broadening their scope and impact in recent years. These cutting-edge technologies are reshaping numerous facets of healthcare delivery, ranging from the meticulous analysis of medical imaging to the intricate processes of drug discovery and the crafting of personalized treatment regimens. In the domain of medical imaging, AI and ML algorithms demonstrate unparalleled precision in deciphering data from a plethora of modalities, including X-rays, MRIs, and CT scans. Their adeptness facilitates the early detection of diseases, notably cancer, thereby enabling timely interventions and treatment stratification. Furthermore, AI and ML play an instrumental role in expediting drug discovery processes by harnessing vast datasets to predict the behavior of molecular compounds. This expedites the identification of potential drug candidates, ultimately streamlining the development pipeline and addressing unmet medical needs. Additionally, predictive analytics underpinned by AI and ML facilitate the forecasting of patient outcomes, thereby enabling healthcare professionals to identify individuals at heightened risk of specific diseases or complications. This proactive stance allows for timely interventions, mitigating disease progression and curbing healthcare expenditure. In healthcare facilities, AI and ML optimize operational efficiency by fine-tuning resource allocation, scheduling, and workflow management. By minimizing inefficiencies, these technologies elevate the overall quality of care delivery, culminating in superior patient experiences and outcomes. As AI and ML technologies continue to evolve, their applications in healthcare are poised for further expansion, offering unparalleled opportunities to revolutionize patient care, drive breakthroughs in pharmaceutical research, and optimize healthcare operations. Through judicious integration and utilization, AI and ML stand to profoundly transform and enhance the healthcare landscape, ushering in a new era of precision medicine and patient-centric care. However, it is imperative to navigate ethical and regulatory considerations meticulously to ensure the seamless integration of these advancements, safeguarding patient safety, privacy, and ethical integrity.

References

Abbas, A. K., & Leonhardt, S. (2014). Intelligent neonatal monitoring based on a virtual thermal sensor. *BMC Medical Imaging, 14*, 1–13.

Adegboro, C. O., Choudhury, A., Asan, O., & Kelly, M. M. (2022). Artificial intelligence to improve health outcomes in the NICU and PICU: A systematic review. *Hospital Pediatrics, 12*(1), 93–110.

Akundi, S., Soujanya, R., & Madhuri, P. M. (2020). Big data analytics in healthcare using machine learning algorithms: A comparative study.

Aleem, S., Huda, N. U., Amin, R., Khalid, S., Alshamrani, S. S., & Alshehri, A. (2022). Machine learning algorithms for depression: Diagnosis, insights, and research directions. *Electronics, 11*(7), 1111.

Aviles-Espinosa, R., Rendon-Morales, E., Luo, Z., Dore, H., Anton, O., Rabe, H., & Prance, R. J. (2019). Neo-SENSE: a non-invasive innovative sensing mattress for cardiac monitoring of babies. In *IEEE sensors applications symposium (SAS)* (pp. 1–5). IEEE.

Babu, A. S., Yashika, K. P., Kushalappa, K. N., Adiga, A., Raj, S. B., Mrudula, N., Jayachandran, R., et al. (2023). NeuroPredict: A comprehensive approach for Alzheimer's disease prediction and development of a medical assisting kit. In *International conference on recent advances in science and engineering technology (ICRASET)* (pp. 1–6). IEEE.

Barragán-Montero, A., Javaid, U., Valdés, G., Nguyen, D., Desbordes, P., Macq, B., & Lee, J. A. (2021). Artificial intelligence and machine learning for medical imaging: A technology review. *Physica Medica, 83*, 242–256.

Castiglioni, I., Rundo, L., Codari, M., Di Leo, G., Salvatore, C., Interlenghi, M., Gallivanone, F., Cozzi, A., D'Amico, N. C., & Sardanelli, F. (2021). AI applications to medical images: From machine learning to deep learning. *Physica Medica, 83*, 9–24.

Cay, G., Solanki, D., Ravichandran, V., Hoffman, L., Laptook, A., Padbury, J. & Mankodiya, K. (2021). Baby-Guard: An IoT-based neonatal monitoring system integrated with smart textiles. In *IEEE international conference on smart computing (SMARTCOMP)* (pp. 129–136). IEEE.

Chandrasekaran, A., Amboiram, P., Balakrishnan, U., Abiramalatha, T., Rao, G., Jan, S. M. S. & Ninan, B. (2021). Disposable low-cost cardboard incubator for thermoregulation of stable preterm infant–a randomized controlled non-inferiority trial. *EClinicalMedicine, 31*.

Datta, S., Barua, R., & Das, J. (2019). Application of artificial intelligence in modern healthcare system. In *Alginates—Recent Uses of this Natural Polymer.* Leonel Pereira, IntechOpen.

Ebrahimi, A., Luo, S., & Disease Neuroimaging Initiative, F. T. A. S. (2021). Convolutional neural networks for Alzheimer's disease detection on MRI images. *Journal of Medical Imaging, 8*(2), 024503–024503.

Fareesha, F., Chandanashree, Y. K., Gowthami, V., Jayachandran, R., & Kalathil, S. (2023). Real-time artificial mood-tracking and health-monitoring system (RAMAHS) for people with mental illness and their caregivers. In *International conference on recent trends in electronics and communication (ICRTEC)* (pp. 1–6). IEEE.

Gopalan, S. S., Raza, A., & Almobaideen, W. (2021). IoT security in healthcare using AI: A survey. In *International conference on communications, signal processing, and their applications (ICCSPA)* (pp. 1–6). IEEE.
Graham, S., Depp, C., Lee, E. E., Nebeker, C., Tu, X., Kim, H. C., & Jeste, D. V. (2019). Artificial intelligence for mental health and mental illnesses: An overview. *Current Psychiatry Reports, 21*, 1–18.
Gulzar Ahmad, S., Iqbal, T., Javaid, A., Ullah Munir, E., Kirn, N., Ullah Jan, S., & Ramzan, N. (2022). Sensing and artificial intelligent maternal-infant health care systems: A review. *Sensors, 22*(12), 4362.
Hamacher, V., Fischer, E., Kornagel, U., Puder, H., Hänsler, E., & Schmidt, G. (2006). Applications of adaptive signal processing methods in high-end hearing aids. To*pics Acoustic Echo and Noise Control*, 599–636.
Henry, C., Saffaran, S., Meeus, M., Bates, D., Van Laere, D., Boylan, G., Sharkey, D., et al. (2022). Application and potential of artificial intelligence in neonatal medicine. *Seminars in Fetal and Neonatal Medicine*, *27*(5), 101346 WB Saunders.
Iliadou, E., Su, Q., Kikidis, D., Bibas, T., & Kloukinas, C. (2022). Profiling hearing aid users through big data explainable artificial intelligence techniques. *Frontiers in Neurology, 13*, 933940.
Jiang, F., Jiang, Y., Zhi, H., Dong, Y., Li, H., Ma, S., & Wang, Y. (2017). Artificial intelligence in healthcare: past, present and future. *Stroke and Vascular Neurology*, *2*(4).
Kelly, C. J., Karthikesalingam, A., Suleyman, M., Corrado, G., & King, D. (2019). Key challenges for delivering clinical impact with artificial intelligence. *BMC Medicine, 17*, 1–9.
Kim, M. J., Yun, J., Cho, Y., Shin, K., Jang, R., Bae, H. J., & Kim, N. (2019). Deep learning in medical imaging. *Neurospine, 16*(4), 657.
Krishna, G., Carnahan, M., Shamapant, S., Surendranath, Y., Jain, S., Ghosh, A., ... & Tewfik, A. H. (2021, November). Brain signals to rescue aphasia, apraxia and dysarthria speech recognition. In *2021 43rd annual international conference of the IEEE engineering in medicine & biology society (EMBC)* (pp. 6008–6014). IEEE.
Kumar, L., Kumar, S., Manoj, M., Kulkarni, A., Dantapur, B., & Jayachandran, R. (2023). WARM WOMB: A wireless real-time monitoring system for neonatal care in resource-limited settings. In *International conference on recent advances in science and engineering technology (ICRASET)* (pp. 1–6). IEEE.
Lin, W., Tong, T., Gao, Q., Di, G., Du, X., Yang, Y., Guo, G., et al. (2018). Alzheimer's Disease Neuroimaging Initiative, Convolutional neural networks-based MRI image analysis for the Alzheimer's disease prediction from mild cognitive impairment. *Frontiers in Neuroscience, 12*, 412254.
Malak, J. S., Zeraati, H., Nayeri, F. S., Safdari, R., & Shahraki, A. D. (2019). Neonatal intensive care decision support systems using artificial intelligence techniques: A systematic review. *Artificial Intelligence Review, 52*, 2685–2704.
Nagendran, M., Chen, Y., Lovejoy, C. A., Gordon, A. C., Komorowski, M., Harvey, H., & Maruthappu, M. (2020). Artificial intelligence versus clinicians: systematic review of design, reporting standards, and claims of deep learning studies. *bmj*, 368.
Nayak, P. A., Shayana, S. A., Raksha, A. R., HS, S. S., & Jayachandran, R. (2024, July). A Compact and Low Power Memristor Based 8-Bit Digital to Analog Converter for Hearing Aid Devices. In *2024 IEEE international conference on electronics, computing and communication technologies (CONECCT)* (pp. 1–6). IEEE.
Patil, V., Madgi, M., & Kiran, A. (2022). Early prediction of Alzheimer's disease using convolutional neural network: A review. *The Egyptian Journal of Neurology, Psychiatry and Neurosurgery, 58*(1), 130.
Qian, Z., & Xiao, K. (2023). A survey of automatic speech recognition for dysarthric speech. *Electronics, 12*(20), 4278.
Rampello, L., Rampello, L., Patti, F., & Zappia, M. (2016). When the word doesn't come out: A synthetic overview of dysarthria. *Journal of the Neurological Sciences, 369*, 354–360.

Rizk, D. K., Hosny, H. M., Voskoglou, M. G., El-Horbaty, E. S. M., & Salem, A. B. M. (2023). Artificial Intelligence Algorithms for Healthcare Services. *American Journal of Applied Mathematics and Statistics, 11*(2), 70–76.

Su, C., Xu, Z., Pathak, J., & Wang, F. (2020). Deep learning in mental health outcome research: A scoping review. *Translational Psychiatry, 10*(1), 116.

Tartarisco, G., Bruschetta, R., Summa, S., Ruta, L., Favetta, M., Busa, M., ... & Pioggia, G. (2021). Artificial intelligence for dysarthria assessment in children with ataxia: A hierarchical approach. *IEEE access, 9*, 166720–166735.

Tutun, S., Johnson, M. E., Ahmed, A., Albizri, A., Irgil, S., Yesilkaya, I., Ucar, E. N., Sengun, T., & Harfouche, A. (2023). An AI-based decision support system for predicting mental health disorders. *Information Systems Frontiers, 25*(3), 1261–1276.

Vadwala, A. Y., Suthar, K. A., Karmakar, Y. A., Pandya, N., & Patel, B. (2017). Survey paper on different speech recognition algorithm: Challenges and techniques. *International Journal of Computers and Applications, 175*(1), 31–36.

Vaishnavi, K., Kamath, U. N., Rao, B. A., & Reddy, N. S. (2022). Predicting mental health illness using machine learning algorithms. *Journal of Physics: Conference Series, 2161*(1), 012021. IOP Publishing.

Venkatesh, D. A. N. (2019). Reimagining the future of healthcare industry through internet of medical things (IoMT), artificial intelligence (AI), machine learning (ML), big data, mobile apps and advanced sensors. Artificial Intelligence (AI), Machine Learning (ML), Big Data, Mobile Apps and Advanced Sensors (October 28, 2019).

Zhang, N., Wood, O., Yang, Z., & Xie, J. (2023). AI-guided computing insights into a thermostat monitoring neonatal intensive care unit (NICU). *Sensors, 23*(9), 4492.

Chapter 7
Exploring Ethical Aspects and Regulatory Framework for India's Sustainable Healthcare Sector in the Era of Artificial Intelligence

Mrinalini Choudhary, Akash Gupta, Arushi Bajpai, and Rubee Singh

Abstract Artificial intelligence (AI) tools like Chat GPT have gained popularity all over the world. The potential of AI has been embraced in some sectors, such as the legal industry, education, and marketing. However, the integration of AI in the healthcare sector has sparked a debate. In jurisdictions like the EU, USA, and Canada, AI is being utilised for medical purposes, accompanied by regulations and guidelines. Conversely, in countries like India, where regulations for AI in healthcare are still developing, there exists a lack of clarity on regulatory frameworks. This chapter analyses regulations from other jurisdictions and identifies challenges India may face in formulating its own. Drawing on an empirical survey involving participants from diverse sectors, the chapter highlights ethical concerns surrounding AI integration in healthcare, including data misuse and patient privacy breaches. The survey underscores the need for robust regulations to govern AI deployment in India's healthcare sector. Additionally, the chapter emphasises the importance of having a sustainable healthcare sector, which ensures sustainability practices are embedded within AI-driven healthcare services. While existing legislation, like the Information Technology Act of 2000 and the National Health Policy of 2017, offer some safeguards, the absence of specific laws poses challenges. Taking inspiration from regulations like the General Data Protection Regulation (GDPR) of the EU, India can develop laws empowering patients to control their data usage, ensuring transparency and accountability in AI-driven healthcare services.

Keywords Artificial intelligence · Healthcare · Regulations · Ethical concerns · Data privacy · Security

---

M. Choudhary
School of Business, Shenandoah University Virginia, Winchester, VA 22601, USA

A. Gupta · A. Bajpai
Jindal Global Law School, Jindal Global University, Sonipat, Haryana, India

R. Singh (✉)
Institute of Business Management, GLA University, Mathura, India
e-mail: dr.rubeerajput@gmail.com

R. Singh et al. (eds.), *Transforming Healthcare Sector Through Artificial Intelligence and Environmental Sustainability*, Approaches to Global Sustainability, Markets, and Governance, https://doi.org/10.1007/978-981-97-9555-0_7

7.1 Introduction

Artificial Intelligence is a word-of-mouth for people across continents. It has permeated into different walks of life of the global population. The most recent example is Chat GPT, which has become extremely popular in the past few years due to its benefits in different sectors of the economy. In the legal industry, it has immense benefits by reducing the workload and increasing the efficiency of the employees. Multiple global law firms have adopted artificial intelligence tools. Even though there are problems associated with these tools, their judicious use is still beneficial in the long run from the perspective of employers because they can use these tools instead of hiring candidates and paying them salaries. Even though AI tools have been accepted by some industries, such as the legal industry and education industry, to some extent, their adoption in healthcare has sparked significant debates. In some jurisdictions like the EU, USA, and Canada, AI is used for medical purposes. Due to its utilisation, certain guidelines and regulations have been formulated. However, in some developing countries like India, there is not an omnipresent knowledge about employing AI tools in the healthcare sector. As a result, India is still in the nascent stages of framing regulations for the adoption of AI in the healthcare sector. With a lack of laws, legislations, and regulations specifically catering to this pertinent issue there is a lack of clarity about how the usage of AI tools is to be regulated. This present chapter seeks to analyse the rules that have been developed by other jurisdictions and attempts to highlight the issues that India will be facing in making such rules by drawing on an empirical study with participants engaged in a wide array of sectors.

7.2 Use of AI in Healthcare Services

The Health Team of DeepMind and Google Health partnered to launch AI-driven tools for the determination of the presence of breast cancer in 2019. It was a landmark occasion for using AI-based solutions in the healthcare sector (Reid, 2020). This incident was highly instrumental and revolutionary in the delivery of healthcare services because AI algorithms produced better results than human radiologists. AI can significantly help in the early diagnosis of diseases, quick decision-making, and research and training by analysing data from a vast pool of patients. Moreover, it can lead to the transformation of medical care rendered to the patients (Davenport & Kalakota, 2019). Moreover, the technology can sift through and analyse a large chunk of clinical data collected from patients and help medical professionals understand various trends in the spread of diseases (Foresee Medical, 2024). Upgradation and development in Machine Learning algorithms, lower cost of hardware, and an increase in internet connectivity have expedited the integration of AI software in medical care (IBM, 2024). Multiple products are still at their developmental stage.

It is essential to follow an approach where the strategy for the development is formulated keeping in mind the interests of the various stakeholders involved (Bajwa et al., 2021). Moreover, AI can help in saving a multitude of lives. Some people urgently require medical care, but they are underprivileged and do not possess the financial resources to get access to healthcare. AI in healthcare has the potential to treat such people and save them from life-taking diseases (World Economic Forum, 2024).

The healthcare sector is one of the most highly valued and is experiencing exponential growth, especially in light of the post-pandemic world. The increasing evolution of digital technology and machine learning is transforming medicinal research (Precedence Research).

However, with the integration of AI-driven tools and algorithms in healthcare services, there are concerns regarding the protection of patient data from unauthorised access. Moreover, there are ethical considerations, such as giving the patient express consent to use their data for the purpose of feeding it into the machine to create an algorithm. It is also pertinent to tackle the controversy that may arise when there is a conflicting opinion between AI-driven technological advice and the advice rendered by the human practitioner. There are pertinent social, political, and ethical considerations that have been flagged by multiple stakeholders such as private hospitals, patients, governmental organisations, doctors, etc.

A report titled 'Ethics and Governance of Artificial Intelligence in Healthcare' was published by WHO in 2021. It highlighted the need for ethical principles to shape the actions of governmental organisations, companies, developers of AI tools, civil society, etc., in the application of such AI-driven tools and algorithms for medical advice (WHO, 2021).

7.2.1 Autonomy of Human Medical Experts

There can be a situation where there is a difference between the medical advice rendered by artificial intelligence technology and the opinion of a human medical practitioner. Since diagnosis and treatment can be life-threatening in some instances, it is essential to ensure that a greater degree is given to the human medical practitioner to reject the opinion of the AI technology and give priority to his opinion. The oversight of a human practitioner is quintessential in making judicious use of technological advancement. This is because it might happen the algorithm is biased towards a specific community due to the data that has been fed into the system. Such prejudices and stigmatisation that may occur will scuttle the entire purpose of using highly state-of-the-art technology for better medical care. Therefore, continuous monitoring and supervision of the algorithms that are utilised for rendering medical opinions have been conducted. Constant evaluation ensures that the algorithm does not damage the interest of any community or any individual (WHO).

Moreover, the integration of AI in clinical care will pose particular challenges that need to be overcome. If the policies do not hold the human medical practitioner liable, then such a human practitioner would be stimulated to give priority to the opinion of the AI technology rather than giving superiority to his analysis of the patient. However, if the human practitioner is penalised for using AI-driven technology to provide healthcare services, then the human practitioner will refrain from taking advantage of the technology. Such a practitioner would constantly challenge the advice rendered by AI technology. He would only take the advice of the AI technology when the opinion of the machine is in line with his professional judgement. Both the extremes are equally troublesome. Human practitioners will not be held liable in a situation when harm is caused to an individual if the practitioner follows the standard of care. In that case, even if the AI-driven technology suggests a better course of action but the human practitioner refuses to adhere to that, then also he would not be held legally liable. However, in instances where the medical advice rendered by technology is germinating from a comprehensive set of data and can be superior from the judgment of the human professional because he lacks a vast dataset, then in that case, the human practitioner would be rejecting the opinion provided by technology in the trepidation of being held legally liable (WHO). The method of treatment prescribed by AI technology may deviate from the standard of care. The human practitioner, in order to avoid any chance of being held liable, may overlook the capacity of technology to positively contribute to patient care by suggesting dosages of medicines (Price et al., 2019). Even institutions such as hospitals can be held liable for the ill effects of AI recommendations. Hospitals can be legally held to be vicariously liable for the actions of their doctors in recommending the treatment to the patient by relying on AI-driven technology (WHO, 2021). Hospitals can be held liable for introducing, using or misusing the AI-driven technology. Therefore, it would be advantageous if hospitals conduct their due diligence on the selection of AI machines and give clear-cut guidelines to their staff to ensure the use of machine learning technology.

Therefore, the regulations that need to be established must be adaptive and strike a delicate balance. The guidelines should ensure that medical practitioners are able to utilise and integrate AI technologies in their medical treatment, and they should hold practitioners liable for the mistakes they commit during the process. Even though the task seems prima facie challenging, it is still pertinent to ensure that the regulations are drafted comprehensively. Moreover, it is quintessential that legal regulations are constantly developing as technology advances. The aim of regulations, laws, and legislations governing the use of AI in healthcare should be to foster innovation and guarantee that the data collected from the patients is kept safe and secure.

7.2.2 Ensuring Transparency

To make the adoption of AI a smooth process, the information that is used to create the software can be published in the public domain. This approach to publication will ensure that there can be a discussion about the development of such technologies. Moreover, even after the AI tool is developed and has received the necessary approval from the relevant authorities, the data that has been fed into the system can also be placed for the public to discuss. Such an approach in the pre-development and post-development stages will increase the trust of the consumers that the technology that they are using for curing their diseases is safe. Moreover, their trust in the company that built the particular software would increase as they would have access to the information. The entire process needs to be transparent as the technology relies on a vast mass of patient data that may vary across demographic regions.

7.2.3 Ensuring Responsibility and Accountability

Even when there is human supervision of the development of algorithms, there is still a slight possibility that the AI technology does not give results as intended. Due to some errors committed in the collection of Big Data, there can be a situation when a specific group of people are not adequately treated or are, in fact, injured by the medical care provided by AI technology. The regulatory bodies, while formulating laws and guidelines on the use of AI in the healthcare sector, need to ensure that the companies who are developing such software are held accountable for their product. The government or the developer company should give the group that has faced the harm some relief in the form of rehabilitation and compensation.

One way to secure compensation for the affected group is to make the manufacturers of the AI-driven products liable for any injury caused by their products. Placing liability on the manufacturers would mean that even in instances when there is no negligence on their end, they can still be held liable and mandated to pay compensation when an individual or a group suffers harm by using their product. An exception that is carved from the principle of product liability is when the product is used for an object that is not the primary object for the product created (Minssen et al., 2020). This will even compel companies to observe a high standard of care when manufacturing this product. Another benefit of extending the liability to the manufacturers is that the affected individuals or groups can claim compensation from them. Otherwise, if the developers of the technology are not held liable, it would be excruciatingly difficult to claim compensation.

The liability of manufacturers depends a great deal on the amount of control they exercise over the technology post-sale. Manufacturers can also avail of the 'development risk defence' that is followed in EU countries, which limits their liability based on the degree of control they possess. Manufacturers are not strictly liable for their products in these EU member states (European Parliament, 2020).

There is a doctrine of 'learned intermediary' that the developers can take advantage of to avoid being held liable. This doctrine stipulates that if appropriate warnings and cautions are conveyed to the medical service provider and such medical service provider, in its professional capacity, prescribes the diagnosis using AI technology, then the human medical service provider will be held liable. This is because the chain of causation between the patient who has been harmed and the developer has been broken by the doctor, who acts as an intermediary in prescribing a treatment using the technology. The doctor makes a professional decision to render a specific treatment to the patient. Therefore, this doctrine holds the doctor liable for their poor judgement in using an AI technology (Price, 2021).

Another crucial doctrine of 'pre-emption' merits in-depth analysis. When the regulatory body approves AI-driven technology, civil liability claims are trumped. It is assumed that since the regulatory body has granted its validation, it is safe to use. However, in cases when there is a medical emergency, regulatory approval might have been granted as a measure to counter that medical emergency. It is also equally possible that the approval was granted without an in-depth examination of the algorithm and analysis of Big Data that has been fed into the system. Such a possibility may arise because of highly intricate AI algorithms and the lack of available data for analysis. Such regulatory approval without scrupulous examination would not be sufficient to shield the manufacturers from lawsuits pertaining to civil liability claims.

In low and middle-income countries, there are pressing concerns regarding the regulatory approval of technologies that are dependent on algorithms, machine learning, etc. Since such countries do not possess the technical expertise to assess and examine the potential of these technologies, it isn't easy to ensure that the technologies are safe for use in such regions of the world. Certain sections of the population in these low and middle-income countries are the most vulnerable to the dangers of AI tools. This is because if they are negatively impacted, then it would be tough for them to claim compensation against influential developers. There has to be careful calibration in formulating policies, taking into consideration the risk–benefit ratio of employing AI technologies in low- and middle-income countries and creating a system for claiming compensation. These policies regarding liability need to be designed together by the local governments of these countries along with international organisations that are suitable to the peculiar situation of these regions.

It is crucial to ensure that the data that is fed into the machine learning program is diverse and collected from multiple backgrounds. This would make it more probable that there is inclusivity in the analysis and designing of data. The algorithm that is created must be designed by accounting for different demographics, multiple age groups, along with disparity in income among a pool of patients. This can be achieved by having a workforce that belongs to a multitude of backgrounds so that they can bring varied viewpoints at every stage of the development of AI technology, such as production, scrutiny, and deployment.

7.2.4 Regulatory Framework of USA

The relevant authority for verifying and determining the security and efficaciousness of AI technologies in the healthcare sector is the Food and Drug Administration (FDA) of the USA. Software that is used for medicine and has the potential of working independently without any integration with hardware medical equipment has been called by the International Medical Device Regulators Forum as Software as a Medical Device. The FDA categorises a medical objective as one with an aim to diagnose, treat, cure, mitigate or prevent disease. For example, software which aids in diagnosing breast cancer by using MRI image analysis or computer-aided detection would be defined as software with a medical objective (FDA, 2017).

Since compartmentalising software and AI-driven technologies as medical devices is difficult, it is hard to determine and document the total number of AI devices that the FDA considers medical devices. Since October 2023, the FDA has granted approval to 692 medical devices powered by AI and Machine Learning. These AI-powered devices have been marketed after requiring the necessary approvals from the FDA (FDA, 2022).

FDA undertook a pilot project to determine the efficiency and feasibility of data collection at each developmental phase of an AI-driven software product. Through such collection, FDA sought to examine and scrutinise the medical device for its safety. The data had to be collected and monitored at each stage of the production cycle (Montanez, 2022). The results of the pilot project showed that the method was not efficacious.

7.2.5 Legal Framework in Canada

Canada follows a licensing system where it is legally necessary for the developers of the company to acquire a license to sell their AI technology in the country. Some of the products that integrate machine learning technology have already obtained a license. For example, products such as diagnostic imaging and radiology have secured the appropriate license (Dumouchel, 2019). The developers are mandated to abide by the established Medical Devices Regulations, which provide a two-step system. Manufacturers must identify if the software they are creating acts as a medical device under the regulations. However, there is a problem with such identification enshrined in the regulations (Regis & Flood, 2021). There are specific software that merely interconnects the symptoms in the patients to the experimental conclusions and international high-standard practices to render medical advice. Since these applications only facilitate and give an analysis that is taken into account in the decision-making of human medical professionals without playing a direct part in diagnosing the patient's condition such applications do not fall under the supervision of the regulations (Regis & Flood, 2021). Moreover, the regulations do not include applications that may use AI technology for administrative purposes under its ambit (Ministry of Canada, 2019). Moreover, a major problem in regulating the software that works on machine learning is that since it evolves with the algorithm it is difficult to monitor it constantly.

7.2.6 Legal Framework in India

The provisions of the Digital Information Security in Healthcare Act (DISHA) Bill and Personal Data Protection would regulate the deployment of AI technologies in healthcare services. Moreover, the Indian Council for Medical Research has formulated guidelines namely, 'Guidelines for Application of Artificial Intelligence in Biomedical Research and Healthcare' for curating an ethical system for producing, developing and deploying AI-driven technology in the healthcare sector (ICMR). There are certain guiding principles that need to be followed while developing the software. The hospitals must possess an awareness of how the data that is collected by them will be utilised. Moreover, there can be a system in place through which the human practitioner can provide feedback to the developers of the software. Moreover, if data of the patients are collected then their consent must be obtained along with a mechanism through which patients can remove their data if they so desire. Moreover, the patients should have the autonomy to decide if their data can be used or not be used in the development of the algorithm. It is quintessential to ensure that the patients are given full knowledge about the risks along benefits of the use of their data and its integration into healthcare services. ICMR has also suggested that if there is any MOU between two companies developing the software then they ensure that the data of patients remain confidential and secure. There should not be any breach of privacy of the data of the patient. It has been emphasised that the software needs to be tested at multiple stages scrupulously to prevent ethical lapses and discrimination against the marginalised sections of the population. In case there is any bias that the algorithm shows during the testing stage then the program should be paused. When the necessary corrections are made in the algorithm then only the program should be restarted to ensure that there is optimal performance without any health hazards.

7.3 WHO Recommendations on the Use of AI in Healthcare

WHO came with a certain set of guidelines titled, 'Ethics and governance of artificial intelligence for health' (Jasarevic, 2021). It outlined six crucial principles of ethics that should govern the development, deployment and integration of machine learning technology in medical healthcare. WHO highlighted the need to ensure that human medical practitioners are able to exercise autonomy over their own decisions while at the same time ensuring that the patient data remains confidential. Moreover, the applications that are developed should be compliant with the regulatory and efficacy requirements. The data that is utilised in the algorithm should be available to public for its scrutiny before actually deploying it in the medical infrastructure. WHO strongly recommends through its guideline that there should be proper training of the stakeholders such as healthcare personnel who use this technology in practice.

7.4 Results of Empirical Study Undertaken

An empirical study was conducted with approximately 70 participants selected randomly who are engaged in varied industries. Their views have been documented and analysed to understand the view of the public on the use of AI in the healthcare sector and the regulations surrounding it. The consumers of the healthcare services are one of the primal stakeholders because they are the ones that receive the services. The data which is collected is the information provided by the consumers. The healthcare services based on the opinion of artificial intelligence has its benefits. However, in certain cases, there can be grievous causes to the patient.

In Fig. 7.1, almost 41% of the service participants were from the hospitality sector. Another 11% belong to the legal sector. A significant proportion of the participants in this survey were engaged in the service sector. Around 14% of the participants came from the legal sector.

A humungous proportion of the participants more than three-fourths affirmed that they have used AI tools in their professional practice at some point in time as is evident in Fig. 7.2. This shows the entrenchment and penetration of AI-driven technology in a multitude of sectors. AI is integrated into multiple tools that are in use in different sectors.

Moreover, 63% of participants agreed that they are familiar with the integration of AI in the healthcare industry as is clear from Fig. 7.3.

The participants were slightly perplexed about whether AI technologies can be incorporated into medical diagnosis. Even though 52% of participants in Fig. 7.4 agreed that AI should be integrated into medical diagnosis, almost 40% were unsure of the answer. Approximately 8% of participants disagreed with integrating AI-driven technologies in medicine. These percentages clearly demonstrate that people are not very accepting of using AI technologies based on algorithms and machine learning for the healthcare sector (Fig. 7.5).

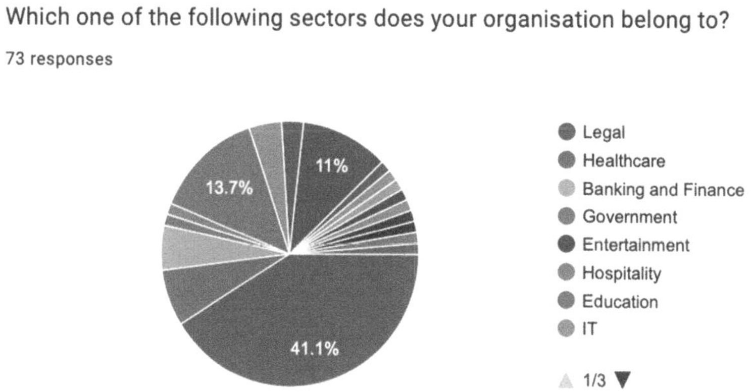

Fig. 7.1 Industry from which the participants belong (AI Empirical Survey, 2024)

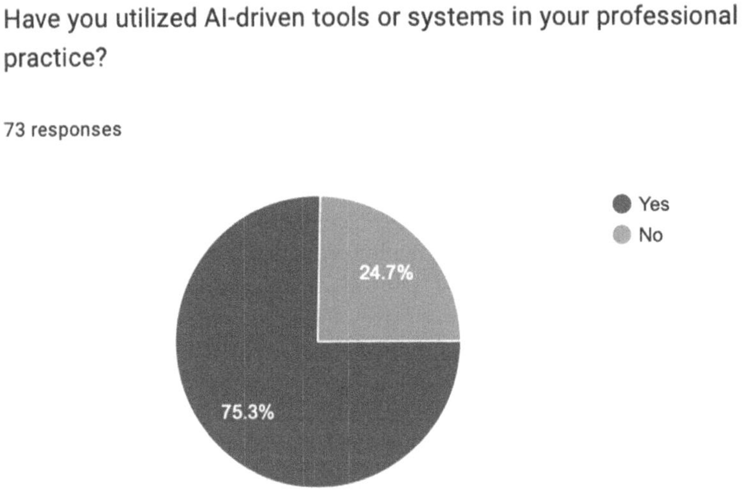

Fig. 7.2 Usage of AI tools (AI Empirical Survey, 2024)

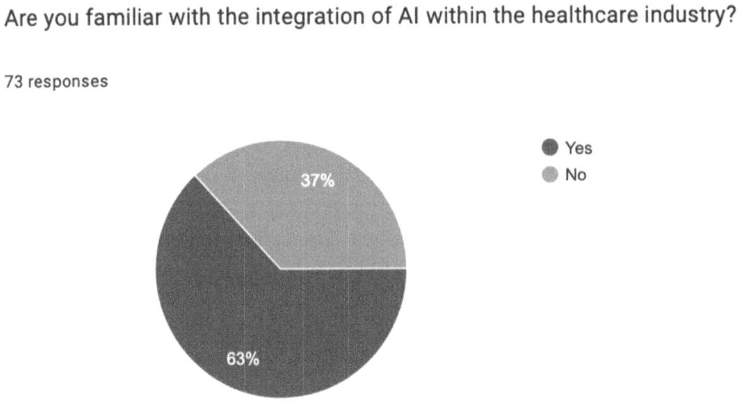

Fig. 7.3 Familiarity with AI in healthcare (AI Empirical Survey, 2024)

Almost 66% of participants felt that patients should have autonomy over the treatment they undertake. The patients should have a mechanism to make a conscious decision between getting treatment from a human medical practitioner or opting for a treatment that is rendered by technology powered by Artificial Intelligence. Another aspect that needs to be highlighted is that a large number of individuals choose that patients should not be given the option of choosing between treatment rendered by human practitioners and diagnosis by AI.

More than three-fourths of participants in Fig. 7.6 denied it when asked if they would consent to be treated by AI diagnosis when there is a possibility of life-threatening consequences. The likelihood of life-threatening consequences was a

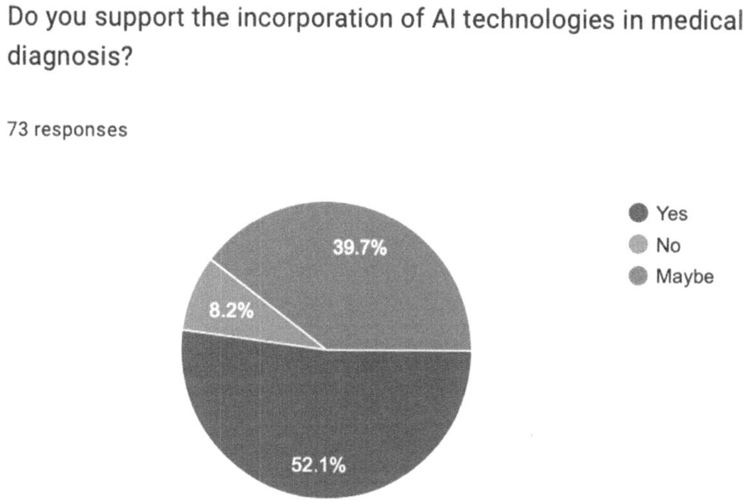

Fig. 7.4 Opinion on AI in healthcare (AI Empirical Survey, 2024)

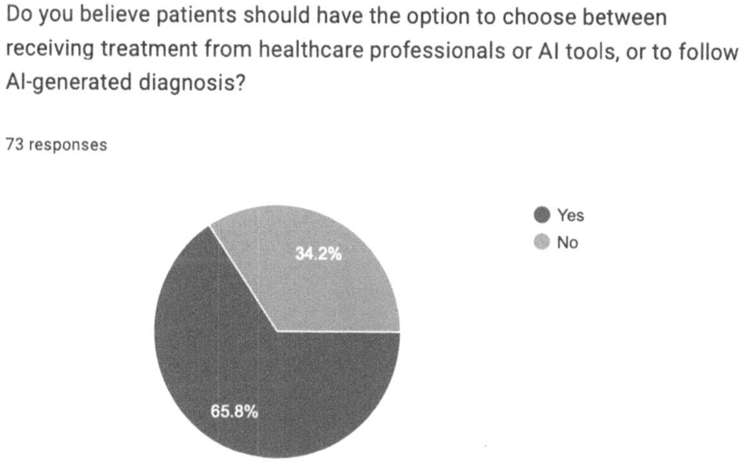

Fig. 7.5 Opinion on different types of treatment (AI Empirical Survey, 2024)

major reason for a majority of participants rejecting the diagnosis using AI and machine learning. However, a significant proportion of participants almost one-fourth of the total still agreed to be treated by AI-driven technology demonstrating their trust in technological advancement (Fig. 7.7).

A majority of participants, approximately 67%, agreed to store their personal data to help diagnose and treat patients. However, a large number of participants expressed discomfort with the storage and management of their data.

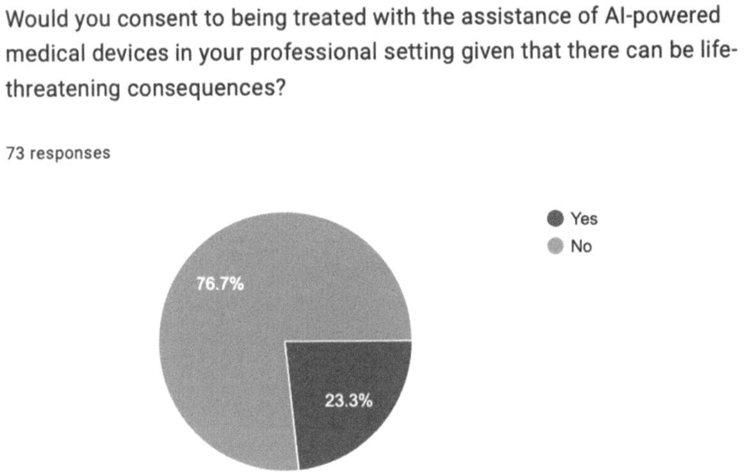

Fig. 7.6 Willingness to undertake medical services by AI (AI Empirical Survey, 2024)

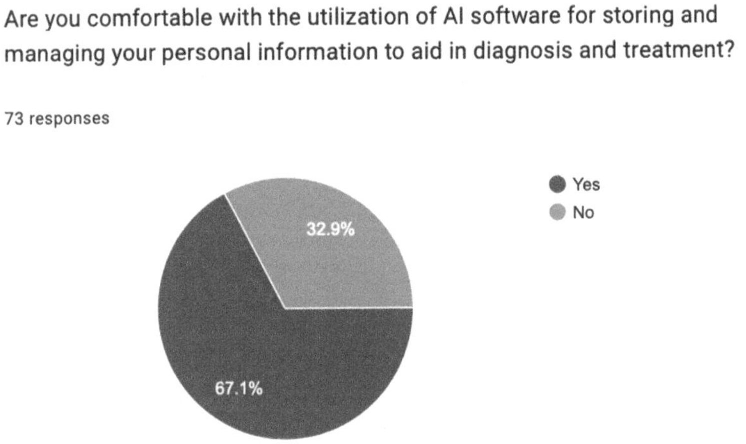

Fig. 7.7 Utilisation of patient's data by AI (AI Empirical Survey, 2024)

There was a razor-thin majority in the total number of participants on the question that they would consent to use their data in developing an algorithm for technology powered by AI as is evident by Fig. 7.8. Participants were almost equally divided on this question. While half of them would give their consent for the usage of their medical data for the creation of AI software, the other half were apprehensive about giving access to their data (Fig. 7.9).

A large number of participants were convinced that they would not recommend undertaking AI-diagnosed treatment to their family members and loved ones due to the possibility of life-threatening harm that may be caused. But even a substantial

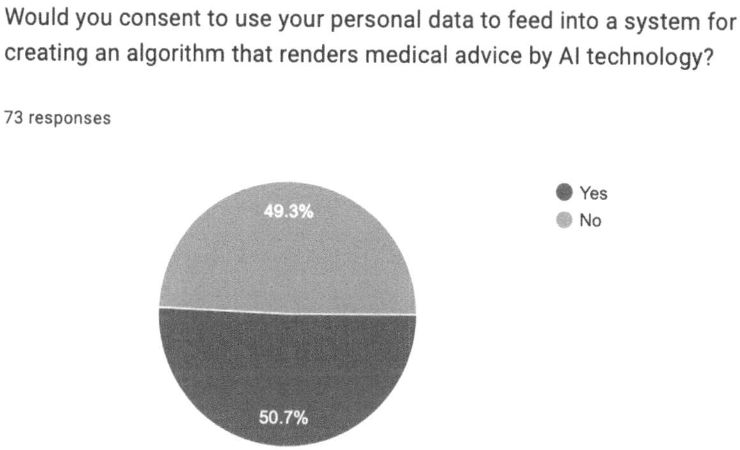

Fig. 7.8 Data for algorithm (AI Empirical Survey, 2024)

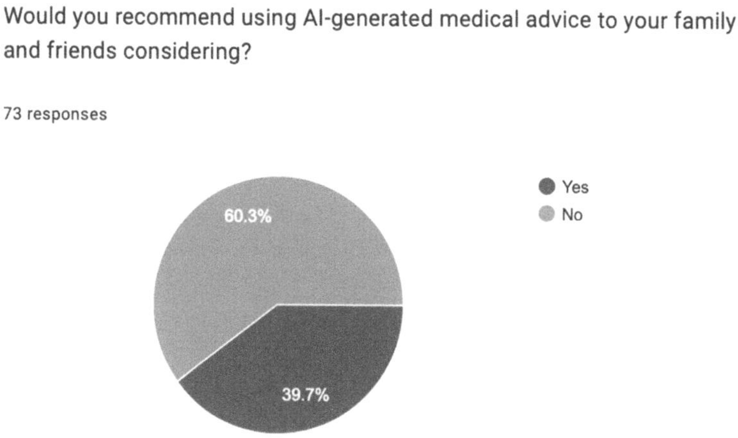

Fig. 7.9 Recommendation of AI-generated medical care (AI Empirical Survey, 2024)

number of participants amounting to approximately 40% would still recommend using the treatment which is rendered by AI-powered technology.

An overwhelming percentage of the participants showed a disinclination towards choosing the opinion of AI-powered technology over a human medical practitioner in Fig. 7.10. This disinclination is a clear indication of the fact that even though AI technologies may have benefits, still, they cannot replace human practitioners' opinions. AI technology can aid and advise human medical service providers in the decision-making process. However, AI cannot completely overhaul human professional judgement. Moreover, people still have more trust and belief in the judgement of a human doctor than a machine that functions on algorithms (Fig. 7.11).

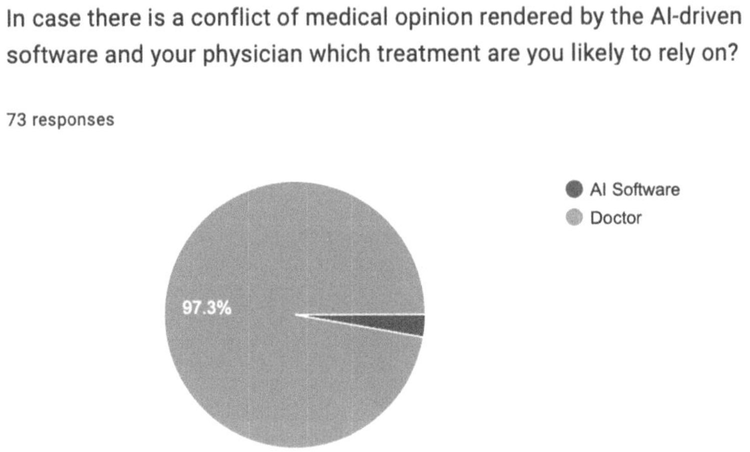

Fig. 7.10 Conflict of opinion (AI Empirical Survey, 2024)

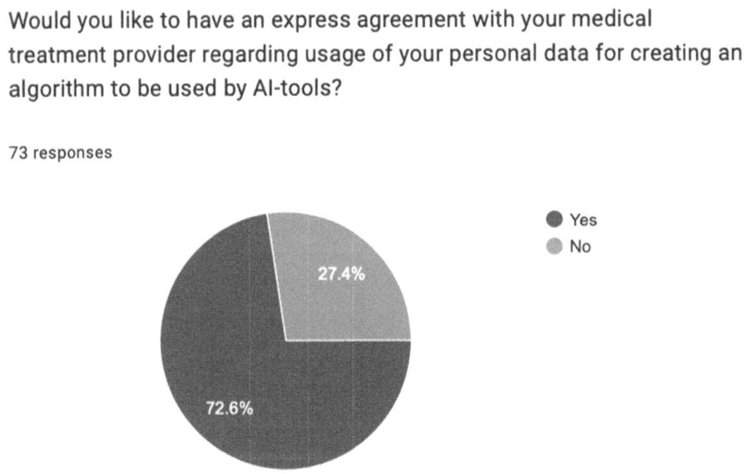

Fig. 7.11 Express agreement between patient and AI company (AI Empirical Survey, 2024)

Almost three-fourths of the total number of participants were of the view that they would appreciate if there was an agreement which expressly states that the data provided by the patient would be utilised in creating an algorithm. Only one-fourth of the total participants were comfortable even without an express agreement about the usage of their personal data for the development of the software. There is a concern among the people that their data should be used only when the patient's consent is obtained. It is an indication of the fact that the patients need to be made aware of the potential use of their by taking express consent from them.

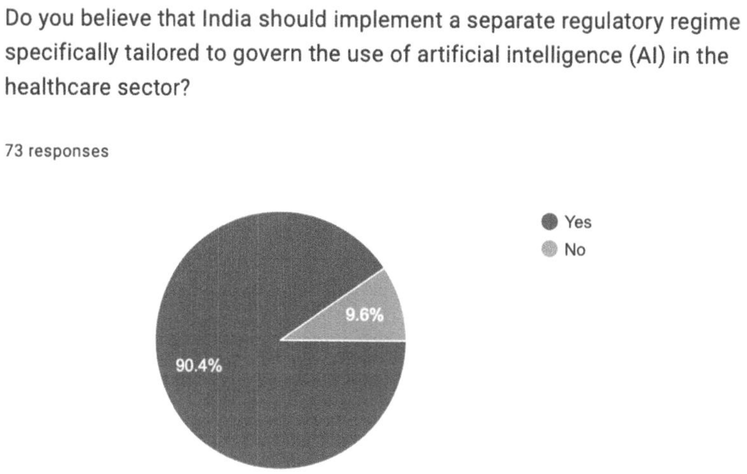

Fig. 7.12 Formulation of regulation (AI Empirical Survey, 2024)

The above-mentioned pie chart in Fig. 7.12 demonstrates that almost 90% of the participants of the survey favoured that India should implement a separate regulatory regime that particularly governs the implementation and deployment of Artificial Intelligence technology in the healthcare sector. Regulation is considered essential by the participants to ensure that there is constant monitoring of the algorithm to ensure the efficacy of the AI-driven tool.

A large proportion of the participants in the survey would be key stakeholders in the healthcare sector because they would be consuming the medical services. Therefore, an overwhelming majority of the participants amounting to almost 53% strongly agreed that healthcare organisations should be held accountable for misusing big data and AI in healthcare as is evident by Fig. 7.13. This survey demonstrates that patients are concerned about their usage of data. The data is provided for medical purposes and treatment by the patients. To ensure that the data is secure and there is no breach of privacy and confidentiality, strict legal provisions should be established so that the data is not misused. The developers of the AI software have a huge mass of data collected from thousands of patients and it is necessary to hold them strictly liable for any misuse that may be detected. The developers need to act responsibly in using such Big Data.

7.5 Discussion

The empirical study undertaken shows uneasiness among the patients about recommending the usage of AI in healthcare to their families. This may be because there are no specific regulations about the deployment of AI and machine learning in healthcare services. Moreover, there is a possibility of life-threatening injury that

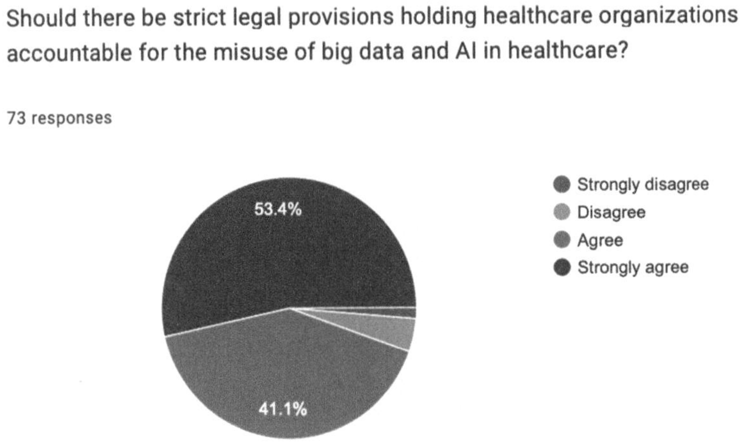

Fig. 7.13 Accountability of healthcare organisations (AI Empirical Survey, 2024)

can be caused to the patient. This is one of the major reasons patients are likely to give primacy to the opinion of a human practitioner rather than AI technology. There is also an alarming lack of knowledge about the deployment and usage of AI in healthcare services among the masses in India. Moreover, there are serious concerns regarding the information that is collected from the patient. When a patient gives his personal information to a hospital then it is not necessary that the patient has consented to use his data for creating an algorithm that is fed into AI technology. There are two ways one can resolve this issue. One is to obtain express consent from the patient through an agreement between the institution and the patient. The other way is to include a term in the terms and conditions that a patient signs when he receives treatment from a hospital. There are different laws pertaining to electronic records and protection of personal data however, there is no standalone law as to this date regarding the health records of the patient.

For fostering a sustainable healthcare sector, where sustainability practices are deeply integrated into AI-driven healthcare services, it is required that long-term environmental, social, and economic impacts are taken into account. By prioritising sustainability in the deployment of AI technologies, healthcare systems can strive towards achieving equitable access to healthcare services while minimising negative environmental footprints and promoting resilience in the face of future challenges.

7.6 Implication and Scope of Future Work

The present chapter highlights the dire need for India to develop laws, legislations and regulations regarding the deployment of AI in the healthcare industry. The regulations must provide clarity on the liability that will be imposed on companies who make AI software using patient data. The companies need to ensure that the data is kept confidential and is protected from unauthorised access. Contemporarily, if a company makes software using an algorithm derived from patient data then there are no laws regulating the utilisation of health data in AI technology.

There can be further research on what specific laws need to be drafted to regulate the usage of health data for creating algorithms. Moreover, methods and regulations that mandate continuous monitoring of Big Data need to be delved into in more detail. There has to continuous consultation among the stakeholders while formulating the regulations so that the concerns of all are addressed. Moreover, proper checks must be developed to ensure that the regulations are effectively implemented and complied with. It is crucial to understand the challenges faced by other jurisdictions in the implementation of such regulations and learn from them. Research in the direction of ethical conduct of companies developing the software is quintessential.

7.7 Conclusion

The survey suggests that even though people are not against the integration of AI, there are ethical concerns like misuse of data, breach of confidentiality and safety of the product. People are sceptical of fully integrating AI technology in healthcare services. Moreover, patients are not in favour of giving primacy to the opinion of AI algorithms. The patients feel that the human medical practitioner should have the final decision-making power in providing health services. AI technology can suggest in the decision-making process by providing its analysis of the Big Data. There must be regulations that need to be established for the implementation and deployment of AI technologies in India's healthcare sector. The Information Technology Act, 2000 has provisions for protecting the sensitive electronic data of individuals. Another mechanism has been provided in National Health Policy, 2017. Even though it is not binding in nature still it underscores the need for using technological advancement including AI in the delivery of healthcare services to patients. The Digital Personal Data Protection Act, of 2023 is legislation which secures the personal data of individuals from unauthorised access and places responsibility on the institutions to ensure that the confidentiality of data is maintained. Since India does not have a standalone law on the usage of AI and machine learning in medical industry it can take a cue from the General Data Protection Regulation, 2018 (GDPR) of the European Union. GDPR gives the command over the data to the patient itself. The patients can manage the usage of such data including disclosure, erasure, processing etc.

References

AI Empirical Survey. (2024). https://docs.google.com/forms/d/1A7AoFkbnJIwldXGHSRZRCvpMX9JfjvM_LeoKT-ePQyI/edit#responses

Bajwa, J., Munir, U., Nori, A., & Williams, B. (2021). Artificial intelligence in healthcare: Transforming the practice of medicine. *Future Healthcare Journal, 8*(2), 188–194. https://doi.org/10.7861/fhj.2021-0095.PMID:34286183;PMCID:PMC8285156

Davenport, T., & Kalakota, R. (2019). The potential for artificial intelligence in healthcare. *Future Healthcare Journal, 6*(2), 94–98. https://doi.org/10.7861/futurehosp.6-2-94.PMID:31363513; PMCID:PMC6616181

Dumouchel, T. (2019). *Regulatory challenges of AI products A pre-market* perspective. Retrieved from https://www.cadth.ca/sites/default/files/symp-2019/presentations/april15-2019/A3-presentation-tdumouchel.pdf

European Commission. (2020). *Report from the commission to the European Parliament, the council, and the European economic and social committee: Report on the safety and liability implications of artificial intelligence, the Internet of Things and robotics.* Retrieved from https://eur-lex.europa.eu/legal-content/en/TXT/?qid=1593079180383&uri=CELEX%3A52020DC0064

Foresee Medical. (2024). *AI in Healthcare.* Retrieved April 5, 2024, from https://www.foreseemed.com/artificial-intelligence-in-healthcare

Hospital Services Market Size to Surpass Around USD 22.57 Trillion by 2033. (n.d.). Retrieved March 20, 2024, from www.precedenceresearch.com

https://www.fda.gov/medical-devices/software-medical-device-samd/artificial-intelligence-and-machine-learning-aiml-enabled-medical-devices

IBM. (2024). *The benefits of AI in healthcare.* Retrieved April 5, 2024, from https://www.ibm.com/blog/the-benefits-of-ai-in-healthcare/

Indian Council for Medical Research. (2023). *Ethical guidelines for application of Artificial Intelligence in Biomedical Research and Healthcare.* Retrieved from https://main.icmr.nic.in/sites/default/files/upload_documents/Ethical_Guidelines_AI_Healthcare_2023.pdf

Jasarevic, T. (2021). *WHO issues first global report on Artificial Intelligence (AI) in health and six guiding principles for its design and use.* Retrieved from https://www.who.int/news/item/28-06-2021-who-issues-first-global-report-on-ai-in-health-and-six-guiding-principles-for-its-design-and-use

Ministry of Canada. Health Canada. (2019). *Guidance document: Software as a medical device (SaMD): Definition and classification.* Retrieved from guidance document: Software as a medical device (SaMD): Definition and classification—Canada.ca

Minssen, T., Mimler, M., & Mak, V. (2020). When does stand-alone software qualify as a medical device in the European Union? The Cjeu's decision in Snitem and what it implies for the next generation of medical devices. *Medical Law Review, 28*(3), 615–624. https://doi.org/10.1093/medlaw/fwaa012

Montanez, M. (2022). *Digital health software pre-certification update: Final FDA report revealed.* Retrieved from https://namsa.com/digital-health-pre-cert-update-fda-final-report/

Price II, W. N. (2021). Medical malpractice and black-box medicine. University of Michigan Public Law Research Paper, No. 536. Retrieved February 4, 2017, from https://papers.ssrn.com/sol3/papers.cfm?abstract_id=2910417#paper-references-widget

Price, W. N., Gerke, S., & Cohen, I. G. (2019). Potential liability for physicians using artificial intelligence. *JAMA, 322*(18), 1765–1766. https://doi.org/10.1001/jama.2019.15064

Régis, C., & Flood, C. M. (2021). AI and health law. *Florian Martin-Bariteau & Teresa Scassa.* Retrieved from SSRN https://ssrn.com/abstract=3733964

Reid, D. (2020). Google's DeepMind A.I. beats doctors in breast cancer screening trial. *CNBC.* Retrieved from https://www.cnbc.com/2020/01/02/googles-deepmind-ai-beats-doctors-in-breast-cancer-screening-trial.html

U.S. Food and Drug Administration. (2017). *What are examples of software as a medical device?* Retrieved from https://www.fda.gov/medical-devices/software-medical-device-samd/what-are-examples-software-medical-device

US Food and Drug Administration. (2022). *Artificial intelligence & medical product.* Retrieved from https://www.google.com/url?sa=t&source=web&rct=j&opi=89978449; https://www.fda.gov/science-research/science-and-research-special-topics/artificial-intelligence-and-medical-products&ved=2ahUKEwjh-Zum06uFAxXFa2wGHWP-C6kQFnoECB8QAQ&usg=AOvVaw2h0H2ea1xUvKKcY-FTI9Nk

World Economic Forum. (2024). *AI in healthcare: Buckle up for change, but read this before takeoff*. Retrieved April 5, 2024, from https://www.weforum.org/agenda/2024/01/ai-in-healthcare-buckle-up-for-big-change-but-read-this-before-takeoff/

World Health Organisation. (2021). *Ethics and governance of artificial intelligence for health.* Retrieved from https://www.who.int/publications/i/item/9789240029200

Chapter 8
Ethical Implications of Artificial Intelligence: Ensuring Patient Data Security

Azamat Ali

Abstract The incorporation of artificial intelligence (AI) in healthcare holds massive potential for improving patient care and clinical consequences. However, as AI technologies develop gradually and integrate into healthcare systems, it is imperative to address the ethical implications, particularly concerning patient data security. This chapter explores the ethical considerations characteristic of the practice of AI in healthcare, with a specific focus on safeguarding the security of patient data. Key ethical concerns include issues surrounding confidentiality, data privacy, consent, and trustworthiness in AI-driven healthcare systems. Safeguarding patient data security is paramount to upholding patient autonomy, confidentiality, and trust in healthcare providers and systems. The chapter discusses various strategies and frameworks for addressing these ethical challenges responsibly, emphasising the significance of robust data governance policies, transparency, and accountability measures. Additionally, it studies the role of governing frameworks and ethical guidelines in ensuring the ethical use of AI in healthcare. By acknowledging and actively addressing these ethical concerns, stakeholders can ensure that AI technologies enhance patient care while respecting patient rights and privacy. Ultimately, prioritising patient data security in AI-driven healthcare is essential to upholding ethical standards and maintaining public trust in healthcare systems.

Keywords Artificial intelligence (AI) · Healthcare · Ethical implications · Patient data security · Data governance

A. Ali (✉)
School of Liberal and Creative Arts, Lovely Professional University, Punjab, India
e-mail: azamatcug@gmail.com

R. Singh et al. (eds.), *Transforming Healthcare Sector Through Artificial Intelligence and Environmental Sustainability*, Approaches to Global Sustainability, Markets, and Governance, https://doi.org/10.1007/978-981-97-9555-0_8

8.1 Introduction

The integration of artificial intelligence (AI) into healthcare systems marks a transformative shift in the delivery of medical services, promising substantial benefits for patients, healthcare providers, and society as a whole (Ramírez, 2024). AI technologies encompass a wide range of applications, including diagnostic support, personalised treatment planning, predictive analytics, and administrative automation. By harnessing the power of machine learning algorithms, data analytics, and natural language processing, AI has the potential to revolutionise medical diagnosis, improve treatment outcomes, optimise resource allocation, and enhance patient experiences (Khalifa et al., 2024).

The adoption of AI in healthcare is determined by the growing availability of digital health data, advancements in computing power, and the increasing demand for efficient and personalised healthcare solutions (Senbekov et al., 2020). In recent years, AI has demonstrated remarkable capabilities in various medical fields, ranging from radiology and pathology to genomics and drug discovery. For instance, AI-powered diagnostic algorithms have shown comparable or even superior performance to human experts in interpreting medical imaging studies, such as X-rays, CT scans, and MRIs. AI-powered prediction models can look at vast amounts of data to find patients who are likely to get certain diseases or have lousy health events. This lets doctors take preventative and early action (Szilágyi & Kovács, 2024).

A significant benefit of AI in healthcare is its capacity to swiftly process and interpret vast amounts of intricate data (Khalifa & Albadawy, 2024). Medical data sources, such as genetic sequences, electronic health records (EHRs), medical imaging archives, wearable device readings, and patient-generated health data, include significant information that can enhance clinical decision-making and enhance patient outcomes. Artificial intelligence algorithms can analyse various data streams in order to detect patterns, correlations, and insights that may not be readily noticeable to human practitioners. This may help in making more precise diagnoses, creating tailored treatment plans, and implementing therapies that are based on evidence (Bekbolatova et al., 2024).

AI has the potential to streamline administrative processes, reduce healthcare costs, and alleviate workforce shortages by automating routine tasks and optimising resource utilisation. Administrative functions such as appointment scheduling, billing and coding, inventory management, and insurance claim processing can benefit from AI-driven solutions, freeing up time for healthcare specialists to focus on patient care and clinical decision-making. AI-powered telemedicine systems and virtual health assistants provide comfortable and easily accessible healthcare services, especially in underprivileged or distant locations with limited access to traditional healthcare institutions (Amjad et al., 2023).

While the incorporation of AI in healthcare holds tremendous promise, it also raises significant ethical concerns that must be carefully addressed to ensure patient safety, privacy, and autonomy. One of the foremost ethical considerations associated with AI in healthcare pertains to patient data security (Williamson & Prybutok, 2024).

Healthcare companies gather and retain extensive quantities of confidential patient data, encompassing medical backgrounds, diagnostic test outcomes, treatment strategies, and demographic particulars. Ensuring the secrecy, accuracy, and accessibility of this data is crucial in order to protect patient confidentiality and avoid illegal entry, misuse, or exploitation (Keshta & Odeh, 2021).

The process of converting health information into digital format and the widespread use of networked medical equipment and systems have increased the vulnerability to cyber-attacks and data breaches in the healthcare industry. Malicious actors may target healthcare networks to steal patient data for identity theft, financial fraud, or other nefarious purposes (Javaid et al., 2023). Unintentional data breaches or weaknesses in systems might jeopardise the security of patient information and undermine trust in healthcare practitioners and institutions. Thus, it is imperative to implement robust cybersecurity measures, encryption protocols, access restrictions, and data encryption to provide a reliable patient data security strategy in AI-driven healthcare settings (Yazid, 2024).

Another ethical concern related to AI in healthcare is the possibility of algorithmic bias and discrimination. AI algorithms are taught on historical data, which might contain inherent biases, inequities, and inequalities in the provision of healthcare (Norori et al., 2021). If these biases are perpetuated or amplified by AI systems, they could exacerbate disparities in access to care, diagnostic accuracy, treatment outcomes, and healthcare resource allocation. For example, AI algorithms used in predictive modelling or decision support may inadvertently prioritise certain demographic groups or exhibit differential performance across population subgroups, leading to unequal treatment or outcomes. Addressing algorithmic bias requires ongoing monitoring, validation, and recalibration of AI models to ensure fairness, equity, and inclusivity in healthcare delivery (Chen et al., 2021).

The ethical practice of AI in healthcare necessitates transparent and accountable decision-making processes to uphold patient trust and autonomy (Bjerring & Busch, 2021). Patients have the right to know how AI tools are being used in their care, including the purposes, limitations, and potential risks associated with AI-driven interventions. Informed consent, shared decision-making, and patient engagement are essential principles that should guide the deployment of AI in clinical practice. Healthcare providers must communicate efficiently with patients, educate them about the role of AI in their treatment, and solicit their preferences and feedback to ensure that AI technologies align with patient values, preferences, and goals (Sarella & Mangam, 2024).

8.2 Ethical Considerations in AI Healthcare

The incorporation of artificial intelligence into healthcare systems presents several ethical problems that require careful navigation to guarantee the appropriate and ethical utilisation of these technologies. With the growing integration of AI in medical practice, it is crucial to thoroughly analyse the ethical considerations that arise from its use. This section explores into the primary ethical concerns linked to the utilisation of AI in healthcare, with a specific focus on secrecy, data privacy, permission, and trustworthiness.

8.2.1 *Confidentiality: Addressing the Risk of Unauthorised Access to Patient Data*

Confidentiality is a core ethical concept in healthcare, which guarantees the safeguarding of patient information from illegal access, utilisation, or disclosure (Mikuletic et al., 2024). The implementation of AI technology increases the likelihood of patient confidentiality breaches as a result of the extensive processing and storage of sensitive health data. Healthcare institutions should use strong security measures, including encryption, access restrictions, and secure authentication methods, to protect patient information from unwanted access or cyber-attacks (Newaz et al., 2021).

Healthcare providers and AI developers are required to comply with rigorous data handling methods and privacy requirements, such as the "Health Insurance Portability and Accountability Act (HIPAA)" in the United States, in order to meet legal and ethical standards. Establishing explicit rules and processes for data exchange, storage, and disposal is crucial in order to mitigate the potential for data breaches and safeguard patient confidentiality (Saeed et al., 2024).

Transparency and accountability are critical in maintaining patient trust and confidence in AI-driven healthcare systems. Prior to data collection, patients should be provided with comprehensive information on the utilisation, accessibility, and privacy safeguards of their data (Moore et al., 2024). By prioritising confidentiality and implementing robust security measures, healthcare organisations can uphold ethical standards and preserve patient trust in the era of AI-enabled healthcare.

8.2.2 *Data Privacy: Ensuring the Protection of Sensitive Patient Information*

Data privacy encompasses safeguarding individuals' personal information against unlawful access, utilisation, or disclosure. Preserving patient trust and upholding their autonomy is of the utmost importance in AI healthcare, particularly when it comes to safeguarding the privacy of patient data (Saputra & Aminah, 2024). AI algorithms depend on extensive datasets to train and enhance their performance, which may include confidential health data like medical records, diagnostic test outcomes, and treatment strategies (Alowais et al., 2023).

To safeguard patient privacy, healthcare organisations must adhere to stringent data protection regulations and industry best practices. This includes implementing data anonymisation techniques, such as de-identification and encryption, to prevent the identification of individual patients from their data (Hadi & Jawad, 2024). Organisations should adopt privacy-preserving AI techniques, such as federated learning and differential privacy, which enable AI models to learn from distributed datasets without compromising individual privacy.

Prior to utilising patients' data for AI-driven healthcare treatments, healthcare professionals are required to get express consent from them. Patients are entitled to comprehend the intended use, authorised access, and possible advantages and drawbacks of AI-powered healthcare services in relation to their data. Healthcare businesses may cultivate a climate of trust and responsibility in AI-powered healthcare by giving importance to data protection and honouring patient autonomy (Wahab & Nor, 2023).

8.2.3 Consent: Issues in AI-Driven Healthcare Interventions

Obtaining informed consent is a fundamental aspect of ethical medical treatment since it guarantees that patients possess the independence to make well-informed choices regarding their healthcare (Pallocci et al., 2023). The complexity and opacity of AI algorithms provide distinct obstacles in obtaining informed consent for AI-driven healthcare interventions. Patients may lack a comprehensive understanding of the mechanics of AI technologies or the potential consequences of their application in clinical decision-making (Asan et al., 2020).

Healthcare practitioners should make a concerted effort to inform patients about the function of artificial intelligence (AI) in their treatment and obtain their permission in a meaningful and transparent way. This entails offering unambiguous and understandable information on the objectives, hazards, and constraints of AI-powered therapies, along with alternate courses of therapy. Patients should be afforded the chance to inquire, articulate apprehensions, and exercise informed decisions on their healthcare preferences (Pruski, 2024).

Continuously obtaining consent is crucial in AI healthcare, as AI algorithms may undergo changes or adjustments over time and necessitate updates or alterations. It is important to notify patients about any modifications to the AI system that might impact their treatment and provide them with the chance to review and adjust their permission accordingly (Wang & Ma, 2024). Patients should be granted the prerogative to revoke their permission at any given moment and demand the elimination of their data from artificial intelligence systems.

By prioritising informed consent and patient engagement, healthcare providers can empower patients to actively participate in their care and ensure that AI-driven interventions align with their values and preferences. Ethical considerations surrounding consent in AI healthcare underscore the importance of respecting patient autonomy and promoting shared decision-making in the delivery of healthcare services.

8.2.4 Trustworthiness: Building Trust in AI Among Patients and Healthcare Providers

Trust is a foundational element of the patient-provider relationship, underpinning effective communication, collaboration, and decision-making (Kohut et al., 2024). From the perspective of AI healthcare, building trust in AI systems is essential to foster acceptance and adoption among patients and healthcare providers. Trustworthiness encompasses various dimensions, including reliability, transparency, accountability, and fairness (Kim et al., 2024).

Patients must have assurance in the accuracy, reliability, and safety of AI-driven healthcare interventions to feel comfortable entrusting their care to these technologies. Healthcare organisations and AI developers must demonstrate the effectiveness and performance of AI algorithms through rigorous testing, validation, and real-world deployment (Habbal et al., 2024). Transparency is critical to building trust in AI systems, as patients should have visibility into how AI technologies are used in their care, including the data inputs, decision-making processes, and potential biases (Sindiramutty et al., 2024).

Accountability is another critical aspect of trustworthiness in AI healthcare, as patients must have recourse in the event of errors, biases, or adverse outcomes resulting from AI-driven interventions (Williamson & Prybutok, 2024). Healthcare practitioners must be held responsible for the choices made utilising AI technology, and there should be established channels to remedy cases of injury or unfairness. Ensuring fairness and equality is crucial for establishing trust in AI systems. Patients need to have confidence that AI algorithms will deliver impartial and just treatment to all individuals, regardless of their various backgrounds (Kerasidou, 2021).

Establishing confidence in AI systems among healthcare practitioners is crucial to encourage their acceptance and incorporation into clinical practice. Healthcare workers should undergo comprehensive training and instruction to ensure they are proficient in utilising AI technologies and possess a strong belief in their dependability, security, and effectiveness (Kim et al., 2024). Healthcare companies may cultivate a culture of trust and cooperation in AI-enabled healthcare by including healthcare providers as partners in the creation and execution of AI-driven solutions.

8.3 Safeguarding Patient Data Security in AI-Driven Healthcare Systems

The application of artificial intelligence technology in healthcare has unique prospects to enhance patient care, improve clinical outcomes, and maximise operational efficiency in a continuously changing environment (Haleem et al., 2022). These AI-driven healthcare systems bring essential advantages, but they also raise serious ethical and security issues, especially when it comes to safeguarding patient data. This section provides an in-depth examination of methods for protecting the security of patient data. It specifically emphasises the adoption of solid data governance regulations, the transparency of AI algorithms, and steps to hold individuals accountable for ethical breaches.

8.3.1 Robust Data Governance Policies

Strong data governance standards are crucial for guaranteeing the appropriate and ethical utilisation of patient data in AI-driven healthcare systems. These policies consist of a comprehensive set of regulations, protocols, and processes that dictate the gathering, retention, entry, utilisation, and dissemination of patient data inside healthcare institutions (Arigbabu et al., 2024). Healthcare providers may reduce the likelihood of data breaches, illegal access, and abuse of patient data by implementing explicit norms and standards for data management.

One key aspect of data governance is data classification, which involves categorising patient data based on its sensitivity, confidentiality, and regulatory requirements (Mahanti and Mahanti, 2021). Healthcare organisations should classify patient data into different tiers or levels of protection, such as public, internal, confidential, and restricted, to determine the appropriate security measures and access controls (Nissar et al., 2024). For instance, confidential data, including medical records, diagnostic test outcomes, and treatment strategies, may necessitate more stringent security protocols, such as access restrictions, encryption, and auditing methods.

Data supremacy policies should address data access and authorisation, specifying who has permission to access patient data and under what circumstances. Role-based access controls (RBAC) can be implemented to restrict access to patient data based on users' roles, responsibilities, and levels of authorisation within the healthcare organisation (de Carvalho Junior & Bandiera-Paiva, 2018). Audit trails and logging mechanisms should be implemented to track and monitor access to patient data, allowing organisations to detect and investigate possible security breaches or unauthorised activities.

Healthcare firms must guarantee adherence to statutory mandates and industry benchmarks for safeguarding data, such as the "Health Insurance Portability and Accountability Act (HIPAA)" in the United States or the "General Data Protection Regulation (GDPR)" in the European Union. These policies enforce strict criteria for protecting patient data, which include mandates for data encryption, access limits, data breach reporting, and patient permission (Bakare et al., 2024). Healthcare firms may exhibit their dedication to safeguarding patient confidentiality and adhering to legal and ethical norms by harmonising data governance policies with regulatory mandates.

8.3.2 Transparency in AI Algorithms

Transparency is paramount in AI-driven healthcare systems to ensure accountability, trustworthiness, and ethical decision-making. Transparency refers to the clarity and openness of AI algorithms and decision-making processes, enabling stakeholders to understand how AI technologies work, how decisions are made, and the potential implications for patient care. Transparency is crucial in patient data security to provide patients with a clear understanding of how AI systems utilise, handle, and safeguard their data (Kiseleva et al., 2022).

One approach to promoting transparency in AI algorithms is through explainable AI (XAI) techniques, which aim to make AI models and decisions interpretable and understandable to humans. XAI techniques include methods for visualising model outputs, generating explanations for model predictions, and identifying factors contributing to decision-making. By providing insights into how AI algorithms arrive at their conclusions, XAI techniques enable healthcare providers and patients to assess the reliability, accuracy, and fairness of AI-driven recommendations and interventions (Hassija et al., 2024).

Transparency in AI algorithms can facilitate accountability and oversight, enabling stakeholders to monitor and evaluate the performance of AI systems and identify potential biases, errors, or ethical concerns (Ansarullah et al., 2024). Healthcare organisations should establish mechanisms for auditing, testing, and validating AI algorithms to ensure compliance with ethical standards and regulatory requirements. By promoting transparency and accountability in AI-driven healthcare, stakeholders can mitigate the risk of unintended consequences, enhance patient safety, and uphold ethical principles (Nasir et al., 2024).

8.3.3 Accountability Measures

Accountability is a cornerstone of ethical governance in AI-driven healthcare systems, ensuring that stakeholders are held responsible for their actions and decisions. Accountability measures encompass mechanisms for identifying, addressing, and remedying ethical lapses, breaches of patient data security, or other wrongdoing in the use of AI technologies (Norman et al., 2024). Healthcare organisations need to establish explicit lines of responsibility, clearly defining the roles, duties, and expectations for all stakeholders participating in AI-driven healthcare.

One key accountability measure is the establishment of governance structures and oversight mechanisms to monitor and evaluate the use of AI technologies in healthcare. This may include the creation of ethics committees, review boards, or regulatory bodies responsible for reviewing AI-driven healthcare interventions, assessing their ethical implications, and ensuring compliance with legal and ethical standards (Kazeem et al., 2024). These oversight bodies should include multidisciplinary expertise, including healthcare professionals, data scientists, ethicists, and patient representatives, to provide diverse perspectives and insights.

Healthcare organisations should implement mechanisms for reporting and addressing ethical concerns or violations related to patient data security in AI-driven healthcare. This may include whistleblower policies, anonymous reporting channels, or incident response protocols for reporting breaches, misconduct, or unethical behaviour (West & Bowman, 2020). Healthcare providers and AI developers should be encouraged to report any ethical lapses or violations they encounter and cooperate with investigations to address and remedy the situation.

Accountability measures should encompass consequences for non-compliance or unethical behaviour in the use of AI technologies in healthcare. Healthcare organisations should establish disciplinary procedures, sanctions, or penalties for individuals or entities found to have violated ethical standards, breached patient data security, or engaged in misconduct (Demura et al., 2020). By holding stakeholders accountable for their actions and decisions, healthcare organisations can foster a culture of ethical responsibility and ensure the integrity and trustworthiness of AI-driven healthcare systems.

8.4 The Role of Governing Frameworks and Ethical Guidelines in AI-Driven Healthcare

Adhering to legislative frameworks and ethical principles is necessary to ensure the ethical use of artificial intelligence in healthcare since it protects patient rights, privacy, and safety. As AI technologies progress and become more widespread in healthcare, it is crucial to carefully examine current frameworks and rules to ensure they effectively handle ethical concerns, including the protection of patient data (Kasula, 2024). This section offers an examination of governing frameworks and ethical guidelines that are pertinent to the ethical utilisation of artificial intelligence (AI) in healthcare. It also reviews regulations and guidelines related to the security of patient data and evaluates the efficacy of existing frameworks in addressing ethical concerns in AI-driven healthcare.

8.4.1 Analysis of Existing Governing Frameworks and Ethical Guidelines

Various regulatory frameworks and ethical principles have been established to guide the responsible and ethical application of artificial intelligence (AI) in the healthcare sector. These frameworks consist of principles, standards, and best practices that attempt to enhance transparency, accountability, justice, and patient-centricity in AI-driven healthcare systems. An exemplary instance is the Ethical Guidelines for Trustworthy AI issued by the European Commission, which delineates seven fundamental prerequisites for guaranteeing the ethical advancement and implementation of AI technology. The standards outlined by Laux et al. (2024) include openness, accountability, fairness, dependability, safety, data protection, and respect for human autonomy.

AI in healthcare is governed by ethical principles and codes of behaviour that professional organisations and industry consortia have produced. The American Medical Association (AMA) has played a crucial role in establishing ethical standards for the utilisation of artificial intelligence (AI) in clinical practice. These recommendations prioritise openness, patient consent, and professional accountability. Transparency in AI algorithms is crucial for healthcare practitioners and patients as it facilitates comprehension and guides patient treatment. Obtaining patient consent facilitates the exercise of autonomy in making decisions regarding AI treatments. Healthcare practitioners have a professional duty to utilise AI technology in accordance with established guidelines, with a focus on ensuring the safety and privacy of patients. The AMA's objective is to advance the ethical utilisation of AI in healthcare by giving importance to these principles. This will help build trust among patients, providers, and society (Bekbolatova et al., 2024).

The International Medical Informatics Association (IMIA) has developed ethical standards for practitioners in the field of health informatics, explicitly addressing ethical considerations in the development, implementation, and evaluation of healthcare systems that include artificial intelligence (AI). These recommendations emphasise the significance of ethical concerns at every stage of AI-driven healthcare solutions, guaranteeing that they adhere to values such as patient autonomy, data privacy, fairness, and responsibility. The IMIA seeks to encourage the appropriate and ethical use of AI in healthcare by offering clear guidelines on ethical best practices. This aims to protect patient rights and welfare while also encouraging innovation and improvement in healthcare delivery (Dadabhoy et al., 2023).

Regulatory entities and government agencies have a vital function in setting and implementing guidelines for the ethical utilisation of AI in the healthcare sector. The U.S. Food and Drug Administration (FDA) has a crucial responsibility in monitoring the safety and effectiveness of medical devices and AI-powered software applications used in healthcare environments. The FDA ensures that these technologies comply with established safety measures, operate efficiently, and satisfy regulatory criteria by implementing strict regulatory standards. The FDA ensures patient welfare and enhances public trust in the quality and dependability of AI-driven healthcare solutions through rigorous review processes and continuous monitoring. The regulatory control in place promotes the development of new ideas and methods while also reducing the possible dangers that may arise from the application of cutting-edge technology in the field of medicine (Stern, 2022).

"The General Data Protection Regulation (GDPR)" of the European Union sets strict duties for data protection on companies that handle personal data, including healthcare providers and developers of artificial intelligence (AI). Organisations are required to treat personal data in a lawful, transparent, and secure manner under

GDPR. They must also get express consent from individuals. The GDPR requires the implementation of solid safeguards for data security, such as encryption, access limits, and the obligation to notify authorities in case of a data breach promptly. Healthcare providers and AI developers must ensure compliance with GDPR in order to safeguard patient privacy, maintain data security standards, and prevent severe fines for failure to comply. This policy enhances confidence in AI-powered healthcare systems by giving priority to the privacy and rights of persons (Pathak, 2024).

8.4.2 Regulations and Guidelines for Patient Data Security in Healthcare

Ensuring the security of patient data is of utmost importance in AI-driven healthcare, considering the sensitive and personal nature of health information. Various legislation and standards exist to ensure the safeguarding of patient data in healthcare settings, including those specifically tailored to handle the use of AI technology. The Health Insurance Portability and Accountability Act (HIPAA) in the United States sets guidelines for safeguarding the confidentiality and integrity of protected health information (PHI). These guidelines include mandates for encrypting data, implementing access controls, and reporting data breaches (Tertulino et al., 2024).

The GDPR in the European Union imposes strict standards for the protection of personal data, including health information, necessitating robust data security measures. Compliance entails obtaining explicit patient consent, anonymising data, and promptly reporting data breaches. Healthcare providers and AI developers must adhere to GDPR guidelines during data collection, processing, and storage to ensure patient privacy and security. By implementing appropriate technical and organisational measures, organisations uphold GDPR requirements, fostering trust and accountability in the handling of sensitive healthcare data (Rupp & von Grafenstein, 2024).

International standards organisations, like the "International Organization for Standardization (ISO)", have established guidelines for "Information Security Management Systems (ISMS)" in healthcare. ISO/IEC 27,001 outlines a framework for creating, implementing, and enhancing ISMS to safeguard sensitive information, such as patient data, from potential security breaches and threats. By adhering to ISO/IEC 27,001 standards, healthcare organisations can systematically manage risks, establish security controls, and ensure the confidentiality, integrity, and availability of patient data. This framework promotes a proactive approach to information security, enhancing trust and confidence in the safety of healthcare data on an international scale (Gaivéo, 2016).

8.4.3 Effectiveness of Current Frameworks in AI-Driven Healthcare Ethical Concerns

Although current legislative frameworks and ethical guidelines offer significant direction for the ethical use of AI in healthcare, some several problems and constraints need to be resolved to ensure their efficacy in resolving ethical concerns. An obstacle arises from the fast advancement of technology and the ever-changing nature of AI algorithms, which may surpass the establishment of regulatory and ethical frameworks (Walter, 2024). Healthcare organisations and regulatory authorities need to modify and develop their rules and standards to stay up to date with breakthroughs in AI technology and developing ethical concerns.

Complying with current standards and norms can be difficult, especially for healthcare companies and AI developers who work in several regulatory settings. Differences in compliance standards between countries might result in discrepancies and uncertainties when interpreting and implementing them (de Almeida et al., 2021). Healthcare firms should allocate resources and acquire knowledge to effectively traverse intricate regulatory environments and guarantee adherence to legal and ethical norms.

Enhanced collaboration and coordination across stakeholders, such as healthcare providers, AI developers, regulatory agencies, and professional organisations, are necessary to tackle ethical challenges and foster responsible utilisation of AI in healthcare. Collaboration among healthcare practitioners, data scientists, ethicists, legislators, and patient advocates from different fields may help create thorough and inclusive frameworks that effectively balance innovation with ethical considerations (Anyanwu et al., 2024).

Regular and systematic monitoring, evaluation, and assessment of AI-driven healthcare systems are crucial in order to detect and address ethical risks and difficulties (Mennella et al., 2024). Healthcare institutions should establish protocols for continuous ethical review, risk evaluation, and stakeholder involvement to guarantee the ethical and responsible use of AI technology. Stakeholders may encourage the ethical use of AI in healthcare by cultivating a culture that values thoughtful consideration of ethical issues and taking responsibility for one's actions. This approach ensures the protection of patient rights, privacy, and safety.

8.5 Conclusion

Prioritising patient data security in AI-driven healthcare is paramount to upholding ethical standards and ensuring patient trust and safety. Patient data comprises sensitive information that must be protected from unauthorised access, misuse, or exploitation, necessitating robust security measures and adherence to ethical guidelines. Critical plans for addressing ethical concerns in AI healthcare while safeguarding patient data security include implementing robust data governance policies, promoting transparency in AI algorithms, and fostering accountability measures for ethical lapses. By prioritising these strategies, healthcare organisations can mitigate the risk of data breaches, ensure patient confidentiality, and uphold ethical principles in the use of AI technologies. Ultimately, maintaining public trust in healthcare systems requires a commitment to ethical AI practices that prioritise patient welfare, respect patient rights, and promote transparency and accountability. By cultivating an environment of ethical accountability and reliability, stakeholders can leverage the revolutionary capabilities of AI in healthcare while ensuring ethical integrity and preserving public trust in healthcare systems.

References

Alowais, S. A., Alghamdi, S. S., Alsuhebany, N., Alqahtani, T., Alshaya, A. I., Almohareb, S. N., & Albekairy, A. M. (2023). Revolutionising healthcare: The role of artificial intelligence in clinical practice. *BMC Medical Education, 23*(1), 689. https://doi.org/10.1186/s12909-023-04698-z

Amjad, A., Kordel, P., & Fernandes, G. (2023). A review of innovation in the healthcare sector (telehealth) through artificial intelligence. *Sustainability, 15*(8), 6655. https://doi.org/10.3390/su15086655

Ansarullah, S. I., Kirmani, M. M., Alshmrany, S., & Firdous, A. (2024). Ethical issues around artificial intelligence. In *A biologist's guide to artificial intelligence* (pp. 301–314). Academic Press. https://doi.org/10.1016/B978-0-443-24001-0.00018-X

Anyanwu, E. C., Okongwu, C. C., Olorunsogo, T. O., Ayo-Farai, O., Osasona, F., & Daraojimba, O. D. (2024). Artificial intelligence in healthcare: A review of ethical dilemmas and practical applications. *International Medical Science Research Journal, 4*(2), 126–140. https://doi.org/10.51594/imsrj.v4i2.755

Arigbabu, A. T., Olaniyi, O. O., Adigwe, C. S., Adebiyi, O. O., & Ajayi, S. A. (2024). Data governance in AI-enabled healthcare systems: A case of the project nightingale. *Asian Journal of Research in Computer Science, 17*(5), 85–107. https://doi.org/10.9734/ajrcos/2024/v17i5441

Asan, O., Bayrak, A. E., & Choudhury, A. (2020). Artificial intelligence and human trust in healthcare: Focus on clinicians. *Journal of Medical Internet Research, 22*(6), e15154. https://doi.org/10.2196/15154

Bakare, S. S., Adeniyi, A. O., Akpuokwe, C. U., & Eneh, N. E. (2024). Data privacy laws and compliance: A comparative review of the EU GDPR and USA regulations. *Computer Science & IT Research Journal, 5*(3), 528–543. https://doi.org/10.51594/csitrj.v5i3.859

Bekbolatova, M., Mayer, J., Ong, C. W., & Toma, M. (2024). The transformative potential of AI in healthcare: Definitions, applications, and navigating the ethical landscape and public perspectives. *Healthcare, 12*(2), 125. https://doi.org/10.3390/healthcare12020125

Bjerring, J. C., & Busch, J. (2021). Artificial intelligence and patient-centered decision-making. *Philosophy & Technology, 34*, 349–371.

de Carvalho Junior, M. A., & Bandiera-Paiva, P. (2018). Health information system role-based access control current security trends and challenges. *Journal of Healthcare Engineering*, 2018. https://doi.org/10.1155/2018/6510249

Chen, R. J., Chen, T. Y., Lipkova, J., Wang, J. J., Williamson, D. F., Lu, M. Y., Mahmood, F. (2021). Algorithm fairness in AI for medicine and healthcare. arXiv preprint arXiv:2110.00603. https://doi.org/10.48550/arXiv.2110.00603

Dadabhoy, F. Z., Marshall, A., & Landman, A. (2023). Informatics. In R.P. Olympia, E.B. Werley, J.S. Lubin, & K. Yoon-Flannery (Eds.). *An emergency physician's path* (p. 51). Springer, Cham. https://doi.org/10.1007/978-3-031-47873-4_51

de Almeida, P. G. R., dos Santos, C. D., & Farias, J. S. (2021). Artificial intelligence regulation: A framework for governance. *Ethics and Information Technology, 23*(3), 505–525.

Demura, M. I., Kononenko, V. A., & Fedosenko, N. A. (2020). Challenges of health care professionals' disciplinary and criminal prosecution. *Wiadomosci Lekarskie (Warsaw, Poland: 1960), 73*(12 cz 2), 2827–2832. https://doi.org/10.36740/WLek202012221

Gaivéo, J. (2016). Information security standards in healthcare activities. *International Journal of Reliable and Quality E-Healthcare (IJRQEH), 5*(4), 15–33. https://doi.org/10.4018/IJRQEH.2016100102

Habbal, A., Ali, M. K., & Abuzaraida, M. A. (2024). Artificial intelligence trust, risk and security management (AI TRiSM): Frameworks, applications, challenges and future research directions. *Expert Systems with Applications, 240*, 122442. https://doi.org/10.1016/j.eswa.2023.122442

Hadi, M. J. (2024). Safeguarding critical infrastructures through data protection laws: A comparative study with a focus on Pakistan. *SSRN*. https://doi.org/10.2139/ssrn.4730720

Haleem, A., Javaid, M., Singh, R. P., & Suman, R. (2022). Medical 4.0 technologies for healthcare: Features, capabilities, and applications. *Internet of Things and Cyber-Physical Systems, 2*, 12–30. https://doi.org/10.1016/j.iotcps.2022.04.001

Hassija, V., Chamola, V., Mahapatra, A., et al. (2024). Interpreting black-box models: A review on explainable artificial intelligence. *Cognitive Computation, 16*, 45–74

Javaid, M., Haleem, A., Singh, R. P., & Suman, R. (2023). Towards insight cybersecurity for healthcare domains: A comprehensive review of recent practices and trends. *Cyber Security and Applications*, 100016. https://doi.org/10.1016/j.csa.2023.100016

Kasula, B. Y. (2024). Advancements in AI-driven healthcare: A comprehensive review of diagnostics, treatment, and patient care integration. *International Journal of Machine Learning for Sustainable Development, 1*(1), 1–5.

Kerasidou, A. (2021). Ethics of artificial intelligence in global health: Explainability, algorithmic bias and trust. *Journal of Oral Biology and Craniofacial Research, 11*(4), 612–614. https://doi.org/10.1016/j.jobcr.2021.09.004

Keshta, I., & Odeh, A. (2021). Security and privacy of electronic health records: Concerns and challenges. *Egyptian Informatics Journal, 22*(2), 177–183. https://doi.org/10.1016/j.eij.2020.07.003

Khalifa, M., & Albadawy, M. (2024). AI in diagnostic imaging: Revolutionising accuracy and efficiency. *Computer Methods and Programs in Biomedicine Update*. https://doi.org/10.1016/j.cmpbup.2024.100146

Khalifa, M., Albadawy, M., & Iqbal, U. (2024). Advancing clinical decision support: The role of artificial intelligence across six domains. *Computer Methods and Programs in Biomedicine Update*. https://doi.org/10.1016/j.cmpbup.2024.100142

Kim, Y. J., Choi, J. H., & Fotso, G. M. N. (2024). Medical professionals' adoption of AI-based medical devices: UTAUT model with trust mediation. *Journal of Open Innovation: Technology, Market, and Complexity, 10*(1), 100220. https://doi.org/10.1016/j.joitmc.2024.100220

Kiseleva, A., Kotzinos, D., & De Hert, P. (2022). Transparency of AI in healthcare as a multilayered system of accountabilities: Between legal requirements and technical limitations. *Frontiers in Artificial Intelligence, 5*, 879603. https://doi.org/10.3389/frai.2022.879603

Kohut, K., Morton, K., Hurley, K., Turner, L., CanGene-CanVar Patient Reference Panel, Dale, C., Foster, C. (2024). A good decision is the one that feels right for me: Codesign with patients to inform theoretical underpinning of a decision aid website. *Health Expectations, 27*(1), e13844. https://doi.org/10.1111/hex.13844

Laux, J., Wachter, S., & Mittelstadt, B. (2024). Trustworthy artificial intelligence and the European Union AI Act: On the conflation of trustworthiness and acceptability of risk. *Regulation & Governance, 18*(1), 3–32. https://doi.org/10.1111/rego.12512

Mahanti, R., & Mahanti, R. (2021). Data and its governance. In *Data governance and data management: Contextualizing data governance drivers, technologies, and tools* (pp. 5–82). https://doi.org/10.1007/978-981-16-3583-0_2

Mennella, C., Maniscalco, U., De Pietro, G., & Esposito, M. (2024). Ethical and regulatory challenges of AI technologies in healthcare: A narrative review. *Heliyon*. https://doi.org/10.1016/j.heliyon.2024.e26297

Mikuletic, S., Vrhovec, S., Skela-Savič, B., & Žvanut, B. (2024). Security and privacy-oriented information security culture (ISC): Explaining unauthorized access to healthcare data by nursing employees. *Computers & Security, 136*, 103489. https://doi.org/10.1016/j.cose.2023.103489

Moore, J., McMeekin, P., Parkes, T., et al. (2024). Contextualizing remote fall risk: Video data capture and implementing ethical AI. *npj Digital Medicine, 7*, 61. https://doi.org/10.1038/s41746-024-01050-7

Nasir, S., Khan, R. A., & Bai, S. (2024). Ethical framework for harnessing the power of AI in healthcare and beyond. *IEEE Access, 12*, 31014–31035. https://doi.org/10.1109/ACCESS.2024.3369912

Newaz, A. I., Sikder, A. K., Rahman, M. A., & Uluagac, A. S. (2021). A survey on security and privacy issues in modern healthcare systems: Attacks and defenses. *ACM Transactions on Computing for Healthcare, 2*(3), 1–44. https://doi.org/10.1145/3453176

Nissar, G., Khan, R. A., Mushtaq, S., et al. (2024). IoT in healthcare: A review of services, applications, key technologies, security concerns, and emerging trends. *Multimedia Tools and Applications*. https://doi.org/10.1007/s11042-024-18580-7

Norman, A. A., Eke, C. I., Phiri, M., et al. (2024). Digital health and deep learning: Opportunities, challenges, and the imperative of data protection in wearable devices. *SSRN*. https://doi.org/10.2139/ssrn.4703449

Norori, N., Hu, Q., Aellen, F. M., Faraci, F. D., & Tzovara, A. (2021). Addressing bias in big data and AI for healthcare: A call for open science. *Patterns, 2*(10). https://doi.org/10.1016/j.patter.2021.100347

Pallocci, M., Treglia, M., Passalacqua, P., et al. (2023). Informed consent: Legal obligation or cornerstone of the care relationship? *International Journal of Environmental Research and Public Health, 20*(3), 2118. https://doi.org/10.3390/ijerph20032118

Pathak, M. (2024). *Data governance redefined: The evolution of EU data regulations from the GDPR to the DMA, DSA, DGA, Data Act and AI Act*. SSRN. https://doi.org/10.2139/ssrn.4718891

Pruski, M. (2024). AI-enhanced healthcare: Not a new paradigm for informed consent. *Bioethical Inquiry*. https://doi.org/10.1007/s11673-023-10320-0

Ramírez, J. G. C. (2024). AI in healthcare: Revolutionizing patient care with predictive analytics and decision support systems. *Journal of Artificial Intelligence General Science (JAIGS), 1*(1), 31–37. https://doi.org/10.60087/jaigs.v1i1.p37

Rupp, V., & von Grafenstein, M. (2024). Clarifying 'Personal Data' and the role of anonymisation in data protection law including and excluding data from the scope of the GDPR (more clearly) through refining the concept of data protection. *Computer Law & Security Review, 52*, 105932. https://doi.org/10.1016/j.clsr.2023.105932

Saeed, M. M. A., Saeed, R. A., & Ahmed, Z. E. (2024). Data security and privacy in the age of AI and digital twins. In *Digital twin technology and AI implementations in future-focused businesses* (pp. 99–124). IGI Global. https://doi.org/10.4018/979

Saputra, A., & Aminah, S. (2024). Telehealth and AI: An ethical examination of remote healthcare services and the implications for patient care and privacy. *Quarterly Journal of Computational Technologies for Healthcare, 9*(1), 1 10. https://vectoral.org/index.php/QJCTH/article/view/72

Sarella, P. N. K., & Mangam, V. T. (2024). AI-driven natural language processing in healthcare: Transforming patient-provider communication. *Indian Journal of Pharmacy Practice, 17*(1). https://doi.org/10.5530/ijopp.17.1.4

Senbekov, M., Saliev, T., Bukeyeva, Z., Almabayeva, A., Zhanaliyeva, M., Aitenova, N., & Fakhradiyev, I. (2020). The recent progress and applications of digital technologies in healthcare: A review. *International Journal of Telemedicine and Applications*. https://doi.org/10.1155/2020/8830200

Sindiramutty, S. R., Tee, W. J., Balakrishnan, S., Kaur, S., Thangaveloo, R., Jazri, H., & Manchuri, A. R. (2024). Explainable AI in healthcare application. In *Advances in Explainable AI Applications for Smart Cities* (pp. 123–176). IGI Global. https://doi.org/10.4018/978-1-6684-6361-1.ch005

Stern, A. D. (2022). The regulation of medical AI: Policy approaches, data, and innovation incentives. *National Bureau of Economic Research*. https://doi.org/10.3386/w30639

Szilágyi, L., & Kovács, L. (2024). Special issue: Artificial intelligence technology in medical image analysis. *Applied Sciences, 14*(5), 2180. https://doi.org/10.3390/app14052180

Tertulino, R., Antunes, N., & Morais, H. (2024). Privacy in electronic health records: A systematic mapping study. *Journal of Public Health (Berl.), 32*, 435–454. https://doi.org/10.1007/s10389-022-01795-z

Wahab, N. A. B. A., & Nor, R. B. M. (2023). Challenges and strategies in data management and governance for AI-based healthcare models: Balancing innovation and ethical responsibilities. *AI, IoT and the Fourth Industrial Revolution Review, 13*(12), 24–32.

Walter, Y. (2024). Managing the race to the moon: Global policy and governance in artificial intelligence regulation—A contemporary overview and an analysis of socioeconomic consequences. *Discovery of Artificial Intelligence, 4*, 14. https://doi.org/10.1007/s44163-024-00109-4

Wang, Y., & Ma, Z. (2024). Ethical and legal challenges of medical AI on informed consent: China as an example. *Developing World Bioethics*. https://doi.org/10.1111/dewb.12442

West, J. P., & Bowman, J. S. (2020). Whistleblowing policies in American states: A nationwide analysis. *The American Review of Public Administration, 50*(2), 119–132. https://doi.org/10.1177/0275074019885629

Williamson, S. M., & Prybutok, V. (2024). Balancing privacy and progress: A review of privacy challenges, systemic oversight, and patient perceptions in AI-driven healthcare. *Applied Sciences, 14*(2), 675. https://doi.org/10.3390/app14020675

Yazid, A. (2024). The intricacies of data privacy in AI-enhanced healthcare systems: A critical examination of challenges and potential solutions. *Eigenpub Review of Science and Technology, 8*(1), 1–8. https://studies.eigenpub.com/index.php/erst/article/view/63

Chapter 9
Artificial Intelligence Transparency and Explainability in Sustainable Healthcare

Wasswa Shafik, Rubee Singh, and Vikas Kumar

Abstract In the period of fast technical innovation, artificial intelligence (AI) has become a transformative force within the medical care market. This study explores the crucial measurements of AI application in healthcare, with a primary concentrate on cultivating openness, interpretability, and cooperation to make specific lasting methods. It establishes the stage by highlighting the crucial function of AI in healthcare. It highlights the necessity of incorporating concepts of openness and interpretability right into AI systems' materials. This structure is essential for developing trust funds amongst stakeholders and advertising liable AI implementation within the healthcare environment. Furthermore, it illuminates the nuanced meanings of openness within the healthcare context, browsing regulative factors to consider and providing studies that brighten effective executions of clear AI in healthcare decision-making procedures. It better explores the details of interpretability and explainability, highlighting their value in boosting the human understanding of AI-driven healthcare choices. Methods and approaches for providing AI choices that are understandable to medical care experts are talked about thoroughly. Human–AI partnership becomes a critical motif in the story, diving right into the collaborating connection between healthcare specialists and AI systems. Techniques for reliable cooperation exist, showcasing exactly how human-in-the-loop techniques boost the total performance and dependability of AI applications in medical care. It checks out the intricacies related to releasing transparent and interpretable AI, describes the instructional

W. Shafik (✉)
School of Digital Science, Universiti Brunei Darussalam, Tungku Link, Gadong, Bandar Seri Begawan 1410, Brunei Darussalam
e-mail: wasswashafik@ieee.org

Dig Connectivity Research Laboratory (DCRLab), Kampala 600040, Uganda

R. Singh
Institute of Business Management, GLA University, Mathura, U.P, India

Department of Management Studies, Kumaun University, Nainital, Uttarakhand, India

V. Kumar
Faculty of Business, Law and Social Sciences, Birmingham City University, Birmingham, United Kingdom

R. Singh et al. (eds.), *Transforming Healthcare Sector Through Artificial Intelligence and Environmental Sustainability*, Approaches to Global Sustainability, Markets, and Governance, https://doi.org/10.1007/978-981-97-9555-0_9

requirements of medical care experts involved with AI, and challenges dominating resistance and apprehension within the market. Preparing for the future, the conversation reaches possible study instructions, discovering arising patterns and innovations positioned to form the future landscape of medical care AI. The chapter wraps up with the ongoing study, advancement, and authentic factors to consider, leading the way for a future where AI perfectly incorporates medical care, guaranteeing both development and sustainability.

Keywords Artificial intelligence · Transparency · Interpretability · Explainability artificial intelligence · Sustainable healthcare · Sustainable cities and communities

9.1 Introduction

Artificial intelligence[1] (AI) stands at the forefront of a healthcare revolution, essentially transforming the way medical diseases are detected and diagnosed in a timely manner due to its transparency and explainability. Using the computational expertise of AI formulas, medical care experts examine substantial datasets with extraordinary rate and precision (Dindorf et al., 2020). The innovation's capacity to look at large databases of clinical info, consisting of individual documents, imaging scans, and hereditary information, allows very early and accurate recognition of illness. This innovative analysis ability holds an enormous pledge, possibly changing the efficiency of medical care treatments. (Taylor & Taylor, 2021). From forecasting illness threats to promoting prompt therapy strategies, AI's analysis abilities not only enhance clinical procedures but substantially boost individual results.

Personalised treatment plans unleashed by AI through joining together specific client information, hereditary details, and historical documents, AI technologies develop customised therapy strategies that resolve everyone's one-of-a-kind features (Goodell et al., 2023). This tailored strategy substantially boosts therapy efficiency while lessening unfavourable impacts. Essentially, AI's function in therapy customisation notes a standard change from the conventional one-size-fits-all version to an extra nuanced and patient-centric medical care shipment system (Huang et al., 2023). As AI formulas continue to progress, the possibility of therapies precisely customised to a person's hereditary makeup holds an excellent pledge for the future of medication. Check out techniques to improve cross-domain explainability, permitting AI versions to learn one medical care domain name to supply vibrant understandings in associated yet various clinical specialists (Goodell et al., 2023).

Anticipating analytics, equipped by AI, has become a game-changer for inpatient administration within the medical care landscape. Leveraging historical personal information, these formulas anticipate illness development, recognise prospective problems, and maximise source allotment (Chauhan & Sonawane, 2022). A doctor equipped with anticipating understanding can proactively handle patient treatment,

[1] https://cloud.google.com/learn/what-is-artificial-intelligence.

expecting and resolving problems prior to the rise. This not only improves patient results but also adds to the total performance of medical care distribution. The combination of AI and anticipating analytics changes the healthcare market right into a much more preemptive and receptive ecological community, where sources are purposefully released to supply the very best feasible individual treatment. (Speith, 2022). Resolve the honest ramifications of explainable AI in healthcare, especially worrying about personal privacy, permission, and the accountable use of delicate wellness information. A need to explore the growth of moral standards, administration frameworks, and regulative structures that stabilise the advantages of AI with honest factors to consider to ensure a lasting and accountable application (Huang et al., 2023).

In the pharmaceutical research arena, AI has increased medication exploration procedures, using an effective device for recognising prospective prospects and anticipating their efficiency. Assessing substantial datasets associated with organic features and chemical communications, AI formulas speed up the recognition of unique medical substances. (Hepworth et al., 2021). This not only minimises the moment and expense generally related to medicine exploration but additionally opens methods for creating therapies for conditions that were, as soon as taken into consideration, unbending, different AI domains are illustrated in Fig. 9.1. The marital relationship of AI and medication exploration is positioned to redefine the pharmaceutical landscape, introducing a period of even more targeted, reliable, and cutting-edge restorative treatments (Lozano-Murcia et al., 2023).

While AI's effect on diagnostics and therapy is extensive, its impact prolongs much past these worlds, penetrating numerous aspects of medical care systems. From enhancing functional jobs and enhancing source monitoring to boosting personal involvement via individualised health and wellness details, AI is improving the whole healthcare environment (Dindorf et al., 2020). Natural Language Processing[2] (NLP) modern technologies, as an example, help with the removal of useful understandings from disorganised healthcare information, using a much deeper understanding of personal problems (Taylor & Taylor, 2021). As AI continues to develop, it holds the pledge of not just attending to specific medical care difficulties but likewise promoting an alternative improvement in the method healthcare is provided, handled, and experienced around the world.

The incorporation of AI in healthcare is revolutionising the sector by enhancing decision-making, improving patient outcomes, and optimising operational efficiencies. AI technologies are increasingly being employed in areas such as diagnosis, treatment planning, personalised medicine, and healthcare management. These advancements are not only transforming traditional healthcare practices but are also paving the way for a more sustainable and efficient healthcare system (Chauhan & Sonawane, 2022). However, as AI becomes more embedded in healthcare, concerns about the transparency and explainability of these technologies have come to the forefront. The complexity and often opaque nature of AI algorithms, particularly in

[2] https://www.ibm.com/topics/natural-language-processing.

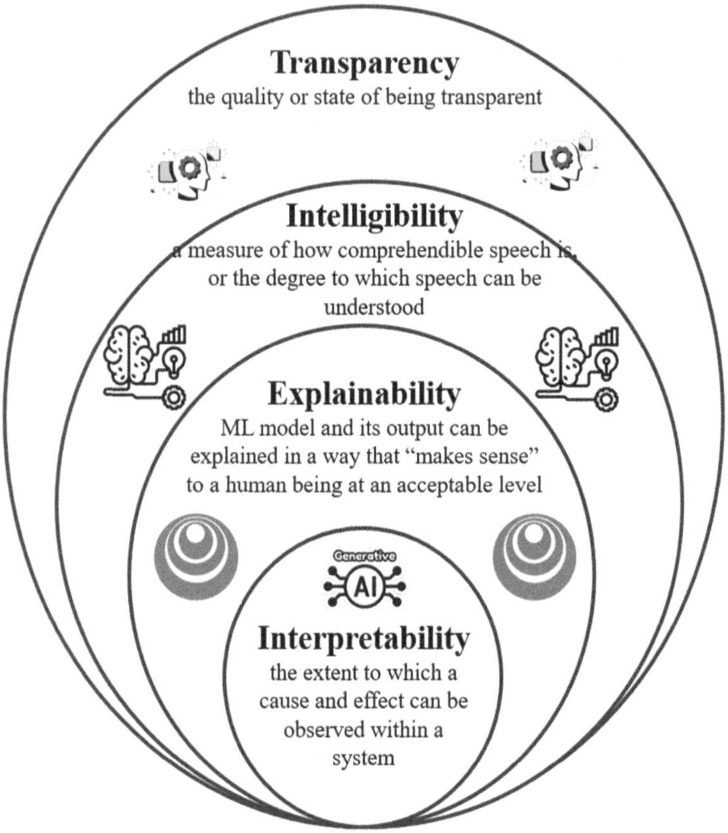

Fig. 9.1 Artificial intelligence domains

machine learning and deep learning models, pose significant challenges for stakeholders who need to understand, trust, and effectively utilise these tools in critical healthcare decisions (Speith, 2022). Thus, transparency and explainability are not just technical issues but fundamental requirements for ensuring the ethical, safe, and practical application of AI in healthcare.

To summarise these technical terms, transparency and explainability in AI refer to the ability to understand and interpret how AI systems make decisions, particularly in complex environments similar to healthcare. In this context, transparency involves making the processes and data used by AI systems visible and understandable to users. At the same time, explainability focuses on providing clear and understandable reasons for the AI's outputs (Huang et al., 2023). These concepts are critical in healthcare, where the implications of AI-driven decisions can be life-altering. The lack of transparency in AI systems can lead to mistrust among clinicians, patients, and regulators, potentially hindering the adoption of AI technologies. Furthermore, without clear explanations of how AI systems arrive at their conclusions, healthcare providers may struggle to validate the results, leading to ethical and legal challenges (Huang et al., 2023). Thus, ensuring that AI systems in healthcare are both transparent and explainable is essential for fostering trust, improving clinical outcomes, and promoting the responsible use of technology in the sector.

The intersection of AI transparency, explainability, and sustainability in healthcare presents both challenges and opportunities. On the one hand, the demand for transparency and explainability can complicate the design and deployment of AI systems, especially in the context of highly complex models like deep learning networks. These models, while powerful, have complex operations making it difficult for users to understand how decisions are made (Goodell et al., 2023). On the other hand, the drive for transparency can lead to the development of more robust, interpretable models that not only improve decision-making but also align with broader sustainability goals. For example, transparent AI systems can help optimise resource allocation, reduce waste, and minimise the environmental impact of healthcare operations. By making the processes and decisions of AI systems visible and understandable, stakeholders can ensure that AI-driven innovations contribute to a more sustainable healthcare system without compromising on quality or safety (Taylor & Taylor, 2021).

The growing need for ethical governance and regulatory oversight further underscores the importance of transparency and explainability in AI-driven healthcare. As AI systems become more autonomous and influential in clinical settings, ensuring that these systems operate in a transparent and explainable manner is crucial for maintaining accountability and protecting patient rights. Regulatory bodies are increasingly emphasising the need for transparency in AI algorithms to ensure that they meet safety and efficacy standards (Joyce et al., 2023). Moreover, ethical considerations, such as the need to avoid bias and ensure fair access to AI-driven healthcare, are closely linked to the transparency and explainability of these systems. In this context, future research must focus on developing AI models and frameworks that prioritise transparency and explainability, thereby ensuring that AI technologies in healthcare are not only innovative but also ethical, trustworthy, and aligned with the goals of sustainable healthcare (Dindorf et al., 2020).

Explainable AI systems give presence right into their decision-making procedures, enabling medical care experts and stakeholders to recognise just how verdicts are gotten. This openness is vital, specifically in important medical care choices where lives go at risk. Interpretability matches openness by making certain that the reasoning behind AI-driven choices is not just noticeable but also understandable to human beings (Huang et al., 2023). This human-centric method cultivates count on and approval of modern AI technologies in medical care, as medical professionals can, with confidence, incorporate AI understandings right into their decision-making procedures. In addition, in the complicated landscape of clinical diagnostics and therapy preparation, the interpretability of AI outputs makes it possible for healthcare experts to fine-tune and boost their very own experience (Dindorf et al., 2020).

9.1.1 Chapter Contribution

This chapter presents the following contributions as summarised below:

- The chapter explores the nuanced meaning of openness within medical care, browses regulatory factors to consider, and showcases studies to illuminate effective executions of transparent AI decision-making.
- The chapter compares interpretability and explainability, highlighting their essential duty in improving human understanding of AI-driven healthcare choices and providing strategies for making AI choices understandable.
- The chapter presents the collaborating connection between medical care specialists and AI systems, providing approaches for reliable cooperation and showcasing the enhancement of performance and integrity with human-in-the-loop strategies.
- The chapter addresses technological and human difficulties connected with applying AI in healthcare, discovering intricacies in releasing transparent and interpretable AI, detailing academic requirements, and facing market resistance and suspicion.
- The chapter illustrates the arising fads and innovations in healthcare AI, expects innovations in openness and interpretability, and details possible research study instructions to form the future landscape of lasting medical care AI.

9.1.2 Chapter Structure

Section 9.2 presents transparency in the context of healthcare AI, illustrating how the regulatory factors are to be considered, as well as standards and studies highlighting transparency in AI applications in healthcare. Section 9.3 distinguishes between interpretability and explainability, showing the significance of interpretable AI in healthcare decision-making. Section 9.4 explores the collaborative relationship between healthcare professionals and some identified strategies for effective cooperation and synergy. Section 9.5 presents the technical challenges in deploying transparent and interpretable AI in healthcare and defeating resistance and scepticism within the healthcare industry. Section 9.6 investigates the emerging research trends and technology directions in AI for sustainable healthcare. Finally, Sect. 9.7 illustrates the lessons learned from the chapter and the conclusion.

9.2 Transparency in AI for a Sustainable HealthCare

Transparency in the context of healthcare, AI implies the clarity and openness with which AI-enabled systems present their decision-making processes. In the medical care domain, where choices have straight ramifications for individual health, it is vital to debunk the "black box" nature of AI systems (Joyce et al., 2023). Explainable and transparent AI makes certain that medical care specialists and stakeholders understand the inputs, reasoning, and elements affecting AI-driven referrals. This quality promotes dependency and trust and encourages experts to seriously assess and confirm the result of AI systems, adding to a joint and liable healthcare atmosphere (Dindorf et al., 2020). The assimilation of AI in healthcare requires a durable structure of governing factors to consider and standards to make certain honest, risk-free, and answerable techniques. Research studies ought to concentrate on establishing flexible systems that are efficient in clarifying choices in real-time, taking into consideration modifications in in-person information, professional standards, and clinical understanding. This makes certain that AI stays interpretable and pertinent as medical care techniques breakthrough (Joyce et al., 2023).

Governing bodies around the world are coming to grips with the difficulty of creating plans that strike an equilibrium between promoting development and securing a person's well-being. Openness demands are progressively becoming important parts of these guidelines, engaging designers and medical care organisations to reveal just how AI formulas run (Taylor & Taylor, 2021). Standards additionally deal with concerns of prejudice, justness, and personal information privacy, recognising the possible dangers connected with AI in healthcare. Striking this fragile equilibrium needs recurring discussion and flexible approaches to equal technical developments while promoting honest requirements (Goodell et al., 2023). To illustrate the current regulatory considerations and guidelines properly, we present some selected efforts demonstrating transparent AI applications in contemporary healthcare systems; some AI data qualities are presented in Fig. 9.2.

9.2.1 IBM Watson for Oncology

IBM Watson for Oncology is a front-runner instance of clear AI in medical care. Made to help oncologists in therapy decision-making, this system discloses openness by offering an in-depth understanding of its logical procedures. By bringing into play a vast data source of clinical literary works, scientific test information, and personal documents, Watson creates therapy referrals based upon one of the most appropriate and current info (Huang et al., 2023). The openness of Watson's decision-making enables healthcare specialists to comprehend exactly how AI comes to certain therapy alternatives, promoting reliance on innovation. By integrating the power of AI with the knowledge of oncologists, IBM Watson for Oncology exhibits a joint method that boosts the top quality and performance of cancer cell treatment (Chauhan & Sonawane, 2022).

Fig. 9.2 Artificial intelligence data qualities criteria

9.2.2 Google's DeepMind in Ophthalmology

DeepMind's cooperation with Moorfields Eye Healthcare facility showcases openness in AI applications for clinical imaging, especially in ophthalmology. The AI system evaluates retinal scans to discover indications of eye conditions. Openness is attained by offering medical professionals a presence right into the formulas' decision-making procedure (Dindorf et al., 2020). Medical professionals can recognise the attributes and patterns affecting the AI's analysis pointers, helping with a collective partnership between human specialists and the AI. This openness not only guarantees that eye doctors can rely on the AI's suggestions but likewise encourages them to confirm and fine-tune the analysis results (Taylor & Taylor, 2021). DeepMind's strategy in ophthalmology establishes a criterion for just how openness can be incorporated right into AI applications, improving the interpretability and approval of AI-generated understandings in professional setups.

9.2.3 AIDoc Medical for Radiology Imaging

AIDoc Medical's application in radiology imaging emphasises the value of openness in AI-assisted diagnostics. The system, created to assess clinical pictures, gives openness by providing radiologists with an understanding of the essential attributes and variables affecting their analysis verdicts. Radiologists can check out the AI's analysis of the pictures, confirming and matching their very own evaluations (Goodell et al., 2023). This clear partnership makes certain that the AI operates as a beneficial device in the radiologist's operations, adding to exact and reliable diagnostics. By honestly connecting the decision-making procedure, AIDoc Medical advertises that it depends on an understanding between the AI system and healthcare specialists, showing exactly how openness improves the harmony between human proficiency and equipment knowledge in the field of radiology (Huang et al., 2023).

9.2.4 PathAI in Pathology

PathAI's application in pathology sticks out for its focus on explainable AI, guaranteeing openness in the analysis procedure. This system allows pathologists to comprehend and translate the attributes that affect their analysis forecasts. By supplying quality on just how the AI examines pathology slides and infers, PathAI helps with a collective partnership between the AI system and pathologists (Chauhan & Sonawane, 2022). The openness of AI's decision-making improves the pathologist's capacity to verify and rely on the analysis's understanding, inevitably causing even more exact and dependable pathology medical diagnoses (Speith, 2022). PathAI's dedication to openness in pathology showcases exactly how explainable AI designs can be critical in promoting a collective and educated strategy for the medical diagnosis of illness.

9.2.5 GE Healthcare's Edison AI

GE Healthcare's Edison AI shows openness in its application to clinical imaging evaluation. Via collaborations with medical care establishments, Edison AI freely connects the formulas and techniques made use of in its picture evaluation procedures. This openness is vital for medical care specialists, enabling them to recognise and verify the AI's payments to clinical diagnostics (Hepworth et al., 2021). By supplying a presence right into the decision-making systems, Edison AI advertises a joint strategy in which medical professionals can incorporate the AI understanding right into their professional operations. This collective openness not only improves the integrity of AI applications but also guarantees that they straighten effortlessly with existing medical care techniques (Lozano-Murcia et al., 2023). GE Medical Care's

Edison AI establishes an instance of exactly how a clear partnership between AI designers and healthcare organisations can add to the effective combination of AI right into clinical imaging operations.

9.3 AI Interpretability and Explainability

Interpretability and explainability of AI are two interrelated terminologies yet distinct concepts crucial in understanding the decision-making processes of AI systems. Interpretability focuses on the ease with which humans preserve and comprehend the internal workings of an AI model, essentially grasping its functionality (McClimens, 2011). Explainability, on the other hand, delves into the capacity to coherence these workings in a manner that is fathomable to non-experts. While interpretability concentrates on the openness of the version's framework, explainability areas focus on producing a story or reasoning behind the design's forecasts (Band et al., 2023). In healthcare AI, comparing these two ideas is crucial for designing AI systems that not only produce precise forecasts but also ensure that medical care experts can rely on, confirm, and efficiently utilise these understandings in their decision-making procedures; several applications are demonstrated in Fig. 9.3.

9.3.1 AI Techniques for Explaining AI Decisions in the Healthcare Sector

Five techniques have been developed to enhance the explainability and interpretability of AI decisions in the health sector.

9.3.1.1 Interpretable Models

Interpretable versions, such as choice trees or direct versions, naturally supply understandings right into their choice reasoning. These designs are developed to be clear, permitting healthcare specialists to conveniently recognise exactly how detailed attributes add to the AI's decision-making procedure. In medical care, interpretable versions provide a clear and simple means to translate and verify AI choices, boosting the count among specialists (Priya Dharshini et al., 2023). An interpretable choice tree design is used to anticipate the probability of diabetes mellitus based on client features such as age, BMI, and family history. Each branch of the tree stands for a choice based upon a details function, offering a clear course to the last forecast. This openness permits healthcare experts to conveniently recognise how the version reaches its diabetes mellitus danger analysis (Pintelas et al., 2021). Medical care experts can, with confidence, make use of and confirm the choice tree's forecasts,

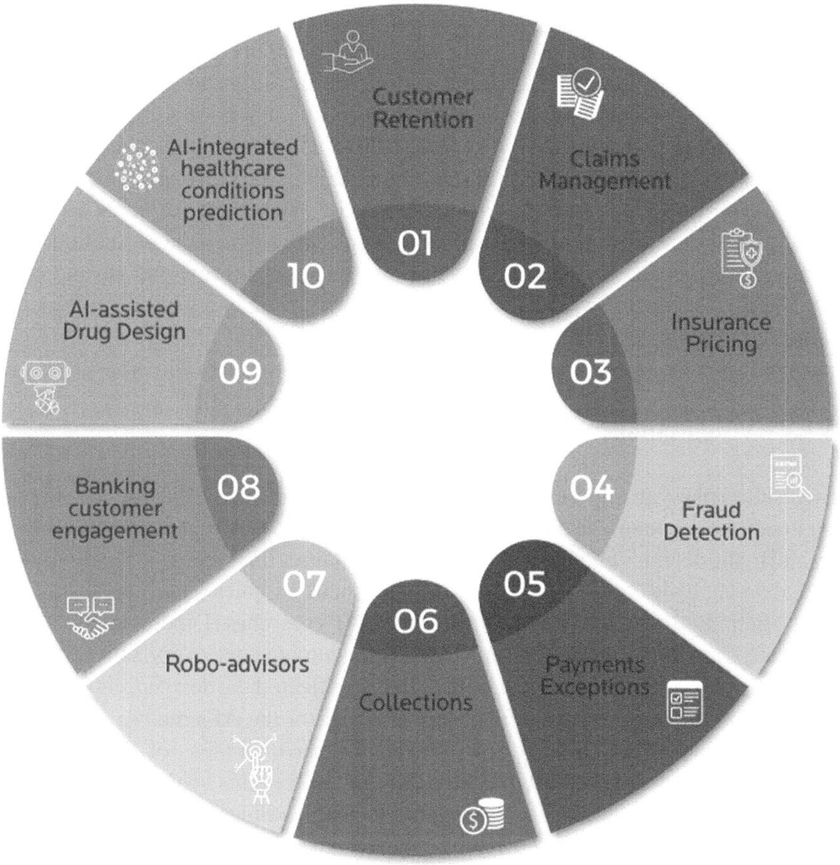

Fig. 9.3 Application of artificial intelligence

comprehending the certain aspects that add to the danger of diabetic issues for every person. This openness boosts the count on the version, making it a beneficial device for threat evaluation and very early treatment (Dindorf et al., 2020).

9.3.1.2 Post-Hoc Interpretability Methods

Post-hoc interpretability techniques are utilised to discuss choices made by complicated, black-box designs. Technologies such as Local interpretable model-agnostic explanations[3] (LIME) and Shapley Additive descriptions[4] (SHAP) produce streamlined, in-your-area estimated designs that illuminate the actions of the initial version

[3] https://christophm.github.io/interpretable-ml-book/lime.html.

[4] https://shap.readthedocs.io/en/latest/.

for detailed circumstances (Goodell et al., 2023). These techniques give important understandings of why a specific choice was made, using openness also for versions with elaborate inner frameworks. In a deep discovering design for growth discovery in clinical images, LIME produces loyal estimations of the version's choice limits for specific photos in your area (Joyce et al., 2023). This permits medical care experts to recognise which pixels added most to the design's choice relating to the existence of a lump. LIME gives an understanding of the certain attributes affecting the design's choice for every picture, making it possible for medical professionals to verify and translate the AI's search. This post-hoc interpretability technique improves the version's dependability, promoting its assimilation right into the analysis process (Vieira & Digiampietri, 2020).

9.3.1.3 Visualisations Models

Visualisation plays an important duty in describing AI choices, particularly in clinical imaging. Saliency maps, as an example, highlight the areas of a photo that considerably add to the design's choice. In radiology, for instance, imagining the locations that affected an analysis choice in an X-ray or MRI check supplies quality to healthcare experts, helping them understand and verify the AI's searches (Dindorf et al., 2020). When an AI version examines breast X-rays for pneumonia discovery, saliency maps highlight locations in the picture that substantially affected the design's choice. Brilliant areas on the map represent functions essential for the AI's medical diagnosis. Radiologists can utilise saliency maps to recognise which areas of the X-ray led the AI to its pneumonia forecast (Taylor & Taylor, 2021). This visualisation helps in confirming the AI's choice and gives important understandings right into the analysis procedure, boosting the interpretability of the design.

9.3.1.4 Natural Language Generation (NLG)

This technique equates the technological and data-driven nature of AI choices right into narrative types, making it extra available to medical care experts who might not be fluent in artificial intelligence terms. NLG makes certain that the reasoning behind AI choices is interacted with understandably, helping with a partnership between the AI system and healthcare experts (Goodell et al., 2023). An AI system giving scientific choice assistance creates human-readable descriptions for its therapy referrals. Rather than offering complicated mathematical information, the system creates natural language descriptions that describe the reasoning behind each suggestion. Medical care experts, no matter their technological know-how, can conveniently understand the factors behind AI's therapy ideas (Huang et al., 2023). NLG presents clear interaction between the AI system and experts, cultivating trust funds and motivating collective decision-making.

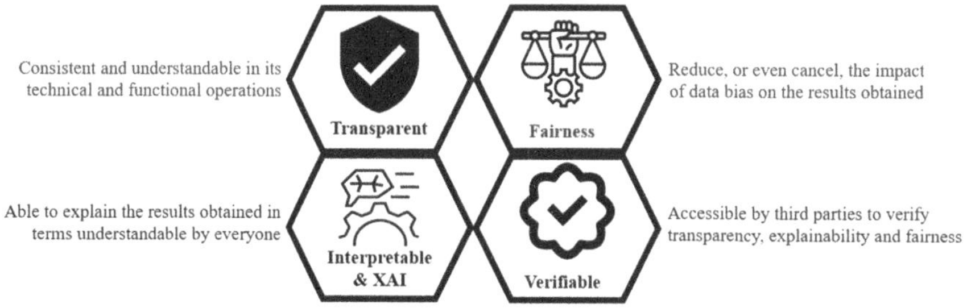

Fig. 9.4 Artificial intelligence and its practical truthfulness

9.3.1.5 Feature Importance Analysis

Examining the value of various functions in the decision-making procedure is an additional strategy for clarifying AI choices. Function relevance evaluation measures the payment of each input variable to the version's outcome. These details permit medical care experts to recognise which elements are most prominent in a specific choice, assisting in the recognition and understanding of the AI's referrals (Chauhan & Sonawane, 2022). In an anticipating design for heart problems, attribute value evaluation measures the effect of variables like cholesterol degrees, high blood pressure, and age on the version's forecasts. It rates these attributes based on their payment to the design's decision-making. Cardiologists can focus on treatments and therapies based on one of the most significant aspects determined by the design (Speith, 2022). Function significance evaluation supplies workable understandings, enabling healthcare experts to concentrate on essential elements of a person's health and wellness and enhance the total precision of therapy techniques, four main factors are considered AI open and trustful as illustrated in Fig. 9.4.

9.3.2 Significance of Interpretability and Explainability AI in Healthcare

Some of the identified significance of these techniques in the healthcare systems are illustrated.

9.3.2.1 Trust and Adoption by Healthcare Professionals

Interpretable and explainable AI in healthcare cultivates count on amongst medical care experts. When specialists can comprehend just how an AI system reaches its verdicts, they are more likely to trust its suggestions (Hepworth et al., 2021). This trust fund is critical for the extensive fostering of AI devices in medical operations, making certain that healthcare specialists welcome and rely upon AI-driven understandings for enhanced individual treatment. Raised depend on causes greater approval and usage of AI innovations by medical care experts (Lozano-Murcia et al., 2023). Medical professionals are most likely to incorporate interpretable AI right into their decision-making procedures, boosting the general performance and efficiency of healthcare distribution.

9.3.2.2 Enhanced Validation and Medical Tool Peer Reviews

The interpretability and explainability of AI systems promote much easier recognition and peer evaluation within the clinical area. Medical care specialists can inspect the decision-making procedures, making sure that the AI straightens with recognised clinical expertise and standards (McClimens, 2011). This peer evaluation procedure is important for preserving the top quality and precision of AI applications in medical care. Boosted recognition and peer evaluation add to the toughness and dependability of AI designs. The openness supplied by interpretable AI permits reliable cooperation and responses amongst medical care experts, lessening the danger of mistakes and making sure that AI innovations follow the highest possible criteria of clinical technique (Priya Dharshini et al., 2023).

9.3.2.3 Ethical and Regulatory Compliance

Interpretable and explainable AI is vital for satisfying honest criteria and regulative needs in medical care. Lots of healthcare guidelines and moral standards demand openness in decision-making procedures, mainly when AI affects individual treatment (Pintelas et al., 2021). Honest factors to consider, such as staying clear of prejudiced results and securing client personal privacy, can be much better dealt with interpretable AI. Adherence to moral and regulative requirements is necessary for the accountable implementation of AI in healthcare (Gurbuz et al., 2023). Interpretable AI guarantees that healthcare establishments and AI designers adhere to standards, promoting an accountable and honest use of AI innovations in individual treatment.

9.3.2.4 Patient Understanding and Informed Decision-Making

Transparent AI adds to the client's understanding of therapy strategies and analysis of results. When AI choices are clarified reasonably, people can much better understand the basis of suggestions. This openness equips people to proactively take part in their medical care choices and therapy strategies (Joyce et al., 2023). Enlightened individuals are most likely to follow suggested therapies and treatments. Interpretable AI sustains everyday decision-making between medical care specialists and clients, boosting personal contentment and adding to a joint medical care atmosphere (Markus et al., 2021). This patient-centered method lines up with the broader objective of boosting total healthcare results and experiences.

9.3.2.5 Mitigation of Bias and Fairness Concerns

Interpretable and explainable AI plays a critical duty in determining and dealing with predispositions in healthcare formulas. By supplying exposure right into the decision-making procedure, interpretable AI permits healthcare specialists to look at possible predispositions and variations in the information utilised to educate the design (Vieira & Digiampietri, 2020). This openness is crucial for advertising justness and equity in healthcare results. Reducing prejudice ensures that AI-driven choices are fair throughout varied person populations. Doctors can proactively analyse and correct any prejudices, bringing about even simpler and more honest healthcare treatments (Farah et al., 2023). Interpretable AI, for that reason, adds to the honest use of AI in medical care by advertising justness and decreasing differences in individual treatment.

9.4 Human-AI Collaboration for a Sustainable HealthCare

The cooperative communication leverages the toughness of both celebrations, producing a medical care community where AI-driven understandings improve the capacities of healthcare specialists, resulting in much more reliable, tailored, and moral individual treatment.

9.4.1 Data Quality and Integration

The challenge of data quality and integration in healthcare arises from the diverse and fragmented nature of healthcare data. Patient information is stored across various systems, often in different formats and with inconsistencies (Dindorf et al., 2020). Integrating and ensuring the quality of this data for AI applications becomes a formidable task. Inconsistent formats, missing values, and errors can undermine the accuracy and reliability of AI algorithms, posing potential risks to patient care (Taylor & Taylor, 2021). Overcoming this challenge demands thorough data cleaning, standardisation, and integration efforts to ensure that AI systems receive high-quality, standardised data for robust performance.

9.4.2 Interoperability and System Integration

The implementation challenge of interoperability and system integration stems from the diverse technology stacks and disparate systems within healthcare institutions. The integration of AI seamlessly into existing workflows and systems faces hurdles due to compatibility issues between AI applications and electronic health record

(EHR) systems (Huang et al., 2023). These compatibility challenges can disrupt clinical workflows and impede the effective deployment and utilisation of AI in healthcare settings. Addressing this challenge requires standardised interfaces, open communication protocols, and collaborative efforts to ensure the seamless integration of AI with existing healthcare infrastructures (Speith, 2022).

9.4.3 Regulatory Compliance and Ethical Concerns

Regulatory compliance and ethical considerations are significant challenges in implementing AI in healthcare. The healthcare industry is subject to stringent regulations, and ensuring AI compliance with frameworks like HIPAA poses a complex task (Taylor & Taylor, 2021). Additionally, ethical concerns related to patient privacy, consent, and responsible AI use further complicate implementation efforts (Huang et al., 2023). Failure to meet regulatory requirements can result in legal consequences and reputational damage. Addressing this challenge involves navigating complex regulatory landscapes, implementing robust data governance frameworks, and incorporating ethical considerations into AI development and deployment.

9.4.4 Resistance to Change and Lack of Trust

Overcoming resistance to change and building trust among healthcare professionals are critical challenges in AI implementation. Concerns about job displacement, a lack of understanding of AI processes, and fear of technology errors impacting patient outcomes can hinder the adoption of AI technologies (Joyce et al., 2023). Resistance and distrust among healthcare providers can impede the successful integration of AI into clinical workflows, limiting the potential benefits of these technologies. Addressing this challenge requires effective communication, comprehensive training programs, and collaborative efforts to demonstrate the value and reliability of AI applications in healthcare (Markus et al., 2021).

9.4.5 Limited Explainability and Interpretability

The challenge of limited explainability and interpretability arises from the complexity of certain AI models, particularly deep learning models. Understanding how these models arrive at specific diagnoses or treatment recommendations is crucial for healthcare professionals (Vieira & Digiampietri, 2020). The lack of interpretability can lead to scepticism, hindering the acceptance and trust in AI-driven decisions. Overcoming this challenge involves developing and integrating explainable AI models and techniques that provide clear insights into the rationale behind AI-driven decisions (Farah et al., 2023; Shafik, 2024). This fosters trust and understanding among healthcare professionals, ensuring that AI recommendations align with their expertise and contribute to informed decision-making.

9.5 AI Implementation Challenges in the Contemporary Healthcare System

Several factors challenge the implementation of healthcare as presented.

9.5.1 Data Quality and Integration

The obstacle to high-quality information and combination in healthcare emerges from the varied and fragmented nature of medical care information. Individual details are kept throughout numerous systems, usually in various layouts and with disparities. Incorporating and guaranteeing the high quality of this information for AI applications ends up being an awesome job (Bogle et al., 2020). Irregular styles, missing out on worths, and mistakes can weaken the precision and integrity of AI formulas, presenting prospective threats to individual treatment. Conquering this difficulty requires detailed information cleansing, standardisation, and assimilation initiatives to guarantee that AI systems obtain top-quality, standard information for durable efficiency (Holzinger, 2021). Explore exactly how healthcare experts and people communicate with AI-driven systems, concentrating on structure, trust fund, and understanding. There has been a raising need of studies need to discover user-centered layout concepts, interaction methods, and academic treatments to improve the approval and cooperation between human beings and AI in healthcare setups (Farah et al., 2023).

9.5.2 Interoperability and System Integration

The application difficulty of interoperability and system assimilation originates from the varied modern technology heaps and inconsonant systems within medical care organisations. The assimilation of AI perfectly right into existing processes and systems encounters difficulties because of compatibility problems between AI applications and EHR systems (Pintelas et al., 2021). These compatibility obstacles can interfere with medical operations and hamper the efficient implementation and usage of AI in medical care setups. Resolving this obstacle needs standard user interfaces, open interaction procedures, and joint initiatives to guarantee the smooth assimilation of AI with existing healthcare frameworks (Band et al., 2023). The study ought to target establishing transferable descriptions that adjust throughout different medical care contexts, advertising the scalability and flexibility of AI designs (Vieira & Digiampietri, 2020). Examine means flawlessly incorporating domain-specific clinical expertise right into AI versions to boost interpretability. This consists of checking out crossbreed designs that incorporate data-driven strategies with rule-based systems, leveraging medical proficiency to enhance the openness and interpretability of AI forecasts (Pintelas et al., 2021).

9.5.3 Regulatory Compliance and Ethical Concerns

Governing conformity and honest factors to consider are substantial obstacles in carrying out AI in medical care. The healthcare market goes through rigorous guidelines, and making sure AI conforms with structures like the Health Insurance Portability and Accountability Act presents a complicated job (McClimens, 2011). In addition, moral issues associated with person personal privacy, authorisation, and accountable AI usage additionally make complex execution initiatives. Failing to satisfy regulative demands can lead to lawful repercussions and reputational damages. Resolving this obstacle entails browsing complicated regulative landscapes, executing durable information administration structures, and including moral factors to consider right into AI growth and release (Hepworth et al., 2021).

9.5.3.1 Resistance to Change and Lack of Trust

Conquering resistance to alteration and construction depends on healthcare care experts, an important difficulty in AI execution. Worries regarding work variation, an absence of understanding of AI procedures, and anxiety about modern technology mistakes influencing individual results can impede the fostering of AI innovations (Dindorf et al., 2020). Resistance and suspicion among doctors can restrain the effective assimilation of AI right into the medical process, restricting the possible advantages of these innovations. Resolving this obstacle needs efficient interaction, extensive training programs, and collective initiatives to show the worth and integrity of AI applications in medical care (Goodell et al., 2023).

9.5.4 Limited Explainability and Interpretability

The difficulty of minimal explainability and interpretability occurs from the intricacy of AI designs, especially deep understanding versions. Recognising just how these versions come to certain medical diagnoses or therapy suggestions is critical for healthcare specialists (Dindorf et al., 2020). The absence of interpretability can cause apprehension, impeding the approval and relying on AI-driven choices. Conquering this obstacle entails establishing and incorporating explainable AI designs and methods that give a clear understanding of the reasoning behind AI-driven choices (Huang et al., 2023). This cultivates trust fund and understanding amongst medical care experts, making certain that AI referrals straighten with their experience and add to educated decision-making. Establish standard structures for assessing and guaranteeing the explainability of AI designs in medical care. This consists of specifying metrics, criteria, and standards to evaluate the degree of interpretability needed for various healthcare applications (Dindorf et al., 2020). Developing a typical language for version explainability will certainly assist in constant examination and contrast throughout varied AI systems.

9.6 Future Research Directions

The future of study in openness, interpretability, and explainability for a lasting health and wellness market includes numerous vital instructions focused on resolving arising obstacles and optimising the influence of modern AI technologies on healthcare.

9.6.1 Advanced Explainable AI Models for Clinical Decision Support

Future research should focus on developing more advanced Explainable AI[5] (XAI) models tailored explicitly to clinical decision support systems[6] (CDSS). Traditional AI models, such as deep neural networks, often operate as "black boxes," making it challenging for healthcare professionals to understand the rationale behind specific recommendations (Dindorf et al., 2020). Future models can offer transparent insights into AI-driven decisions, enabling clinicians to trust and verify the results. Research should also explore the integration of domain-specific knowledge into these models, allowing for more accurate and context-aware explanations (Joyce et al., 2023). The development of hybrid models that combine rule-based systems with machine learning could offer a promising pathway, ensuring that the AI's reasoning aligns with established medical guidelines while also adapting to new data. Furthermore, future research could investigate the scalability of these models across different healthcare systems and regions, ensuring that the transparency and explainability of AI are maintained regardless of the context (Huang et al., 2023). This will be crucial for fostering widespread adoption and ensuring that AI-driven healthcare remains both effective and ethical.

[5] https://www.ibm.com/topics/explainable-ai.

[6] https://en.wikipedia.org/wiki/clinical_decision_support_system.

9.6.2 Transparent AI for Sustainable Healthcare Operations

As AI increasingly drives operational efficiencies in healthcare, future research must prioritise transparency in AI algorithms used for resource allocation, waste management, and energy consumption in healthcare facilities (Speith, 2022). AI models in this domain should not only optimise sustainability but also provide clear, interpretable data on how decisions are made to achieve environmental goals. Developing AI systems that can transparently balance cost, environmental impact, and patient care quality will be essential. Future work should focus on creating frameworks that allow for the auditability of AI systems, ensuring that stakeholders can understand and trust the mechanisms by which sustainability metrics are achieved (Huang et al., 2023). Furthermore, research should explore the integration of real-time data analytics to provide continuous, transparent monitoring of environmental performance. This will enable healthcare administrators to make informed decisions that align with both economic and environmental objectives (Dindorf et al., 2020). The ethical implications of AI-driven sustainability efforts should also be examined, particularly in ensuring that cost-saving measures do not inadvertently compromise patient care.

9.6.3 Ethical and Transparent AI in Personalised Medicine

The future of AI in personalised medicine will hinge on the development of ethical and transparent algorithms that both patients and clinicians can trust. As AI systems increasingly tailor treatments based on individual genetic, environmental, and lifestyle factors, transparency becomes critical in understanding how specific decisions are made (Speith, 2022). Future research should investigate methods for enhancing the explainability of AI-driven personalised treatment plans, particularly in complex cases involving multiple comorbidities. This includes developing techniques to ensure that AI recommendations are not only evidence-based but also clearly communicated to both healthcare providers and patients (Huang et al., 2023). In addition, research should explore the integration of patient preferences and values into AI models, ensuring that personalised medicine is not only technically accurate but also aligned with the ethical principles of autonomy and informed consent. Another important direction is the development of transparent AI systems that can explain the limitations and uncertainties inherent in their recommendations, enabling clinicians to make more informed decisions and engage in shared decision-making with patients (Chauhan & Sonawane, 2022).

9.6.4 Governance Frameworks for AI Transparency in Healthcare

Establishing robust governance frameworks that enforce transparency and explainability in AI systems used within healthcare is a critical future research direction. As AI becomes more integral to healthcare delivery, the development of regulatory standards that ensure transparency without stifling innovation will be essential (Speith, 2022). Future research should focus on identifying the key components of such frameworks, including the establishment of guidelines for algorithmic accountability, the creation of standardised methods for AI model validation, and the development of protocols for continuous monitoring of AI performance in clinical settings. The role of interdisciplinary collaboration in shaping these frameworks should also be explored, bringing together expertise from AI, healthcare, ethics, and law to ensure that governance structures are comprehensive and adaptable (Huang et al., 2023). Additionally, research should examine the implications of transparency on patient trust and the doctor-patient relationship, ensuring that governance measures enhance rather than erode confidence in AI-driven healthcare (Dindorf et al., 2020).

9.6.5 AI Transparency in Predictive Analytics for Public Health

Research into AI transparency in predictive analytics for public health is poised to become increasingly important as these tools are used to forecast disease outbreaks, track the spread of infectious diseases, and manage public health resources. Future directions should focus on the development of transparent predictive models that public health officials, policymakers, and the public can easily understand (Huang et al., 2023). This includes creating algorithms that not only predict health trends accurately but also clearly articulate the factors contributing to these predictions. Research should also explore methods for integrating diverse data sources, such as social determinants of health, into AI models while maintaining transparency (Chauhan & Sonawane, 2022). Ensuring that these models are interpretable will be critical for building public trust and facilitating effective communication during health crises. Furthermore, future research could investigate the ethical implications of transparency in public health AI, particularly in balancing the need for accurate predictions with the potential for public panic or misinterpretation of data (Speith, 2022). The development of standardised protocols for transparent reporting of predictive analytics outcomes and their uncertainties will be crucial for the responsible use of AI in public health.

9.6.6 Cross-Disciplinary Approaches to Explainability in AI-Driven Healthcare

Future research should explore cross-disciplinary approaches to enhance the explainability of AI in healthcare, combining insights from computer science, medicine, ethics, and cognitive psychology (Chauhan & Sonawane, 2022). Understanding how different stakeholders, clinicians, patients, and administrators perceive and interpret AI explanations will be critical to improving the usability and acceptance of AI systems. Research could investigate the design of user-centric interfaces that present AI-driven insights in ways that are intuitive and aligned with the cognitive processes of different users (Huang et al., 2023). Furthermore, studies could explore the role of human factors engineering in the design of explainable AI systems, ensuring that these systems are not only technically sound but also practically applicable in clinical settings. By drawing on cognitive psychology, research can develop strategies to mitigate the potential cognitive overload that might arise from overly complex explanations, instead promoting a balance between detail and clarity (Dindorf et al., 2020). The integration of ethical principles, particularly in explaining the limitations and uncertainties of AI recommendations, will be essential in maintaining transparency while avoiding overreliance on AI systems (Dindorf et al., 2020).

9.6.7 Artificial Intelligence Transparency in Remote and Telehealth Services

As telehealth and remote healthcare services continue to expand, ensuring AI transparency in these settings will be a vital area of research. AI algorithms are increasingly used to monitor patient health remotely, diagnose conditions, and recommend treatments, often without direct human oversight (Holzinger, 2021). Future research should focus on developing transparent AI models that can clearly communicate their reasoning and limitations to patients and healthcare providers in remote settings. This includes designing AI systems that provide easily understandable explanations of their recommendations, allowing patients to make informed decisions about their care (Shafik et al., 2024). Additionally, research should explore how to ensure transparency in AI-driven telehealth applications across different cultural and socioeconomic contexts, recognising that perceptions of transparency may vary. The development of standards for AI transparency in telehealth will also be crucial, particularly in ensuring that AI recommendations are consistent with those that would be made in traditional, in-person healthcare settings (Dindorf et al., 2020). Moreover, research should investigate the role of AI in facilitating patient-clinician communication during remote consultations, ensuring that AI tools enhance rather than hinder the delivery of transparent and effective care (Chauhan & Sonawane, 2022).

9.6.8 *Artificial Intelligence-Driven Environmental Impact Assessments in Healthcare*

The future of sustainable healthcare will likely involve AI-driven environmental impact assessments that can transparently evaluate and mitigate the environmental footprint of healthcare activities. Research should focus on developing AI systems capable of assessing the full lifecycle environmental impact of medical procedures, pharmaceuticals, and healthcare operations. These systems should provide clear, interpretable insights into how healthcare practices affect energy consumption, waste production, and resource use (Chauhan & Sonawane, 2022). Future research could explore the integration of real-time environmental data into AI models, enabling continuous monitoring and immediate feedback on the environmental consequences of healthcare decisions. Additionally, the development of AI tools that can transparently balance environmental impact with patient outcomes will be critical for making sustainable healthcare decisions. This includes the creation of models that can explain the trade-offs between different sustainability metrics, such as carbon emissions and water usage, in a way that healthcare professionals can easily understand (Goodell et al., 2023).

9.6.9 *AI Transparency in Patient Data Privacy and Security*

As AI systems increasingly handle sensitive patient data, ensuring transparency in how these systems manage privacy and security is critical. Future research should explore the development of AI models that can clearly explain how patient data is processed, stored, and protected against breaches or misuse. This includes creating algorithms that offer real-time transparency regarding data access, usage, and potential vulnerabilities (Taylor & Taylor, 2021). Researchers should investigate techniques for enhancing the explainability of AI-driven encryption methods and privacy-preserving technologies, such as differential privacy and federated learning, ensuring that both healthcare providers and patients understand how their data is safeguarded (Chauhan & Sonawane, 2022). Furthermore, research could focus on developing transparent consent mechanisms that allow patients to make informed decisions about their data, with AI systems providing clear explanations of the risks and benefits associated with data sharing. Future work should also examine the ethical implications of transparency in AI-driven data privacy, particularly in balancing the need for robust security measures with the need to maintain patient trust (Joyce et al., 2023). The development of international standards for transparency in AI-driven data security, tailored to the healthcare sector, will be essential for protecting patient privacy in an increasingly digital healthcare landscape.

9.6.10 AI Transparency in Healthcare Innovation and Drug Development

The future of healthcare innovation and drug development will heavily rely on AI, making transparency in these processes a critical area of research. AI models are increasingly used to accelerate drug discovery, design clinical trials, and predict therapeutic outcomes. Future research should focus on developing transparent AI systems that can clearly communicate the rationale behind drug candidate selection, trial design, and the prediction of treatment efficacy (Dindorf et al., 2020). This includes creating models that provide detailed explanations of how AI-driven predictions align with biological mechanisms and clinical evidence. Researchers should also explore methods for making the decision-making process in AI-driven drug development more interpretable for regulators, clinicians, and patients, ensuring that new treatments are both effective and safe (Dindorf et al., 2020). The role of AI in optimising personalised medicine approaches, where treatments are tailored to individual genetic profiles, should also be a key focus, with transparency ensuring that patients and healthcare providers understand the basis for specific therapeutic recommendations (Taylor & Taylor, 2021). Furthermore, research should investigate the ethical considerations of transparency in AI-driven innovation, particularly in ensuring that AI models do not inadvertently introduce biases that could impact the safety or efficacy of new treatments. Establishing transparent frameworks for AI-driven drug development will be essential for fostering trust in new medical innovations and ensuring that they meet rigorous safety and efficacy standards (Joyce et al., 2023).

9.7 Lessons Learned and Conclusion

This section summarises the insights presented on AI, the lessons learned from the chapter, and, finally, the conclusion.

9.7.1 Lessons Learned

Below are the vital lessons discovered:

- Clear and interpretable AI versions are fostered depending on healthcare experts, individuals, and stakeholders. Trust funds are a requirement for an efficient partnership between people and AI systems, highlighting the value of focusing on explainability in medical care AI growth.

- The research highlights the value of resolving moral problems associated with individual personal privacy, approval, and the liable use of AI-driven understandings. Future executions ought to focus on honest structures to guarantee that AI lines up with well-established requirements and safeguards client legal rights.
- Increasing human proficiency with AI-driven understandings boosts analysis accuracy, tailored therapy strategies, and total client treatment. The research study highlights the requirement to grow a collective frame of mind amongst healthcare specialists to make the best use of the advantages of modern AI technologies.
- The vibrant nature of medical care needs AI designs that can adjust and supply clear understandings in time. Future studies and applications need to concentrate on creating systems that continue to be interpretable in developing medical care atmospheres, making sure of continual significance.
- Creating AI systems with a user-centric method is vital for effective fostering. Various stakeholders, consisting of healthcare specialists, clients, and policy-makers, have unique demands for interpretability. Lessons discovered highlight the value of customising descriptions to varied individual teams, considering their cognitive capacities and decision-making procedures.
- Making certain of the top quality, honesty, and personal privacy of healthcare information is essential for reputable AI-driven understandings. Lessons were found to stress the requirement for detailed information administration structures to deal with data-related difficulties in healthcare AI.
- The research study highlights the value of attending to predispositions to make certain reasonable and fair healthcare results. Lessons found emphasise the demand for continual tracking, evaluation, and improvement of AI designs to reduce prejudices and boost the justness of medical care forecasts.
- Comprehending the professional effectiveness, functionality, and socio-economic ramifications of these innovations is essential for forming future executions. Lessons discovered tension in the significance of influence analyses to make sure that AI adds favourably to client treatment and medical care systems.

9.7.2 Conclusion

Finally, the research study on openness, interpretability, and explainability for a lasting health and wellness field has lit up important lessons that chart the training course for the liable combination of AI innovations right into medical care. Count on became the cornerstone, highlighting the requirement of clear and interpretable AI designs to promote self-confidence amongst medical care specialists, individuals, and stakeholders. Moral factors to consider stood apart as vital, stressing the demand for durable structures dealing with personal privacy, approval, and liable AI usage. The research study strengthened the transformative possibility of human-AI cooperation, showing that increasing human know-how with AI-driven understandings improves diagnostics, therapy strategies, and total person treatment. Continual versatility and user-centric layout showed crucial, advising the advancement of AI systems that develop gradually and accommodate varied stakeholder demands. The lessons emphasised the fundamental duty of information administration in making sure of trusted AI-driven understandings and highlighted the importance of clear interaction in developing understanding. Recurring predisposition reduction initiatives and real-world effect analyses were considered necessary, stressing the dedication to justness and favourable payments to medical care results. As we browse the vibrant landscape of healthcare AI, these lessons give a roadmap for lasting, honest, and impactful applications that focus on individual wellness and partnership between human knowledge and expert systems.

References

Band, S., Yarahmadi, A., Hsu, C. C., Biyari, M., Sookhak, M., Ameri, R., Dehzangi, I., Chronopoulos, A. T., & Liang, H. W. (2023). Application of explainable artificial intelligence in medical health: A systematic review of interpretability methods. *Informatics in Medicine Unlocked*, *40*. https://doi.org/10.1016/j.imu.2023.101286

Bogle, G., Silver, J., Lickerman, E., Li, K., & Murickan, T. (2020). The future of artificial intelligence in medicine. In *Intelligence-based medicine: artificial intelligence and human cognition in clinical medicine and healthcare*. https://doi.org/10.1016/B978-0-12-823337-5.00011-1

Chauhan, T., & Sonawane, S. (2022). Contemplation of Explainable Artificial Intelligence Techniques Model Interpretation using Explainable AI. *International Journal on Recent and Innovation Trends in Computing and Communication*, *10*(4). https://doi.org/10.17762/ijritcc.v10i4.5538

Dindorf, C., Teufl, W., Taetz, B., Bleser, G., & Fröhlich, M. (2020). Interpretability of input representations for gait classification in patients after total hip arthroplasty. *Sensors (Switzerland)*, *20*(16). https://doi.org/10.3390/s20164385

Farah, L., Davaze-Schneider, J., Martin, T., Nguyen, P., Borget, I., & Martelli, N. (2023). Are current clinical studies on artificial intelligence-based medical devices comprehensive enough to support a full health technology assessment? A systematic review. In *Artificial intelligence in medicine* (Vol. 140). https://doi.org/10.1016/j.artmed.2023.102547

Goodell, J. W., Ben Jabeur, S., Saâdaoui, F., & Nasir, M. A. (2023). Explainable artificial intelligence modeling to forecast bitcoin prices. *International Review of Financial Analysis*, *88*. https://doi.org/10.1016/j.irfa.2023.102702

Gurbuz, E., Turgut, O., & Kok, I. (2023). Explainable AI-based malicious traffic detection and monitoring system in Next-Gen IoT healthcare. In *2023 international conference on smart applications, communications and networking, smartNets 2023*. https://doi.org/10.1109/SmartNets58706.2023.10215896

Hepworth, A. J., Baxter, D. P., Hussein, A., Yaxley, K. J., Debie, E., & Abbass, H. A. (2021). Human-swarm-teaming transparency and trust architecture. *IEEE/CAA Journal of Automatica Sinica*, *8*(7). https://doi.org/10.1109/JAS.2020.1003545

Holzinger, A. (2021). The next frontier: AI we can really trust. *Communications in Computer and Information Science*, *1524 CCIS*. https://doi.org/10.1007/978-3-030-93736-2_33

Huang, Q., Peng, S., Deng, J., Zeng, H., Zhang, Z., Liu, Y., & Yuan, P. (2023). A review of the application of artificial intelligence to nuclear reactors: Where we are and what's next. In *Heliyon* (Vol. 9, Issue 3). https://doi.org/10.1016/j.heliyon.2023.e13883

Joyce, D. W., Kormilitzin, A., Smith, K. A., & Cipriani, A. (2023). Explainable artificial intelligence for mental health through transparency and interpretability for understandability. In *npj Digital Medicine* (Vol. 6, Issue 1). https://doi.org/10.1038/s41746-023-00751-9

Lozano-Murcia, C., Romero, F. P., Serrano-Guerrero, J., & Olivas, J. A. (2023). A comparison between explainable machine learning methods for classification and regression problems in the actuarial context. *Mathematics, 11*(14). https://doi.org/10.3390/math11143088

Markus, A. F., Kors, J. A., & Rijnbeek, P. R. (2021). The role of explainability in creating trustworthy artificial intelligence for health care: A comprehensive survey of the terminology, design choices, and evaluation strategies. In *Journal of Biomedical Informatics* (Vol. 113). https://doi.org/10.1016/j.jbi.2020.103655

McClimens, A. (2011). Where is the humanity? *Learning Disability Practice, 14*(3). https://doi.org/10.7748/ldp2011.04.14.3.8.p4986

Pintelas, E., Liaskos, M., Livieris, I. E., Kotsiantis, S., & Pintelas, P. (2021). A novel explainable image classification framework: case study on skin cancer and plant disease prediction. *Neural Computing and Applications, 33*(22). https://doi.org/10.1007/s00521-021-06141-0

Priya Dharshini, S., Ram Kumar, K., Venkatesh, S., Narasimhan, K., & Adalarasu, K. (2023). An overview of interpretability techniques for Explainable Artificial Intelligence (XAI) In *Deep learning-based medical image analysis. 2023 9th International Conference on Advanced Computing and Communication Systems, ICACCS 2023*. https://doi.org/10.1109/ICACCS57279.2023.10113001

Shafik, W., Hidayatullah, A. F., Kalinaki, K., & Aslam, M. M. (2024). Artificial Intelligence (AI)-Assisted Computer Vision (CV) in healthcare systems. In *Computer vision and AI-integrated IoT technologies in the medical ecosystem* (pp. 17–36). CRC Press. https://doi.org/10.1201/9781003429609-2

Shafik, W. (2024). Deep learning impacts in the field of artificial intelligence. In *Deep learning concepts in operations research* (pp. 9–26). CRC Press. https://doi.org/10.1201/9781003433309-2

Speith, T. (2022). A review of taxonomies of Explainable Artificial Intelligence (XAI) methods. In *ACM International Conference Proceeding Series*. https://doi.org/10.1145/3531146.3534639

Taylor, J. E. T., & Taylor, G. W. (2021). Artificial cognition: How experimental psychology can help generate explainable artificial intelligence. In *Psychonomic bulletin and review* (Vol. 28, Issue 2). https://doi.org/10.3758/s13423-020-01825-5

Vieira, C. P. R., & Digiampietri, L. A. (2020). A study about explainable articial intelligence: using decision tree to explain SVM. *Revista Brasileira de Computação Aplicada, 12*(1). https://doi.org/10.5335/rbca.v12i1.10247

Part III
Generative AI and Healthcare for Different Sectors

Chapter 10
Perceived Value of Prospect of AI-Enabled Healthcare: View of Z Generation

Soumik Gangopadhyay, Sreeparna Guha, and Poulami Basu

Abstract Artificial intelligence (AI) is changing the way healthcare is accessible in India; healthcare organizations have collected huge datasets from various fields, which includes visual data and information on clinical experimentations and related data. Some of the obvious factors which hinders the wider adoption in healthcare continue to be difficulty in collection and accessibility, privacy concerns, misuse and regulatory ambiguity. The aim of this study is to investigate the perception of this very important group about AI in healthcare and the influence of digital affinity and socio-demographic factors. A cross-sectional study was conducted using a paper survey with patients and their companions. The primary data was collected by a well-structured questionnaire prepared by the authors. The primary data were analyzed using machine learning algorithms on both raw and engineered data to test the potential of the variables in predicting consumer's perceived value of AI in day-to-day healthcare operations. The model predicts the consumer's acceptability with overall accuracy of 85% with high sensitivity of 88%. The result depicts high acceptability of AI for minor health issues, whereas for prolonged healthcare, consumers are still relying on human intervention over AI-enabled services.

Keywords Artificial intelligence · Machine learning · Healthcare · Consumer perception · Predictive analytics

---

S. Gangopadhyay · S. Guha (✉)
Institute of Engineering & Management, University of Engineering and Management, Kolkata, India
e-mail: sreeparna.guha@iem.edu.in

P. Basu
University of Engineering and Management, Kolkata, India

R. Singh et al. (eds.), *Transforming Healthcare Sector Through Artificial Intelligence and Environmental Sustainability*, Approaches to Global Sustainability, Markets, and Governance, https://doi.org/10.1007/978-981-97-9555-0_10

10.1 Introduction

With all its newer forms and advances in past decade, artificial intelligence (AI) aims to automate customer service without explicit human intervention. AI enables one to concentrate on the most essential tasks and enhance their decision-making process by leveraging data pertaining to a specific use case. AI-enabled tools are quick in an emergency scenarios, which is efficient for forecasting, clustering and classification tasks. Simply put, AI can streamline numerous business operations, allowing one to focus on their core business activities. Furthermore, AI is expected to be utilized in numerous innovative ways in the future. Furthermore, the field of AI incorporates knowledge from economical methods, computing science which harnesses mathematical and statistical knowledge.

Artificial intelligence techniques are ideal for exploring data and detect underlying pattern and insights which is ambiguous for plain human sight. With AI, healthcare organizations aim to improve patient care implementing better clinical prognosis (Margam & Rakesh, 2024). In the early 1970s, research grew as AI entered healthcare and was applied to solve biomedical problems. Finally, in the 1980s, AI found its way into clinical fields, where it is enthusiastically embraced today. AI has been making waves in global healthcare for several years now, with new technologies ranging from block chain to artificial intelligence having significant potential to change healthcare forever. The global AI picture covers a wide range of applications, including pharmaceuticals, medical technology, drug development, diagnostics, hospital operations, health insurance operations, clinical decision making and robotic surgery. A deeper dive suggests that use cases include implementing chatbots to interact with patients, population health monitoring, treatment referral and care services, data-driven diagnosis, image-based diagnosis, clinical decision support, compliance monitoring, retention of medical data, fraud prevention, capacity planning, human resource management, quality assurance and training. Z-generation perceives the role of AI in minor healthcare services rather than major one.

10.1.1 Intervention of AI in Healthcare

Novel clinical practices includes design, functional and operational advancements i.e. digital healthcare, AI-based algorithms, targeting high risk patients, medical decision support, telemedicine etc. can provide quality service at reduced cost of therapy (Asif et al., 2023). It can ensure better integration of sub sectors and improve adaptability (Dagan, 2018). In 2017, an USA-based study of Accenture has justified the potential of artificial intelligence in healthcare which indicated an annual savings of $150 billion by 2026 in healthcare (PWC, 2016). Integrating AI can provide effective solution to accessibility and availability problem in rural areas. Many AI-driven medical diagnostic gadgets that are affordable, portable, easy to handle and have low maintenance are already in use. Telemedicine is a by-product of policy implementation of health administration of nations with compromised health status (Cremer & Kasparov, 2021) as it functions at the intersections of 'health', 'healthcare' and 'health policy'. In consonance with sustainable development goals, many underdeveloped nations implemented telemedicine, M-Health to reduce heath inequality among the residents (Kumar et al., 2020; Zhang & Zaman, 2020). Cost, distance or time of travelling influence 'Perceived usefulness' and 'Perceived ease of use' of individuals to adopt telemedicine for self -treatment. Big data analytics has also revolutionized the run time analysis of data on chronic diseases. Robotic Process Automation (RPA) integrated AI has facilitated patient appointment scheduling, data entry, data extraction, claim management, insurance claim management, hospital patient care and has ability to improve patient experience by reducing delivery time, improving efficiency and eliminating errors. Elsa Science app developed in Sweden for sharing experience of rheumatoid patients; AI-embedded software was used to save 30% time of oncologists of Slovakia with the help of electronic cancer data management system of Georgia. CroDiab, a web-based diabetes data registry system of Croatia, is a sophisticated example used in recent time.

Critical challenges in Indian healthcare include low doctor–patient ratios, uneven distribution of knowledge, affordability, inadequately trained personnel, and late detection and diagnostic errors. As a powerful tool, AI can solve many of these challenges. The fact that health technology can improve and accelerate implementation is a time-tested fact. By leveraging AI and digital technologies in these ways, India and other developing countries can overcome the challenges of affordability and accessibility in diagnostics, ultimately improving healthcare outcomes for their populations. Over time, two branches of artificial intelligence have developed in healthcare: physical AI and virtual AI. While physical AI involves the use of physical AI-integrated devices such as robots to help deliver quality healthcare to patients, virtual AI is built using deep learning frameworks, where algorithms are built through recursive training and on-filed experience.

Further, due to the emergence of COVID-19 pandemic, global health crisis landscape has unexpectedly shifted. But, constraints can spur creativity. Recent shift in healthcare services from physical to digital explores the probability of possibilities in healthcare and way forward to digital health (Gangopadhyay et al., 2022). But value gap on versatile areas is an eye opener that vehemently argues the actions for change. Changing attitude, due to mounting awareness towards preventive healthcare, is giving birth of consumerism in healthcare. Moreover, healthcare is a labour intensive service. Other than manpower shortage, variable cost, capital investment, depreciation and cost of maintenance can exaggerate the expenditure burden (Gangopadhyay & Ukil, 2022).

Many AI-driven medical diagnostic gadgets that are affordable, portable, easy to handle and have low maintenance are already in use. Sig Tuple's diagnostic tool is an example of such category that can generate a pathology report by analyzing blood slides without the assistance of a pathologist. Niramai Health Analytix is also a thermal analytics-based device that is bringing revolutionary changes in early-stage breast cancer detection. AI-enabled chatbot such as 'Wyasa' provides mental health support. Identical examples are under application which are providing empathetic support and suggesting a practitioner for consultation (Nautiyal, 2022). ‘Health Hub’ is a One-stop kiosk, developed by Delhi-based CBSE School students as a project of “hackathon”, an IBM initiative, now allow rural residents to avail proper treatment or other medical facilities without the presence of the doctor (The Wire, 2021). ‘Medicine from the Sky’ is an initiative of government of Telengana in association with World Economic Forum, NITI Aayog, and Health Net Global is delivering medicines and other health assistance with the help of drone in remote areas in a quick and safe way (Mohammad, 2021). In consonance with sustainable development goals, many underdeveloped nations implemented telemedicine to reduce heath inequality among the residents (Xiang et al., 2016). M-Health is effective for managing scarce resource at hand (Kumar et al., 2020). Cost, distance or time of travelling influence ‘Perceived usefulness’ and ‘Perceived ease of use’ of individuals to adopt telemedicine for self-treatment (Zhang and Zhaman, 2020). In India, it was initiated as a trial and error method to reduce the patient burden on public clinical establishments.

Medical Buyer (2023) has also revealed that digital consultation allows individual patients to interact with expert professional virtually, breaking the distance barrier as well as beneficial for minor medical conditions that do not necessitate a hospital visit. Moreover, Social Assistive Robots (SARs) are becoming increasingly popular in the healthcare industry. These robots are intelligent and capable of communicating with patients in a manner that conveys a sense of care. They are able to interact, communicate, be companions, and form emotional connections with patients. Utilizing robots in healthcare can lead to improved patient care, streamlined medical procedures, and increased safety for all those in the hospital. By reducing the risk of mistakes and forgetting information, these robots are enabling healthcare professionals to perform their duties more effectively. These robots are revolutionizing healthcare for the benefit of all. Medical Buyer, 2023 have recorded the utilization of AI and smart devices has enabled the development of remote patient monitoring. In addition to traditional health monitoring tools such as glucose monitors and blood pressure gauges, the use of intelligent implants and prosthetic devices has enabled the tracking of post-surgery recovery. These devices are implanted into the body to monitor movement and muscle strength, providing information about a patient’s progress. Additionally, digital pills, nano-robots and intelligent fabrics have been developed to monitor patients and ensure their medication intake, wound management and overall health. By 2025, approximately half of the population in developed countries is expected to use AI-driven monitoring tools and wearable technology. With an estimate growth of $10 million dollars by 2025, AI enabled medical services are projected to replace human presence in at least 80% of different services. AI’s expertise and scalability make it capable of outperforming humans in terms of accuracy and scalability. For example, IBM Watson was able to identify missed treatment options for 30% of cancer cases, and AI triage diagnoses were able to achieve 90.2% accuracy compared to 77.5% for humans. While AI has been a boon in healthcare sector, some of the obvious barriers to wider adoption in healthcare continue to be blind spots in data access and data collection, privacy concerns, data misuse and regulatory ambiguity. India has started its journey in AI-based healthcare and is growing rapidly, but there is still a long way to go. The idealistic goals of universal healthcare and affordability that can be achieved through AI-based healthcare is now a distant reality. However, there is hope on the horizon as more and more innovations are coming in AI-powered healthcare services that could soon make this dream a reality. Application of AI is currently supply driven, which may end up as an act of capital loss. This research contributes to the development of automation psychology by focusing on consumer behavior in medical settings, and expands the knowledge of consumer decisions in healthcare. Therefore, end users perspectives on critical role and importance of AI in healthcare day-to-day operations have to be explored. Basic objectives of the study are

- To study the perceived value of AI in healthcare among the common people.
- To study how AI is changing the game in clinical trials by utilizing patient data.
- To understand the factors influencing user perception and their acceptance predictability.

10.2 Survey of Literature

Healthcare systems effectiveness is a function of resource allocation that leads to high costs of operations and slow processing. Leveraging the support of AI and RPA, healthcare operation can achieve highest precision as it can automate many manual functioning. By leveraging technology innovations to integrate preventive, personalized and precision medicine across the healthcare continuum, healthcare providers are projected to deliver improved patient care quality with much cost efficiency, ultimately improving patient outcomes and enhancing population health (Dhawan et al., 2015). AI has a role to health, health policy and healthcare. Longoni et al. have observed the potential for Artificial Intelligence (AI) to revolutionize healthcare through its applications such as heart disease diagnosis, medical advice, skin cancer detection, and more. Research of Nagy and Sisk have revealed that artificial intelligence (AI) is efficient to reduce the administrative burden of clinicians, potentially providing them with more time to engage with patients in meaningful ways; however, there is uncertainty as to whether this new time will be used to build relationships. Murdoch, 2021 have argued that the proliferation of artificial intelligence (AI) has posed a significant risk to the privacy of patients. Fritsch et al. have critically justified by their observations that the attitude towards AI-enabled healthcare services still coursing through much debate. Contrary to the perception of health professionals, the attitudes of patients and their peers have so far been less interested. Moreover, study of Malek et al. (2022) have identified that the utilization of Artificial Intelligence in healthcare is rapidly growing worldwide. Data about patients is exchanged between medical professionals and staff through AI tools, which can contain a great deal of confidential information. Ali et al. have examined positive and negative impact of AI-integrated healthcare, as well as the various methods and functions associated with its use. It highlights the rapid development of AI systems, which have enabled the early detection of diseases and the development of effective treatments. Khan et al. (2023) have revealed that the utilization of artificial intelligence (AI) can be seen as a significant factor in the development of the healthcare sector provided data privacy issues, social and ethical considerations, hacking of the system and the development of the software are considered before design of the service. Medical Buyer (2023) have observed that healthcare robots are making a positive impact on the lives of healthcare workers. These robots are not limited to the operating room, but can also be employed in clinic and outpatient settings to improve the quality of life for patients. The robotic surgery has revolutionized the minimally invasive procedure, replacing the traditional laparoscopic approach. It is predicted that the majority of surgeries in the future will be robotic. To ensure that surgeons are adequately prepared for this new reality, it is essential to provide realistic training. This includes exposure to new techniques, such as robotic surgical simulation, telementoring, and virtual reality simulators. Additionally, robotic technology is revolutionizing surgeon training, as virtual reality simulators provide a more realistic practice environment, allowing for the easier learning of procedures. Artificial intelligence (AI) is aiding surgeons by providing feedback and suggestions in real-time during procedures. Furthermore, deep learning data allows for the optimization of surgical approaches to individual patients. Finally, robots are bridging the gap between experts from larger cities and smaller towns, providing access to high-quality medical care for all. Robotics is having a significant impact on surgery and healthcare. Underdeveloped nations are using more e-health in adequate allocation of budget of health in GDP that can lead to more dependence on e-health. Disease dynamics of rural India has a complex scenario as it is not only flooded with infectious diseases but also with non-communicable diseases. Following the changing burden of disease, Indian health service approach has shifted its direction from reactive to proactive initiative (The Wire, 2021).

In post COVID-19, situation is more critical which demands advanced facility and diagnostic services (Gangopadhyay et al., 2022). Moreover, poor accessibility and affordability of legitimate healthcare is a function of social disparity (Singh & Badaya, 2014). Lack of awareness, inadequate transportation and shortage in medical and paramedical professionals have potentiated the problem further (Devarakonda, 2016). Existing health facilities are in absolute dilemma to fix several burning issues i.e. shortage in skilled manpower, inadequate infrastructure, zero preventive option, non-availability of guaranteed treatment, high mortality, etc. But solving the issue is a humongous undertaking. Indian rural has only 33% of Indian doctors (Paul et al., 2018). So, it is time for both quality and quantity (Barik & Thorat, 2015). But, it is impossible to transform the primary healthcare of India within a short period (Kelsey, 2021). By harnessing the power of technology and innovation, we can create a more equitable, efficient, and sustainable healthcare system for the benefit of all individuals and communities (Xiang et al., 2016). Rapid advancement in AI is a game changer even in healthcare due to cost saving option (Reddy et al., 2019). Healthcare access and quality score of India has improved its position from 24.7% in 1990 to 41.2% in 2016 due to change in policy. But, replacement of human with AI application is both contextual and conditional (Dick, 2020) as hybrid intelligence has sector specific application (Dellermann et al., 2019). Moreover, emotional intelligence is a missing component in AI-driven applications (Shabbir & Anwer, 2018). Further, introduction of AI is best suitable for system augmentation rather than human replacement (Cremer & Kasparov, 2021; Mohammad, 2021). Rong et al. (2020) have recorded the latest developments in integration of AI into biomedicine is poised to revolutionize healthcare delivery, biomedical research, and drug development, leading to more accurate diagnosis and prognosis, personalized treatment strategies and improved patient outcomes.

10.3 Design of Hypothesis

10.3.1 Methodology

The aim of this study is to investigate the perception of this very important group about AI in healthcare and the influence of digital affinity and socio-demographic factors. A cross-sectional study was conducted using a paper survey with patients and their companions. The present study is based on both secondary and primary data. Secondary sources were from books, magazines, brochures and web sites. Primary data was collected by a well-structured questionnaire prepared by the authors. The entire study was carried out in two stages.

Stage I: A survey was undertaken among people. To collect the primary data, we have approached 200 respondents belonging to the age group of 18–60 years having disparity in profession with the help of a structured questionnaire at Kolkata area. Responses have been collected on a 5-point scale anchoring from strongly disagree to strongly agree. Strongly disagree, disagree, neutral, agree and strongly agree have been coded as 1, 2, 3, 4 and 5, respectively.

Stage II: Multiple predictive models are built using PYTHON IDE on both raw and engineered data to test the potential of the variables in predicting consumer's perceived value of artificial intelligence in day-to-day healthcare operations.

10.4 Data Analysis and Result

10.4.1 Primary EDA

A basic descriptive analysis on the raw, unprocessed data reveals that the distribution contains response majorly from age group 18–25 years, indicating that the results will be partly biased towards younger generation and their perception of AI for healthcare in the near future (Table 10.1).

The primary data collected contains targeted questions which focus on multiple aspects of integrating artificial intelligence in day-to-day healthcare operations. The questions are encoded as follows.

Table 10.1 Demographic distribution of responses

Index	Age	Gender	Educational qualification	Monthly household income
Count	200	200	200	200
Unique	5	6	11	15
Top	18–25	Male	Post graduate	Rs.70,000–100,000
Freq	40	50	47	27

Q6. AI based clinical establishment is user friendly.

Q7. AI can bring more precision in clinical quality.

Q8. AI enhances the quality of clinical performance.

Q9. AI will shift control of clinical supervision from manpower to machine power.

Q10. AI based healthcare sector can provide more effective outcome.

Q11. AI has efficiency to enhance healthcare service quality.

Q12. AI based medical data is more secured.

Q13. AI based training saves time and money.

Q14. AI assisted clinical care is automated, so more focused.

Q15. AI assisted clinical care is more time saving.

Q16. Human assisted clinical care is more flexible.

Q17. AI based healthcare services are more safe and secure.

Q18. Minor medical conditions can be solved faster by using virtual medical support.

Q19. AI is a threat of unemployment for healthcare workers.

Q20. AI will improve healthcare outcomes and reduce cost in the medical procedures.

Q21. AI poses a threat to the information about patients.

Q22. Patients are reluctant to accept AI as a change in the clinical sector.

Q23. AI oriented medical services are cost effective than the traditional healthcare.

The 23 questions are treated as variables to analyze the dependency and variability of the dataset.

Correlation-based similarity index reveals weak to no similarity among the variables showcasing that there are no redundant features in our dataset.

Similarity Index

q14	q10	0.555915
q8	q10	0.454847
q7	q6	0.454132
q11	q10	0.433694
q12	q6	0.430899
q13	q10	0.412727
q7	q11	0.408696
q6	q8	0.401326
q23	q20	0.400951

Similarity Index

q9	q20	-0.141337
q10	q21	-0.093320
q18	q6	-0.092382
	q8	-0.086333
q10	q18	-0.085950
q8	q17	-0.080258
q16	q7	-0.071071
q22	q7	-0.059954
q7	q15	-0.058444
q9	q16	-0.057233

10.4.2 Principal Component Analysis (PCA)

Due to the high dimension of the features, our investigation redirects to dimensionality reduction methods, primarily to identify variables with high contribution towards the variability of the dataset (Fig. 10.1).

We modelled PCA on filtered dataset containing only the relevant variables removing the demographic features. The PCA model contains 10 components enclosing approximately 80% variability of the features.

The sorted loading scores elucidate the features having more weight towards the variance. The top 5 features as follows:

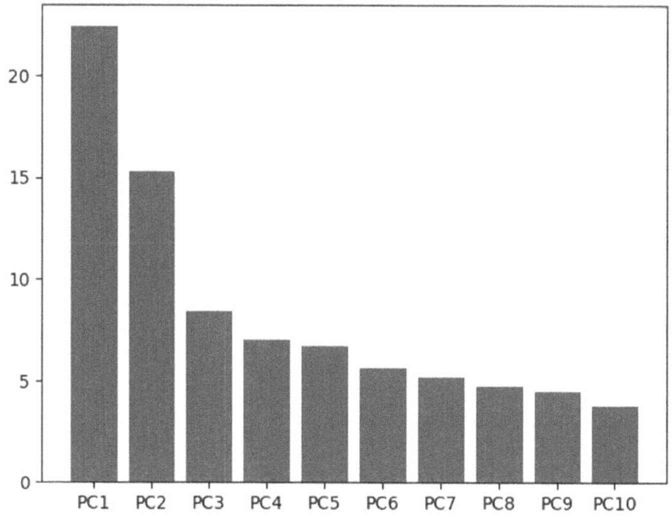

Fig. 10.1 Scree plot for PCA

q10 0.397811.

q13 0.355214.

q6 0.343322.

q14 0.327226.

q8 0.318561.

q11 0.308435.

The above combination reveals that efficiency, cost, ease of use and time management are four primary factors which contributes towards perceptivity. This in turn helped us to identify a target variable to classify respondents.

10.4.3 Building Predictive Model

10.4.3.1 Decision Tree Model

Deriving insight from PCA result, we used Q23 as the target variable for our predictive model. Through Q23, we hypothesized to identify respondent according to their level of acceptance of AI-integrated healthcare.

Hence, in our model we have 17 predictor variables against one target variable. In raw form, the target variable has 5 levels, which makes the problem as multi-class classification problem.

We chose tree-based classification algorithm for the said multi-class classification problem with 7:3 ratio in training and testing subset. In the initial training, the overall prediction accuracy was 0.56 with level 5 (strongly agree) being predicted with the highest precision (Fig. 10.2).

	Precision	recall	f1-score	support
1	0.29	0.40	0.33	5
2	0.71	0.45	0.56	11
3	0.60	0.80	0.69	15
4	0.47	0.54	0.50	13
5	0.73	0.50	0.59	16
accuracy			0.57	60
macro avg	0.56	0.54	0.53	60
weighted avg	0.60	0.57	0.57	60

Fig. 10.2 Classification report decision tree model 1

	precision	recall	f1-score	support
0	0.80	0.67	0.73	36
1	0.60	0.75	0.67	24
accuracy			0.70	60
macro avg	0.70	0.71	0.70	60
weighted avg	0.72	0.70	0.70	60

Fig. 10.3 Classification report decision tree model 2

10.4.3.2 Analysis with Re-Engineered Features

We further re-encoded the target variable to achieve more clarity and interpretability in classification and prediction of consumer's acceptance or rejection of AI-integrated healthcare systems.

The re-encoded target variable is the following:

Q23 = 0, for Neutral or Disagree or Strongly Disagree (1 or 2 or 3).

= 1, for Agree or Strongly Agree (4 or 5).

Here, 0 indicates 'Rejection' and 1 indicates 'Acceptance'.

The model training with re-engineered target variable yielded improved accuracy of 0.7 with better precision of 80% in predicting **Rejection** (Fig. 10.3).

10.4.3.3 Random Forest Model

Due to random subset selection, the decision tree model is often biased towards random feature patterns. Instead of generalization the decision tree model tends to under fit in several occasion. To investigate further potentiality and data diversity of our model, we trained the dataset using ensemble classification algorithm to capture variability of multiple samples and considering the final result using majority voting.

100 Decision Tree models were trained and the result yielded provided improved over all accuracy of 85%. The prediction capability for both rejection and acceptance yielded high value—82 and 88%, respectively (Figs. 10.4 and 10.5).

The feature importance analysis revealed the contribution of the features towards data split. Considering top 3 features elucidates schemes and factors which cam influence users to showcase trust and acceptance towards AI-integrated healthcare operations. The analysis reveals

(a) Users are well accepting use and assistance of AI for minor health conditions which can be solved faster than traditional process which requires human intervention.
(b) Although for patients who are undergoing or will undergo long term medical care are still reluctant to accept AI generated assistance.
(c) Due to its automation and time efficiency, users understand the potentiality of AI to improve cost effectiveness in day-to-day operations.

```
print(classification_report(y_test,y_pred))

              precision    recall  f1-score   support

           0       0.82      0.90      0.86        31
           1       0.88      0.79      0.84        29

    accuracy                           0.85        60
   macro avg       0.85      0.85      0.85        60
weighted avg       0.85      0.85      0.85        60
```

Fig. 10.4 Classification report random forest model

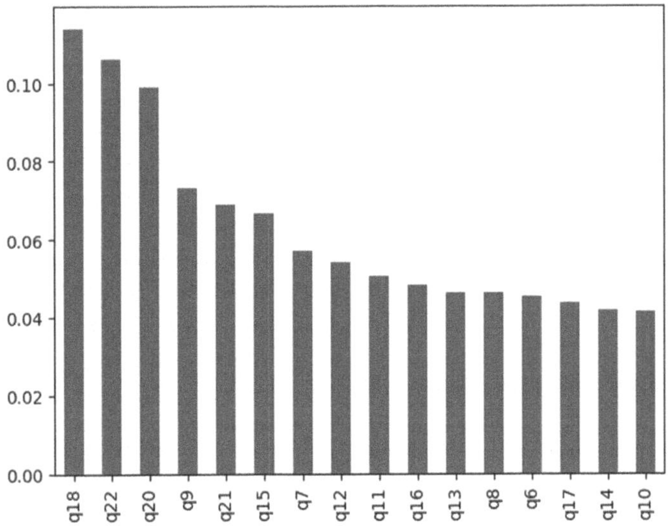

Fig. 10.5 Feature importance analysis

Our predictive model is built with 200 primary responses which is biased towards comparative younger generation. Perception criterion for millennial or older generation is not represented properly in our model. Hence, for future reference more responses targeted for comparative older generation can be investigated to gather similar insight.

10.5 Conclusion

In conclusion, AI has the potential to revolutionize healthcare, offering a wide range of benefits to patients, healthcare providers, and the overall healthcare system.

While there is still some public uncertainty about AI, there is growing recognition of its potential to improve healthcare. Education and awareness campaigns can help people understand the affectability of AI in improving patient care, address concerns about bias and privacy, and foster trust in this technology.

AI is transforming clinical trials by utilizing patient data to identify potential participants, predict patient outcomes, monitor patient safety, and develop new types of trials. This can lead to faster drug development, improved patient outcomes, and reduced costs.

AI algorithms can analyze heavy loads patient data to identify patterns and trends that can inform personalized medicine, predictive analytics, and overall health management. This can lead to more effective treatments, possibility of early detection based on historical records and improved diagnosis outcomes.

AI has tremendous potential to revolutionize healthcare by making it more personalized, precise, predictive, and accessible. Transparent AI algorithms leads to better interpretability, which in turn helps the healthcare providers to manage time, patient load and quality of care giving. Hence, harnessing AI powers within ethical and medical considerations can take the health care services to provide improved facility and in the end save lives.

References

Ali, O., Abdelbaki, W., Shrestha, A., Elbasi, E., Ali, A., Alryalat, M., & Dwivedi, Y. K. (2022). A systematic literature review of artificial intelligence XE "Artificial intelligence" in the healthcare sector: Benefits, challenges, methodologies, and functionalities. *Journal of Innovation and Knowledge, 8*(1), 100333.

Asif, M., Alam, M., & Alam, Q. (2023). Future of pharmaceutical marketing in India-an extensive review. *International Journal of Pharma Professional's Research, 14*(3), 164–185.

Barik, D., & Thorat, A. (2015). Issues of unequal access to public health in India. *Frontiers in Public Health, 3*, 245.

Bocas J. (2022). *AI training for healthcare workers*. LinkedIn. https://www.linkedin.com/pulse/how-ai-changing-medical-field-jo%C3%A3o-bocas

Cremer, D. D., & Kasparov, G. (2021). AI should augment human intelligence, not replace it. *Harvard Business Review*, 18.

Davenport, T., & Kalakota, R. (2019). The potential for artificial intelligence in healthcare. *Future Healthcare Journal, 6*(2), 94–98.

Dellermann, D., Ebel, P., Söllner, M., & Leimeister, J. M. (2019). Hybrid intelligence. *Business and Information Systems Engineering, 61*(5), 637–643.

Devarakonda, S. (2016). Hub and spoke model: Making rural healthcare in India affordable, available and accessible. *Rural and Remote Health, 16*(1), 3476.

Dhawan, P. A., Heetderks, J. W., Pavel, M., Acharya, S., Akay, M., et al. (2015). Current and future challenges in point-of-care technologies: A paradigm-shift in affordable global healthcare with

personalized and preventive medicine. *IEEE Journal of Translational Engineering in Health and Medicine, 3*(1–10), 2800110.
Dick, S. (2020). Artificial intelligence. *Harvard Dta Science Review, 1*, 102–129.
Gangopadhyay, S., Ukil, A., Chatterjee, S. (2022). Emerging paradigm of smart healthcare in the management of COVID-19 pandemic and future health crisis. In C. Chakraborty, M. R. Khosravi (Eds.), *Intelligent healthcare* (pp. 455–478). Springer, Singapore. https://doi.org/10.1007/978-981-16-8150-9_20
Gangopadhyay, S., Ukil, A. (2022). Being resilient to deal with attrition of nurses in private COVID-19 hospitals: Critical analysis with respect to the crisis in Kolkata, India. In L. Garg, C. Chakraborty, S. Mahmoudi, V. S. Sohmen (Eds.), *Healthcare informatics for fighting COVID-19 and future epidemics. EAI/Springer innovations in communication and computing* (pp. 353–363). Springer, Cham. https://doi.org/10.1007/978-3-030-72752-9_18
Hendrix, N., Veenstra, D. L., Cheng, M., Anderson, N. C., & Verguet, V. (2022). Assessing the economic value of clinical artificial intelligence: Challenges and opportunities. *Value in Health, 25*(3), 331–339.
Kelsey, E. A. (2021). Joy in the workplace: The mayo clinic experience. *American Journal of Lifestyle Medicine, 17*(3), 413–417. https://doi.org/10.1177/15598276211036886
Khan, B., Fatima, H., Qureshi, A., Kumar, S., Hanan, A., Hussain, J. (2023). *Drawbacks of artificial intelligence and their potential solutions in the healthcare sector* (pp. 1–8). PMC9908503. https://www.ncbi.nlm.nih.gov/pmc/articles/PMC9908503/
Kumar, A., Nayar, K. R., Koya, S. F. (2020). COVID-19: Challenges and its consequences for rural health care in India. *Public Health in Practice, Science Direct, 1*, 100009.
Lee, D. H., No Yoon, S. (2021). Application of artificial intelligence-based technologies in the healthcare industry: Opportunities and challenges. *International Journal of Environmental Research and Public Health, 18*(1), 271. https://www.mdpi.com/1660-4601/18/1/271
Malek, L. A., Jain, P., Johnson, J. (2022). *Data privacy and artificial intelligence in health care*. Reuters. https://www.reuters.com/legal/litigation/data-privacy-artificial-intelligence-health-care-2022-03-17/
Margam, R. (2024). Mindful machines: How artificial intelligence transforms mental healthcare. *International Journal of Engineering and Techniques., 10*(1), 1–7.
Mohammed, S. (2021). *Telangana launches 'Medicine from the Sky' project to drone-deliver vaccines, medicines to remote areas, The Hindu, 11 September* (p. 1).
Moore, J. (2023). AI in health care: The risks and benefits. *Medical Economics*. https://www.medicaleconomics.com/view/ai-in-health-care-the-risks-and-benefits
Murdoch, B. (2021). Privacy and artificial intelligence: Challenges for protecting health information in a new era. *BMC Medical Ethics, 22*(1), 122. https://doi.org/10.1186/s12910-021-00687-3
Nautiyal, S. (2022). Artificial intelligence can significantly improve radiology reporting. *Financial Express Health Care, 9* February, 1.
Paul, Y., Hickok, E., Sinha, A., Tiwari, U., Mohandas, S., Ray, S., Hickok, E., Bidare, P. M. (2018). Artificial intelligence in the healthcare industry in India. In *The Centre for Internet and Society*. www.news.medgenera.com/12-artificial-intelligence-healthcare-startups-india-ai/
PWC. (2016). *Indian Healthcare on the cusp of a digital transformation*. Tele Medicon
Reddy, S., Fox, J., & Purohit, M. P. (2019). Artificial intelligence-enabled healthcare delivery. *Journal of the Royal Society of Medicine, 112*(1), 22–28.
Shabbir, J., & Anwer, T. (2018). Artificial intelligence and its role in near future. *Journal of Latex Class Files, 14*(8), 1–11.
Singh, S., & Badaya, S. (2014). Health care in rural India: A lack between need and feed. *South Asian Journal of Cancer, 3*(2), 143–144.

The Wire (2021). *Economic survey takes aim at out-of-pocket spend, high private healthcare costs.* https://thewire.in/health/economic-survey-takes-aim-at-out-of-pocket-spend-high-private-healthcare-costs

Xiang, G. Y., Zeng, Z., & Shen, Y. J. (2016). Present situation and development trend of China's intelligent medical construction. *Chin Gen Pract, 19*(24), 2998–3000.

Zhang, X., & Zaman, U. B. (2020). Adoption mechanism of telemedicine in underdeveloped country. *Health Informatics Journal, 26*(2), 1088–1103.

Chapter 11
The Integration of AI and Social Commerce in Healthcare: New Era of Engagement, Personalisation, and Decision Making

Abhay Grover and Nilesh Arora

Abstract Patients can be more actively involved in their healthcare and given greater autonomy through the use of social commerce and artificial intelligence (AI) to improve their health literacy, make more informed decisions, and engage in individualised health education. The combined effects of artificial intelligence and social commerce on healthcare delivery and patient outcomes are the focus of this research. This research aims to examine the effectiveness of social commerce platforms that enable collaborative decision-making, behaviour modification interventions, and AI-driven personalised health education platforms. This study intends to shed light on how these integrated approaches affect patient empowerment and involvement by conducting a thorough literature analysis and analysing current scenarios. In order to improve healthcare engagement and patient empowerment, this study found that utilising AI and social commerce had promising upsides. Successfully delivering individualised health information and personalised health education platforms powered by AI algorithms have improved health literacy and fostered patient empowerment. There is evidence that behavioural change therapies supported by social commerce platforms can encourage patients to make positive changes to their lifestyles and adopt healthier habits. Treatment plans that are in line with patient's values and preferences are achieved through shared decision-making procedures made possible by AI-driven decision support systems. Adopting patient-centric approaches in healthcare delivery is crucial, as highlighted by the theoretical implications of this study. By combining AI with social commerce, we can build a system for individualised healthcare that goes beyond the status quo and puts the patient at the centre of their own health management.

A. Grover (✉) · N. Arora
Management, Lovely Professional University, Punjab, India
e-mail: abhay.16051@lpu.co.in; abhaygrover@yahoo.co.in

N. Arora
Management, Chandigarh University, Chandigarh, India

R. Singh et al. (eds.), *Transforming Healthcare Sector Through Artificial Intelligence and Environmental Sustainability*, Approaches to Global Sustainability, Markets, and Governance, https://doi.org/10.1007/978-981-97-9555-0_11

11.1 Introduction

When applied to healthcare, the term "social commerce" is the practice of utilising social marketing and social enterprises to increase the accessibility, cost, and availability of medical treatment (Todua et al., 2024). With approaches including inclusive business models, collaborative engagement with stakeholders, and flexible payment systems, healthcare social enterprises (HCSEs) have the ability to revolutionise the healthcare industry. These techniques lead to better health outcomes for patients. Open innovation methods like AgorIP have also been tested to help commercialise healthcare opportunities (POWNEY, 2020). Furthermore, healthcare social businesses contribute to expanding access to services, raising standards of care, and strengthening public health infrastructure. Health promotion programmes that use social marketing principles to zero in on specific demographics and meet their unique requirements have a track record of success in improving people's health.

In today's digital age, artificial intelligence (AI) has emerged as a game-changing technology that could revolutionise numerous industries (Barth, 2024). The innovative powers of AI are revolutionising the way businesses engage with customers and tailor their buying experiences. This is especially true in the realm of social commerce, where AI plays a pivotal role in enhancing customer engagement, facilitating better support interactions, and offering personalised recommendations (Akram et al., 2019). Creating technological devices that can perform tasks that normally need a human brain is the most fundamental definition of artificial intelligence (AI). Recent years have seen rapid advancements in artificial intelligence (AI) due to the availability of massive amounts of data and the exponential growth in processing capacity. Thus, AI has made its way into numerous sectors, including as the medical field, financial services, transportation, and the entertainment industry (Laskowski, 2023).

While the AI healthcare business was worth $11 billion in 2021, Statista predicts that number would rise to $187 billion in 2030 (IBM Education, 2023). As a result of this enormous growth, the healthcare industry as a whole, including hospitals, medical providers, pharmaceutical and biotechnology businesses, and others, will certainly undergo more profound transformations. Rapid advancements in machine learning (ML) algorithms, increased data accessibility, more affordable technology, and the advent of 5G have all sped up the use of AI in healthcare, which in turn has quickened the rate of change. Medical records, clinical trials, and genetic data are all part of the massive amounts of health data that AI and ML systems can process at a far faster rate than humans (IBM Education, 2023).

The use of social media platforms that make use of Web 2.0 technologies is greatly enhancing medical education, research, and healthcare through encouraging teamwork and making it easier to share findings. Despite healthcare providers' use of these platforms to raise public health literacy, questions regarding the veracity and integrity of content continue to be raised. Social media sites such as Facebook, YouTube, Instagram, TikTok, and Twitter have grown in importance in the healthcare industry in 2023. These sites facilitate patient-provider communication, professional growth, and information sharing (Jeyaraman et al., 2023) thanks to companies like Meta Platforms, Inc., Menlo Park, California, United States; X Corp., Carson City, Nevada, United States; and ByteDance Ltd., Beijing, China. Numerous health-related subjects, including vaccinations, medications, smoking, NCDs, pandemics, eating disorders, and medical treatments, have become trending on social media due to the large number of patients who use these platforms to research and exchange health-related information. A wide variety of health-related social media uses, such as health interventions, health education, health campaigns, and disease outbreak tracking, have aided medical research. Additionally, there has been a lot of interest in the correlation between social media usage and psychological well-being, suggesting that it could be a useful tool for helping those who are struggling with mental health issues (Jeyaraman et al., 2023).

In 2020, consumers spent $992 billion on social media commerce; by 2026, experts predict that figure will hit $2.9 trillion. Instagram and Facebook accounted for 55% of the buyers, and the majority of them were young adults (18–24). When planning how to make purchasing easier for users, business owners should keep their target demography in mind and tailor their offerings to their tastes. The goal of any new trend and technological development should be to make it easier for customers to complete their purchases through all available channels (Jaros, 2023). India's spending on artificial intelligence jumped 108% in 2018 to 665 million USD, and by 2025, experts predict that number will have risen to 11.78 billion USD. The expenditure on artificial intelligence is anticipated to provide one trillion US dollars to the Indian economy by the year 2035 (Chatterjee et al., 2023). So far, India has taken a cautious approach to integrating AI, despite being a founding member of the Global Partnership on AI (GPAI) coalition 2022–23. Diagnosis algorithms are already incorporating AI to test for a variety of diseases, including cancer, diabetic retinopathy, and cardiovascular disorders. Many people in India struggle with issues related to healthcare affordability, accessibility, and quality, making it one of the country's most dynamic and demanding industries.

India is planning to use AI primarily in the following sectors: healthcare, education, retail/customer engagement, agriculture/food processing, and financial technology. Through the use of AI-powered point-of-care diagnostics, teleconsultation, and e-prescription capabilities, healthcare providers are able to better serve patients in underserved areas. In terms of AI's impact and development, India has seen a shambolic path. As part of its Digital India Initiative, the Government of India has launched a slew of programmes with the goal of making India an AI powerhouse by streamlining the country's legislative and political frameworks to speed up the introduction of AI systems. As part of its objective to promote the adoption of artificial intelligence throughout the nation, India plans to develop a detailed set of guidelines for the next five years that centre on targeted industry and research initiatives (Chatterjee et al., 2023).

Healthcare problems may be amenable to technological solutions, such as social commerce and artificial intelligence (AI) (Fig. 11.1 shows the integration of AI and Social Commerce). Technology advancements in mobile, the Internet of things (IoT), processing power, and data security, along with the growing availability of multi-modal data (genomics, economics, demographics, clinical, and phenotypic), signal a turning point in healthcare and technology that will radically alter healthcare delivery models with the help of AI-augmented healthcare systems (Bajwa et al., 2021). This study steps into a knowledge vacuum in healthcare by investigating how artificial intelligence (AI) and social commerce (SCM) could work together to empower and engage patients. Poor communication and decision-making are common outcomes of the impersonal nature of traditional healthcare. This study seeks to transform healthcare services by investigating the potential of artificial intelligence and social commerce to personalise health education, promote behaviour change, and enable shared decision-making. The ultimate goal is to establish a patient-centric paradigm that encourages individuals to take an active role in their own health.

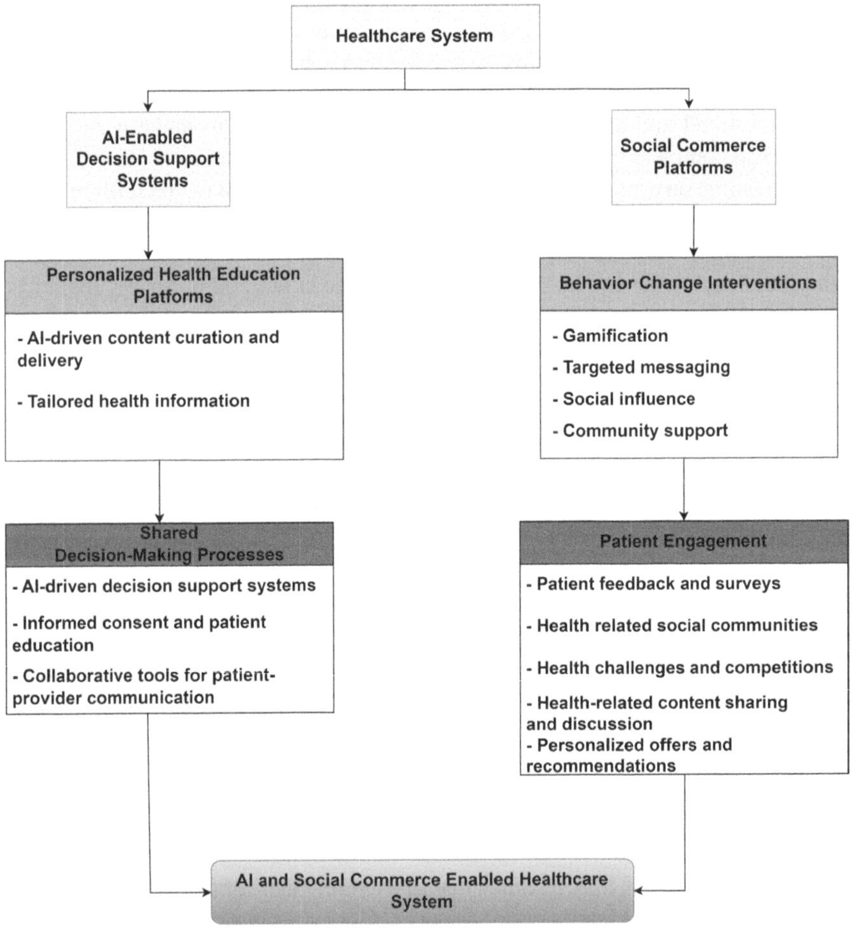

Fig. 11.1 Healthcare system. Author's Creation

11.2 Theoretical Framework

11.2.1 Patient Engagement and Empowerment

Patients' engagement and adherence have been considered the "last mile" barrier in the healthcare industry for a very long time. This refers to the final obstacle that stands between ineffective and good health outcomes. Utilisation, financial performance, and the overall experience of members are all improved when patients take an active role in their own health and treatment (Krist et al., 2017). Big data and artificial intelligence are becoming increasingly commonplace in the process of addressing these difficulties. When it comes to developing a treatment plan for their acute or chronic patients, healthcare practitioners and hospitals commonly rely on their clinical expertise in order to improve the health of their patients. In spite of this, there are situations in which the patient's behaviour does not change as directed. For example, if the patient does not reduce their weight, schedule a follow-up appointment, take their meds as directed, or adhere to their treatment plan, then this may become irrelevant.

Noncompliance, which occurs when patients do not take their medications or treatment programmes as advised, is a significant problem that presents a significant challenge. If patient participation leads to better health outcomes, would it be beneficial to have the power to tailor and contextualise care through the use of artificial intelligence? Within the context of the care continuum, a growing number of researchers are concentrating their efforts on figuring out how to make use of business rules engines and machine learning to power sophisticated interventions. It is an attractive area of study to conduct research on messaging alerts and tailored, relevant material that promotes actions at important intervals of time. An additional topic that is attracting increasing attention in the field of healthcare is the significance of meticulously building the "choice architecture" in order to encourage patients to make more proactive behaviour changes that are supported by evidence. Programmes are able to tailor treatment plans by comparing patient data to known successful treatment paths for comparable cohorts. This comparison is made utilising information obtained from electronic health record systems, biosensors, watches, cellphones, conversational interfaces, and other devices. Anyone, from medical professionals and patients to nurses and others who work in contact centres, is able to receive the guidance (Wen & Huang, 2019).

11.2.2 AI and Social Commerce in Healthcare

As a result of artificial intelligence, social commerce is severely impacted. Customers participate in social commerce, which is a combination of e-commerce and social media, by communicating with businesses on social media platforms and making purchases online (Eriksen, 2023). A combination of factors, including the growing importance of social media and the demand for a buying experience that is free of hassles, has contributed to its meteoric rise in popularity over the past few years. The application of artificial intelligence (AI) is bringing about a transformation in the social commerce landscape by enabling algorithms and data analysts to tailor services and experiences for customers (Arora, 2023). The subject of artificial intelligence known as machine learning enables computers to "learn" from data in order to improve their decision-making and prediction capabilities. In order to deliver individualised recommendations, artificial intelligence systems are able to assess the behaviours, preferences, and purchasing history of users. Using this personalisation makes shopping more enjoyable, increases the number of conversions, and encourages customers to remain loyal to the company.

There are a number of artificial intelligence technologies that are improving customer interactions on social commerce platforms. Two examples of these technologies include natural language processing (NLP) and sentiment analysis (SEA) (Team, 2024). Customer service has been changed by chatbots and virtual assistants that are powered by artificial intelligence. These features enable users to receive individualised advice in a few seconds. Chatbots have the capacity to participate in in-depth conversations with users, provide answers to questions, and even assist in making purchases by providing assistance. By utilising AI, businesses have the opportunity to provide service around the clock, increase the level of satisfaction experienced by their customers, and decrease the amount of time it takes to respond to inquiries.

Recommendation systems that are powered by artificial intelligence are essential to the success of social commerce since they play a crucial part in fostering individuals involvement (Taliwal et al., 2024). These systems do a comprehensive analysis of user data, which includes social interactions, purchase trends, and browser history, with the goal of gaining an understanding of customer preferences and providing appropriate product recommendations. Artificial intelligence has made it possible for organisations to do sentiment analysis and social listening, which provides them with invaluable insights about the preferences of their customers and the tendencies of the market. Systems powered by artificial intelligence are able to examine posts, reviews, and comments made on social media platforms in order to determine the feelings of customers and recognise developing patterns. By utilising this information, businesses are able to modify their marketing tactics, generate more focused advertisements, and satisfy the wants of their customers.

The use of artificial intelligence to social commerce, on the other hand, is not without its difficulties and factors to take into consideration. Artificial intelligence relies heavily on the collecting and processing of enormous amounts of user data, which is why protecting users' privacy is a significant concern (Office of the Victorian Information Commissioner, 2024). When dealing with customer information, businesses have a responsibility to be open and honest about how they use customer data and to ensure that they adhere to all applicable data protection regulations. Ethical factors must be taken into account because artificial intelligence algorithms should not discriminate in any way, shape, or form. If companies want to earn the trust of their clients, they should work towards transparently making algorithmic decisions and implementing ethically sound AI methods.

11.3 Personalised Health Education

11.3.1 AI-Powered Health Content Curation and Delivery

More favourable results for patients and reduced expenses are the primary bets. In terms of speed and accuracy, businesses are adopting machine learning to diagnose patients more quickly than hospitals and medical professionals. Watson, developed by IBM, is one of the most well-known technologies in the field of healthcare. To put it another way, it is able to grasp and respond to questions that are given to it using everyday language. Following the collection of data from patients and other sources, the system will next create a hypothesis and provide a confidence grading scheme to support the hypothesis. The use of chatbots and virtual health assistants that are accessible online can provide assistance to patients and other individuals who receive healthcare with a wide range of administrative activities. These jobs include locating medical information, scheduling appointments, comprehending billing, and more. With the assistance of a wide range of artificial intelligence technologies, pandemics such as COVID-19 are not only being better understood but also fought against and predicted (Laskowski, 2023).

11.3.2 AI-Powered Health Content Curation and Delivery

EdApp is a feature-rich adaptive learning software that allows users to create their courses and implement strategies like spaced repetition (RAPL, 2023). With Brain Boost, a spaced repetition tool, you won't have to worry about forgetting anything. Learners are able to retain more material with Brain Boost's automatic and personalised follow-up examinations. Healthcare providers can personalise their learning experiences with the use of adaptive learning solutions that employ data collected in real-time. This way, students can focus on the material that will help them the most. Personalisation and support for different learning styles are two additional ways that adaptive learning can assist students. Through the use of adaptive learning, you may be able to provide superior care to patients, become more knowledgeable professionals, and satisfy your hunger for knowledge. Furthermore, it has the potential to facilitate the matching of teachers with students in accordance with the particular learning requirements of each student, so opening the door to more meaningful interactions between the two parties. UpToDate and eMedicine are two examples of systems that provide access to expert information. Additionally, there is a growing trend of using apps to gain access to medical journals (Sharma et al., 2017).

11.4 Behaviour Change Interventions

11.4.1 Behavioural Analytics and Prediction Models

There has been a rapid acceleration in patient deterioration, readmissions, mortality, improved documentation, disease identification, end-of-life care, patient movement, chronic care management, and predictive modelling as a result of digital transformation in the healthcare industry (Yang, 2022). Both machine learning and artificial intelligence have a number of potential uses in the field of clinical trials and research in the healthcare industry. In addition to being labor-intensive, expensive, and time-consuming, clinical studies typically take years to complete successfully. Using machine learning-based predictive analytics, researchers are able to construct a list of potential patients to include in clinical trials. These individuals can be identified from a variety of sources, including social media, previous medical records, and other data.

In addition, machine learning is utilised to determine the ideal size of the test sample, ensure that trial participants have access to data in real time, and make use of electronic records in order to decrease errors that are dependent on data. By merging data with insights powered by artificial intelligence, the unified life science platform known as Medidata is able to optimise clinical trials and breakthrough treatments that are focused on patients. For medical purposes, it makes use of artificial intelligence and machine learning. Additionally, Rave EDC, which stands for electrical data capture, enables this platform to make use of enormous amounts of extensive data for the purpose of conducting experiments. The application of predictive analytics to medical data is yet another common application of machine learning in the healthcare industry. Algorithms that are powered by artificial intelligence and machine learning are able to sift through mountains of data in search of patterns and trends, which they can then use to predict what is going to happen in the future. Automation based on artificial intelligence and machine learning has the potential to significantly enhance accuracy and efficiency in this field, which deals with complex data. Through the use of predictive analytics, Kaiser Permanente is able to identify individuals who have a higher risk of developing diabetes or cardiovascular disease. In addition to being an insurance firm, they are a healthcare organisation. Using patient data to discover risk factors through extensive data analysis, KP HealthConnect helps patients take responsibility of their health (Avcontentteam, 2024). In addition, the platform offers individualised actions designed to assist patients in managing their illness further.

When it comes to data science, artificial intelligence, and machine learning, the healthcare business is increasingly adopting these technologies. Having a solid relationship between artificial intelligence and practical clinical and operational skills is absolutely necessary. Patients, physicians, and operational leaders should be involved in data interrogation, AI pipeline reform, and the iterative interaction between change-informed AI and AI-informed change from the start (Lindsell et al.,

2020). It is vital to make improvements to the algorithms used in artificial intelligence. Yet, it is equally important to keep humans involved and involve all key stakeholders in order for AI to be effective in the healthcare industry (Yang, 2022).

11.4.2 Gamification Strategies

A novel form of gamification that is founded on the principles of behavioural economics is currently making its way into the realm of medicine. Prominent healthcare institutions and thinkers are investigating gamification as a potential method to drive patients to engage in better habits, such as taking their medications as prescribed, losing weight, and engaging in more physical exercise (Volpp & Patel, 2019). When referring to the process of employing game elements in order to motivate particular actions, the term "gamification" is used. The goal of gamification in health care is to encourage involvement in healthy behaviours by making them more fascinating, making them more fun, and making them more profitable. The following are some examples of tactics that can be used to implement gamification: objectives, challenges, levels, points, progress, feedback, prizes, and badges that are awarded. The process of adding game mechanics into situations that are not traditionally associated with gaming is referred to as "gamification." According to the notion that was given by Deterding et al. (2011), intervention developers can incorporate the same addictive qualities found in games into their digital technologies. By incorporating game elements, such as the accumulation of points for taking medication, into a monotonous and non-game activity, we may transform it into a game that is both entertaining and engaging (Cugelman, 2013).

11.4.3 Social Media and Peer Support Network

The spread of social media may be beneficial to various mental health initiatives, including online peer-to-peer aid and other forms of mental health care. Because of social media, users are able to connect with people who understand them, establish communities that support one another, and take control of their own behaviour by determining how much to share and engage with others (Rayland & Andrews, 2023). All of these benefits are made possible by social media. Sixty-eight percent of young adults (those between the ages of 18 and 35) who self-report having a mental illness said that they used social media to talk about their experiences, sixty-six percent said that they connected with other people who understood what they were going through, fifty percent said that they learned coping mechanisms and forty-two percent said that they learned about mental illness from them (Rayland & Andrews, 2023).

Facebook is the most popular platform for social networking in the United Kingdom, with 23% of its users falling within the age bracket of 18–24 years old and 31% falling within the age bracket of 25–34 years old. According to recent research (Rayland & Andrews, 2023), approximately ninety per cent of people who self-identify as having a mental illness and are between the ages of twelve and thirty-four use Facebook. On Facebook's platform, which is entirely free to use, you can communicate in either a synchronous or asynchronous manner, depending on your preferences. Users can submit information and participate in debates. The messenger tool enables users to speak with one another in a private setting when they are in a group because they do not wish to disclose everything on the main feed. As a consequence of this, Facebook is rapidly becoming a widespread platform for peer-to-peer support groups. There are two types of groups: public and private. The former allows anybody to read and write to it, while the latter restricts access to only those individuals who an administrator has authorised (Rayland & Andrews, 2023). Moderators sometimes also set up live chat rooms, where members can meet and talk to one another in real time, either in front of or behind the camera. In contrast to peer support groups that are formally organised, run, or regulated by mental health professionals, Facebook peer support groups are typically more informal and self-organised.

11.5 Decision-Making Processes

11.5.1 Behavioural Analytics and Prediction Models

Decision support systems (AI-DSS) are intelligent computer systems that are designed to aid decision-makers (such as doctors) in making better judgements. These systems are also known as artificial intelligence-driven decision support systems. Artificial intelligence-driven decision support systems have the potential to significantly improve the quality of data that is available to medical professionals, which can result in more accurate illness diagnosis and other treatment options. According to Dlugatch et al. (2024), decision-support systems should not be regarded as autonomous decision-makers but rather as extra tools that can assist medical professionals in making well-informed, evidence-based recommendations and decisions.

Viewers of artificial intelligence must always keep in mind the fundamental hierarchy of data, which begins with raw data and progresses upwards through the levels of information, knowledge, and wisdom. An expanding number of traditionally "knowledge-based" fields are being penetrated by modern artificial intelligence as a result of the exponential development in processing power and memory capacity. However, at this point, it is not within our capabilities to either impart or derive intelligence into or from artificial intelligence. To give just one example, a babysitter that is powered by artificial intelligence could "see" a youngster safely putting a spoon or fork in their mouth. It may even "know" better than to put a pocket knife anywhere near a child's face, much less their lips or eyes. This is because it would be hazardous. However, an astute parent will be aware that a child is capable of putting anything in its mouth, including toxic chemicals, animals in the vicinity, and any object that is rolling around. A razor, a shard of pottery, or a nearby worm are all things that they will rapidly add to the list of things that could potentially cause harm to their child. When it comes to a grandmother who is suffering from dementia, that same astute parent will apply the same criteria without much coaxing. There are a number of restrictions that afflict existing AI programmes. At the top of the list is the absence of "awareness." Another limitation of the programme is that it is dependent on firsthand experiences of both success and failure, which restricts its capacity to learn from past outcomes (Sloane & Silva, 2020).

11.5.2 Facilitating Patient-Provider Communication

Social media platforms have shown to be an extremely significant resource for the support networks of consumers in the field of health communication. As stated by Mikal et al. (2020), the concept of social support is based on the premise that it can assist individuals in experiencing a greater sense of connection, improve their capability to deal with challenging circumstances, increase their self-esteem, and increase their levels of competence in their positions within society. According to the findings of a significant number of studies, social support is multi-dimensional, encompassing not only informational but also emotional, instrumental, and companionship forms of aid. There are a number of health outcomes that have been found to be influenced by social support, including emotional support, companionship, and psychological well-being.

One further important aspect of communities that are created on social media is the effectiveness with which social storytelling can be accomplished through the use of these platforms. By way of illustration, healthcare providers routinely make use of social media platforms such as Facebook and Twitter to broadcast information about patients' experiences with various illnesses. Through the utilisation of digital storytelling, interventions that have been designed to increase the number of people who undergo screening for breast, colorectal, and prostate cancers have all been demonstrated to be effective. Through the use of the power of storytelling, social support is typically provided in contexts that are associated with healthcare. On the other hand, this circumstance begs for the facilitation of storytelling by subject matter experts. Significant ethical concerns arise when underprivileged populations are told stories through the use of the Internet. Among the many ethical concerns that the authors brought up in their call for a comprehensive investigation into the ownership and sharing of digital content, harm reduction in group settings (mainly when dealing with potentially antagonistic group dynamics or trauma disclosure) and the possibility of perpetuating stigmatising narratives were among the many concerns that were brought up (Khan et al., 2022).

11.6 Impact of AI and Social Commerce Integration

Based on the data compiled by Hootsuite, healthcare businesses are steering clear of Pinterest, Threads, and Twitter/X, but they have complete faith in the return on investment (ROI) that LinkedIn, Instagram, and Facebook provide (Schmalfuhs, 2023). It is essential to point out that engagement is the most critical indication that 64% of healthcare marketers use to demonstrate a return on investment. Both the sort of post and the frequency of posts are barriers to engagement. As a result of the fact that 79% of healthcare businesses are active on LinkedIn, forty per cent of those organisations only publish once a week at the most. According to Hootsuite, LinkedIn is set to take over as the social media hub for the healthcare industry. More than eighty percent of members of Generation Z have expressed an interest in using social media to research health-related topics. In addition, they are ready to follow more businesses and organisations on LinkedIn. In light of the fact that forty percent of millennials have the intention of working in the healthcare industry (Schmalfuhs, 2023), Hootsuite forecasts that LinkedIn will be a fantastic resource for the future generation of professionals. For social media users, it is not enough to connect; they also want to be amused. A majority of respondents (over 50%) think that businesses would do well to be more approachable on social media. Interestingly, when businesses are overly concerned with promoting themselves, 34% of customers find it very off-putting.

11.7 Ethical and Privacy Concerns

The recently made available knowledge about artificial intelligence presents further challenges. As an illustration, according to the trends that Hootsuite has identified, it is anticipated that healthcare organisations will significantly increase their utilisation of artificial intelligence for customer support activities by 375% in the coming year. Moreover, it is anticipated that 356% of these organisations intend to expand their use of artificial intelligence for picture editing (Schmalfuhs, 2023). A further 62% of customers are less likely to engage with or trust content that was generated by an artificial intelligence application when they are aware of its origin. A more significant number of individuals have expressed that they are unable to determine whether something they come across on the internet is genuine or manufactured artificially. Hootsuite received feedback from 63% of healthcare social marketers indicating that guidelines do, in fact, make their tasks more difficult. More than half of healthcare firms consider compliance rules to be their top worry when it comes to social media safety. Furthermore, nearly half of these organisations have company policies that dictate what employees are allowed to post and what they are not allowed to post. In addition, the application of AI in the medical field creates a variety of ethical difficulties. The majority of medical care decisions have been made by people in the past. The utilisation of intelligent computers for the purpose of making or assisting with these decisions raises questions about accountability, transparency, permission, and privacy. Considering the current level of technology, transparency may be the most challenging problem to tackle.

A large number of AI algorithms, especially those that deal with picture processing through deep learning, are tough, if not impossible, to understand or describe. A patient may naturally wonder what factors contributed to the cancer diagnosis when told that an imaging study had shown the disease. Even doctors with a general understanding of how deep learning algorithms work might not be able to explain them. When it comes to diagnosing and treating patients, AI systems will inevitably make mistakes, and finding those responsible can be a challenge. Additionally, there will inevitably be cases where people get medical advice from AI systems that they would rather get from a compassionate doctor. Healthcare machine learning systems could potentially be biassed by algorithms, leading to inaccurate illness risk predictions based on gender or ethnicity. The use of AI in healthcare is anticipated to bring about numerous changes in the fields of ethics, medicine, the workplace, and technology. Structures for monitoring critical issues, responsible reactions, and governance measures to limit harmful implications should be established by healthcare institutions, governments, and regulatory agencies. Constant focus and deliberate policymaking will be necessary for a long time because this is a very powerful and impactful technology that will affect human society (Wen & Huang, 2019).

11.8 Future Directions and Opportunities

11.8.1 Emerging Trends in AI and Social Commerce in Healthcare

A number of potential applications of artificial intelligence are discussed, including the decrease of human error, the help of medical staff and professionals, and the provision of patient care around the clock. The utilisation of artificial intelligence (AI) in the fields of medical imaging, diagnosis, and treatment planning has the potential to expand as the capabilities of AI continue to increase. With the assistance of artificial intelligence applications, a variety of tasks, including answering phones and assessing patterns of population health, will become less difficult to perform (and possibly also for other purposes). Future artificial intelligence systems may, for instance, make it possible for clinicians and staff workers to perform their jobs in a more automated or upgraded manner. According to IBM Education (2023), this will make it possible for humans to commit more time to giving face-to-face professional care that is not only effective but also compassionate.

The "quadruple aim" of healthcare systems is to achieve precision medicine, precision diagnostics, precision therapeutics, and networked and AI-augmented care. Artificial intelligence can assist these systems in achieving this "quadruple aim" by standardising and making precise medicine more accessible. There are a multitude of potential uses of artificial intelligence (AI) in the healthcare industry, which is experiencing rapid development. Some examples of these applications include drug discovery, virtual clinical consultation, sickness diagnosis, prognosis, medication management, and health monitoring. Here are just a few examples. A non-exhaustive suite of short-, medium-, and long-term applications in the healthcare industry is provided by us in light of the fact that artificial intelligence has the ability to supplement, automate, and revolutionise medicine. As things stand, artificial intelligence systems are not thinking engines; they are unable to make use of "common sense" or "clinical intuition and experience" in the same way that human physicians can. On the contrary, artificial intelligence is more analogous to a signal translator that extracts patterns from files. AI technologies are beginning to be utilised by healthcare businesses in order to automate operations that are repetitive, high-volume, and require a significant amount of time. In addition, the proof of concept for artificial intelligence in precision diagnostics has come a long way, with examples such as radiation planning and diabetic retinopathy (Bajwa et al., 2021).

The development of precision medicine, which is a much-needed advance in healthcare, is entirely dependent on this capability, which mainly shows itself as machine learning. Even if initial attempts at providing treatment recommendations and diagnoses have been challenging, we believe that artificial intelligence will eventually become an expert in this field. This is something that we predict will happen. As a result of the rapid advancement of artificial intelligence for imaging analysis, robots will likely be able to process an increasing number of images in the domains of radiology and pathology at some time in the future. Speech and text recognition are presently being utilised for a variety of tasks, including the capture of clinical notes and the facilitation of patient communication, and this demand is only going to increase in the future. Ensuring that the technologies are beneficial is not the most significant challenge for artificial intelligence in these healthcare domains; instead, the most significant challenge is getting them to be employed in clinical practice on a daily basis. The need for regulatory approval, interaction with electronic health record systems, standardisation to the point that comparable products perform similarly, clinician education, funding from public or commercial payer groups, and field-based updates are all prerequisites for the widespread adoption of artificial intelligence systems. In spite of the fact that it will take a far more extended period to solve these challenges than it takes for the technologies to advance, they will ultimately be defeated. As a result, we estimate that artificial intelligence will be utilised in clinical practice in a limited capacity during the next five years, followed by a more extensive application within the following decade. Additionally, it is also becoming more evident that artificial intelligence systems will not totally replace human clinicians but rather will collaborate with them in order to deliver superior care to patients. There is a possibility that jobs and responsibilities for human clinicians will grow to involve a greater reliance on uniquely human characteristics, such as the capacity to empathise with others, to exert influence over others, and to see the bigger picture. The only people who would lose their jobs in the long term may be medical professionals, such as nurses and doctors, who are hesitant to work with artificial intelligence (Wen & Huang, 2019).

11.8.2 Potential Applications in Chronic Disease Management

Over the last ten years, there has been an increase in the utilisation of health-related mobile applications (apps) to assist in the treatment of chronic diseases. These applications have been utilised in three different ways: digital teaching, self-monitoring, and feedback. Despite the fact that the World Health Organisation (WHO) defines mobile health (mHealth) as the "spread of mobile technologies as well as advancements in their innovative application to address health priorities", there is no definition that all parties universally accept. It can be challenging for patients, politicians, and physicians to identify which of the many applications that are accessible for managing chronic diseases will be the most beneficial (Agarwal et al., 2021). As a result, a number of applications are available. Conducting experimental trials is the most effective strategy for determining how these applications function; nevertheless, this particular method might be challenging to implement due to the fact that the number of apps is constantly growing, and their features are frequently being updated. Instead, academics are increasingly turning to checklists of quality criteria that are simple and quick to use in order to evaluate mobile applications.

As a result of the expansion of research that makes use of such a preset set of criteria, questions have arisen regarding the scope of the criteria that are utilised in the process of evaluating the quality of mobile applications. As an illustration, Fernández et al. (2023) stated that while evaluating mobile applications for asthma treatment, it is essential to take into consideration the applications' goals, whether or not they conform to care standards, whether or not they adhere to explicit language and usability criteria and the relationship between quality and affordability. One example is the Mobile App Rating Scale (MARS), which was developed as a consequence of an effort to consolidate different frameworks that were already in existence for the purpose of evaluating digital health applications. Despite the fact that it is an excellent tool for determining the overall popularity of an app, it does not always take into account how well the app functions for specific medical objectives.

11.8.3 Innovations in AI-Driven Personalised Medicine and Precision Health

The dynamic field of biomedical research has been revolutionised by the incorporation of cutting-edge technology and multidisciplinary techniques, which has resulted in a significant change in the way complex health issues are addressed and resolved. Through the intersection of medical microbiology, immunology, molecular biology, and artificial intelligence (AI), a substantial force is pushing discoveries in precision medicine, particularly in infectious diseases and beyond. This is especially true in the context of this setting. Artificial intelligence (AI) in healthcare can take various forms, ranging from virtual (e.g., health information management systems powered by deep learning and actively guiding doctors in their treatment decisions) to cyber-physical (e.g., surgical assistant robots and drug delivery nanorobots). Numerous image-based detection and diagnosis systems in healthcare are now able to match or surpass physicians in terms of performance because of the ability of AI technologies to identify complex patterns and underground structures. The use of artificial intelligence in clinical decision-support systems has the potential to lessen diagnostic mistakes, enhance decision-making capabilities, and facilitate the extraction and reporting of EHR data for doctors. According to Johnson et al. (2021), AI will be able to tackle more complex issues with the help of new computational advancements in areas such as pattern recognition, efficient search, prediction, bias-free reasoning, and natural language processing (NLP).

The drug discovery pipeline has been considerably optimised as a result of the application of artificial intelligence in diagnostics for precision medicine. This has led to the identification of possible therapeutic candidates with an accuracy and efficiency that is unparalleled. Significant advancements have been made in the areas of molecular pathology, proteomics, and neurodevelopmental diseases as a result of the contributions made by artificial intelligence. One example of how this integration has prepared the way for a multidisciplinary approach to addressing healthcare issues is the implementation of the One Health initiative in the Arctic, which is constantly evolving. It is important to note that artificial intelligence has substantial implications for clinical oncology as well as COVID-19 diagnosis and treatments. Allami and Yousif (2023) state that artificial intelligence-driven precision vaccinology has the ability to enhance our understanding of mechanisms and design vaccinations that are more specifically targeted to the individual.

11.9 Conclusion

The healthcare industry is undergoing a paradigm change towards patient-centric care models with the integration of social commerce and artificial intelligence (AI). Some important conclusions have been drawn from extensive research on collaborative decision-making processes, behaviour modification treatments, and personalised health education. The first is the impressive effectiveness of AI-powered personalised health education platforms in providing patients with specific, individualised health information. These platforms let patients feel more empowered by using their data and preferences to deliver them knowledge that is both relevant and easy to understand. This improves their health literacy. Secondly, there is hope for the promotion of healthy behaviours and lifestyle alterations through interventions that use social commerce and AI to support behaviour change. These programmes successfully encourage patients to engage in sustainable health practices, which improves their health outcomes, by using gamification tactics, tailored messaging, and social influence dynamics. Finally, one of the most important aspects of patient empowerment and participation is the shared decision-making processes that are made possible through AI-driven decision support systems that are integrated into social commerce platforms. These procedures guarantee that treatment programmes are in line with patients' beliefs, interests, and objectives by giving patients individualised insights and facilitating collaborative conversations with healthcare practitioners.

There are a number of potential future projects that may be suggested to further integrate social commerce with AI in healthcare. First, hospitals and other medical facilities should put money into creating decision support systems powered by artificial intelligence that can communicate with current EHRs. In turn, this will improve the delivery of individualised treatment by allowing for the smooth incorporation of insights enabled by AI into clinical workflows. The second point is that social commerce platforms and healthcare professionals should work together to meet the specific requirements of different patient groups. Partnerships with healthcare organisations to promote health awareness and involvement, the formation of communities focused on health, and the integration of health-related elements into existing social commerce platforms are all possible steps in this direction. And lastly, healthcare providers should be required to participate in continuing education and training programmes so they can use AI and social commerce to their patients' advantage. Data ethics, privacy laws, and the best ways to use AI-driven innovation into healthcare are all part of this curriculum. Finally, there's a lot of hope that patients will be more involved and empowered in their treatment if AI and social commerce are able to work together. Improved patient outcomes and a more sustainable healthcare system can be achieved when healthcare organisations employ shared decision-making procedures, behaviour change interventions, and personalised health education to revolutionise care delivery.

References

Agarwal, P., Gordon, D., Griffith, J., Kithulegoda, N., Witteman, H. O., Sacha Bhatia, R., Kushniruk, A. W., Borycki, E. M., Lamothe, L., Springall, E., Shaw, J. (2021). Assessing the quality of mobile applications in chronic disease management: a scoping review. *Npj Digital Medicine, 4*(1). https://doi.org/10.1038/s41746-021-00410-x

Akram, U., Hui, P., Khan, M. K., Yan, C., Tanveer, Y., & Hashim, M. (2019). Shopping online without thinking : Myth or reality. In J. Xu, F. Cooke, M. Gen, S. Ahmed (Eds.), *Proceedings of the twelfth international conference on management science and engineering management. ICMSEM 2018. Lecture Notes on Multidisciplinary Industrial Engineering*. Springer, Cham. https://doi.org/10.1007/978-3-319-93351-1_2

Allami, R. H., & Yousif, M. G. (2023). Integrative AI-driven strategies for advancing precision medicine in infectious diseases and beyond: A novel multidisciplinary approach. *Medical Advances and Innovations Journal, 1*(2), 1–11.

Arora, S. (2023). *AI analytics explained: How it works and key industry use cases*. ThoughtSpot. https://www.thoughtspot.com/data-trends/ai/ai-analytics

Avcontentteam. (2024). Machine Learning & AI for Healthcare in 2024. Analytics Vidhya. https://www.analyticsvidhya.com/blog/2023/03/machine-learning-ai-for-healthcare/#:~:text=PredictiveAnalytics,make predictions about future events.

Bajwa, J., Munir, U., Nori, A., & Williams, B. (2021). Artificial intelligence in healthcare: Transforming the practice of medicine. *Future Healthcare Journal, 8*(2), e188–e194. https://doi.org/10.7861/fhj.2021-0095

Barth, S. (2024). *AI in healthcare*. ForeSee Medical. https://www.foreseemed.com/artificial-intelligence-in-healthcare

Chatterjee, S., Rawat, C., Ramani, A., Kundu, R., Ganguly, S., Nandwani, R., Ghosh, R., Chandra, D., Mukherjee, S., Chakraborty, S., Srimany, Swarnendu, Sinha, S., Misra, A., Brahma, U., Sawant, P., Jadhav, P., Samantaray, S., Tayade, S., Das, D., Choudhary, A. (2023). Enabling healthcare with technology. In *Pwc* (Issue March).

Cugelman, B. (2013). Gamification: What it is and why it matters to digital health behavior change developers. *JMIR Serious Games*. https://doi.org/10.2196/games.3139

Deterding, S., Dixon, D., Khaled, R., & Nacke, L. (2011). Effect of a comprehensive surgical safety system on patient outcomes. In *MindTrek '11: Proceedings of the 15th International Academic MindTrek Conference: Envisioning Future Media Environments* (pp. 9–15). https://doi.org/10.1145/2181037.2181040

Dlugatch, R., Georgieva, A., & Kerasidou, A. (2024). AI-driven decision support systems and epistemic reliance: A qualitative study on obstetricians' and midwives' perspectives on integrating AI-driven CTG into clinical decision making. *BMC Medical Ethics, 25*(1), 1–11. https://doi.org/10.1186/s12910-023-00990-1

Eriksen, M. (2023). *The use of Artificial Intelligence (AI) in social commerce*. Comment Sold. https://try.commentsold.com/blog/business-operations/use-of-artificial-intelligence-in-social-commerce/

Fernández, C., Vicente, M. A., Guilabert, M., Carrillo, I., & Mira, J. J. (2023). Developing a mobile health app for chronic illness management: Insights from focus groups. *Digital Health, 9*. https://doi.org/10.1177/20552076231210662

IBM Education. (2023). *The benefits of AI in healthcare*. https://www.ibm.com/blog/the-benefits-of-ai-in-healthcare/

Khan, M. I., Rahman, Z. U., Saleh, M. A., & Khan, S. U. Z. (2022). Social media and social support: A framework for patient satisfaction in healthcare. *Informatics, 9*(1), 1–14. https://doi.org/10.3390/informatics9010022

Jaros, D. (2023). *How digital natives are powering social commerce—and what that means for your business. YEXT.* https://www.yext.com/blog/2023/05/how-digital-natives-are-powering-social-commerce-and-what-that-means-for-your-business

Jeyaraman, M., Ramasubramanian, S., Kumar, S., Jeyaraman, N., Selvaraj, P., Nallakumarasamy, A., Bondili, S. K., Yadav, S. (2023). Multifaceted role of social media in healthcare: Opportunities, challenges, and the need for quality control. *Cureus, 15*(5). https://doi.org/10.7759/cureus.39111

Johnson, K. B., Wei, W. Q., Weeraratne, D., Frisse, M. E., Misulis, K., Rhee, K., Zhao, J., & Snowdon, J. L. (2021). Precision medicine, AI, and the future of personalized health care. *Clinical and Translational Science, 14*(1), 86–93. https://doi.org/10.1111/cts.12884

Krist, A. H., Tong, S. T., Aycock, R. A., & Longo, D. R. (2017). Engaging patients in decision-making and behavior change to promote prevention. *Stud Health Technol Inform. 240*, 284–302

Laskowski, N. (2023). *A guide to artificial intelligence in the enterprise*. Tech Target. https://www.techtarget.com/searchenterpriseai/definition/AI-Artificial-Intelligence

Lindsell, C. J., Stead, W. W., & Johnson, K. B. (2020). Action-informed artificial intelligence—matching the algorithm to the problem. *JAMA—Journal of the American Medical Association, 323*(21), 2141–2142. https://doi.org/10.1001/jama.2020.5035

Mikal, J. P., Beckstrand, M. J., Parks, E., Oyenuga, M., Odebunmi, T., Okedele, O., Uchino, B., & Horvath, K. (2020). Online social support among breast cancer patients: Longitudinal changes to Facebook use following breast cancer diagnosis and transition off therapy. *Journal of Cancer Survivorship, 14*(3), 322–330. https://doi.org/10.1007/s11764-019-00847-w

Office of the Victorian Information Commissioner. (2024). Artificial intelligence and privacy—Issues and challenges. https://ovic.vic.gov.au/privacy/resources-for-organisations/artificial-intelligence-and-privacy-issues-and-challenges/

POWNEY, M. (2020). *£20m AgorIP project investment to boost innovation in wales*. Business News Wales. https://businessnewswales.com/20m-agorip-project-investment-to-boost-innovation-in-wales/

RAPL. (2023). *Revolutionize training: Top 10 spaced repetition platforms and their benefits*. https://getrapl.com/blog/revolutionize-training-top-10-spaced-repetition-platforms-and-their-benefits/#:~:text=EdApp also offers a variety,activates the Brain Boost feature.

Rayland, A., & Andrews, J. (2023). From social network to peer support network: Opportunities to explore mechanisms of online peer support for mental health. *JMIR Ment Health*. https://doi.org/10.2196/41855

Schmalfuhs, M. (2023). *Artificial intelligence changing the landscape for healthcare social media marketing*. https://www.pharmexec.com/view/artificial-intelligence-changing-the-landscape-for-healthcare-social-media-marketing

Sharma, N., Doherty, I., & Dong, C. (2017). Adaptive learning in medical education: The final piece of technology enhanced learning? *Ulster Medical Journal, 86*(3), 198–200.

Sloane, E. B., & Silva, R. J. (2020). *Artificial intelligence in medical devices and clinical decision support systems*. Elsevier Inc. https://doi.org/10.1016/B978-0-12-813467-2.00084-5

Taliwal, A., Pathania, M., & Chugh, M. (2024). Social commerce recommendation systems: leveraging user behaviour and preferences. In *Advanced applications of generative AI and natural language processing models*. IGI Global. https://doi.org/10.4018/979-8-3693-0502-7.ch013

Team, MobiDev. (2024). *Natural Language Processing (NLP) use cases for business optimization*. MobiDev. https://mobidev.biz/blog/natural-language-processing-nlp-use-cases-business

Todua, N., Jashi, C., & Todua, N. (2024). The role of social media in healthcare marketing. In *Modern healthcare marketing in the digital era*. IGI Global. https://doi.org/10.4018/979-8-3693-0679-6.ch002

Volpp, Kevin, G., & Patel, M. S. (2019). How gamification can change health behaviors. https://hbr.org/webinar/2019/05/how-gamification-can-change-health-behaviors

Wen, Z., & Huang, H. (2019). The potential for artificial intelligence in healthcare. *Future Healthcare Journal, 27*(4), 217–224. https://doi.org/10.5912/jcb1327

Yang, C. C. (2022). Explainable artificial intelligence for predictive modeling in healthcare. *Journal of Healthcare Informatics Research, 6*(2), 228–239. https://doi.org/10.1007/s41666-022-00114-1

Chapter 12
Artificial Intelligence Adoption for Sustainable HRM in HealthCare Industry: A Conceptual Review

Rubee Singh, Toh Guat Guan, Kamalesh Ravesangar, and Amit Joshi

Abstract An efficient workforce is essential to managing customers, as evidenced by the growing demand for competitive and skilled HR professionals on a global scale. The HR division stands out as a crucial component that elevates organizational performance in the fiercely competitive business world of today. Artificial intelligence (AI) and its related technologies are becoming more and more prevalent in many different sectors of the economy, including healthcare. With its ability to handle administrative duties, decision-making, big data, diagnosis, and other tasks, AI offers a promising answer to the HR problems facing the healthcare industry. The goal of this book chapter is to provide readers with more in-depth understanding of how AI is being adopted by sustainable HRM practices in the healthcare industry. Along with outlining the benefits of incorporating AI into HRM and healthcare, it also projects what the future holds for this sector of the economy. This chapter is a goldmine for academics and students studying subjects related to information, computer science, healthcare, organizational innovation, and human resource management. It provides insightful advice to HR specialists, businesses, and workers alike.

Keywords Artificial intelligence · Sustainable · HRM · Healthcare

---

R. Singh (✉) · A. Joshi
Department of Management Studies, Kumaun University, Nainital, Uttarakhand, India
e-mail: dr.rubeerajput@gmail.com

R. Singh
Institute of Business Management, GLA University, Mathura, India

T. G. Guan · K. Ravesangar
Department of Accountancy and Business, Tunku Abdul Rahman University of Management and Technology, Kuala Lampur, Malaysia

R. Singh et al. (eds.), *Transforming Healthcare Sector Through Artificial Intelligence and Environmental Sustainability*, Approaches to Global Sustainability, Markets, and Governance, https://doi.org/10.1007/978-981-97-9555-0_12

12.1 Introduction

Human resource management (HRM) has a deep-rooted historical foundation. HRM originated in the 1960s and applies to individuals in organizational positions (Marler & Parry, 2016). The human resources (HR) department oversees various organizational units linked directly or indirectly to personnel (Jainh, 2021). Within organizations, HR performs two essential roles: administrative duties and employee management. The former is typically clear-cut, given the well-defined regulations. Nonetheless, even rule-based HR functions demand continuous assessment and fine-tuning to avoid employee discontent. This requirement places a substantial responsibility on HRM to consistently gauge employee sentiments. Consequently, HR should prioritize activities essential fostering a sustainable business model rather than getting entangled in labor-intensive processes. The rapid advancement of technology has had a profound impact on individuals' lives, forcing businesses to adjust to these changes. Consequently, human resource management (HRM) has experienced significant evolution. The primary goals of HRM include attracting and retaining talent, enhancing employee motivation, and selecting the suitable candidates. Technological advancements have fundamentally transformed HR, employee management strategies, and the creation of job profiles. The wide availability and expertise have played a pivotal role in driving numerous enhancements in HR operations. García-Arroyo and Osca Segovia (2021) emphasize that information technologies promote decentralized decision-making, allowing a broader range of decision-makers to. This shift results in increased capabilities and improved of inputs (data) into outputs. Moreover, García-Arroyo and Osca highlight the revolutionary influence of leveraging data and big data in HR functions. They identify the main challenges as the acquisition, processing, and interpretation of data to align with organizational goals, while also raising critical questions regarding data quantity, structure accessibility, and distribution. The complex nature, extensive volume, and rapid flow of data empower machine learning algorithms to analyze and forecast outcomes based on this information. Hong notes that contemporary HRM systems can collect, analyze, and generate statistical data; however, the integration of machine learning (ML) algorithms is essential for unraveling the connections within the data.

The phrase artificial intelligence (AI) was introduced in 1956 by John McCarthy, a computer scientist, to refer to computer programs exhibiting human-like. These programs can autonomously and effectively complete tasks, thereby decreasing reliance on human effort (Rajaraman, 2014). It is also defined as the formulation of algorithms that designed to train machines by feeding them with large datasets and applying diverse techniques like deep learning and Bayesian modelling (Esteva et al., 2019; and Zarringhalam et al., 2014), aimed to mimic cognitive functions of human (He et al., 2019; Dreyer & Geis, 2017, Goodfellow et al., 2016; and Ghahramani, 2015). Once the machines start operating, their algorithms will be updated and improvised according to the feedback given on the mistakes they have made. At present, it has reached remarkable levels of performance in which it is capable of executing multiple and complicated tasks that closely match or even surpass what the human brains are capable of doing (Van Hartskamp et al., 2019). AI is leading the way in advancing and researching healthcare with a new approach. Recent progress in AI, including improved access to extensive and intricate health data, faster computer processing speed, a larger pool of AI experts, and the availability of AI tools, has made AI an essential component of modern healthcare (Bohr & Memarzadeh, 2020). The healthcare sector has been traditionally hesitant in adopting AI technology, but now its implementation is rapidly gaining momentum. The industry acknowledges the importance of AI systems and is integrating to address healthcare professional shortages, meet rising patient expectations, keep with advancements, improve quality, and achieve optimal outcomes. It is widely recognized that AI holds the potential to revolutionize patient care through precision medicine and patient-centered services. The integration and utilization of AI have extended across various sectors such as manufacturing, agriculture, transportation, and healthcare. Despite its impact, AI remains an emotionless entity with an operating system fundamentally distinct from human cognitive functions (Klien et al., 2004; Korteling et al., 2018, 2021; Moravec, 1998; Shneiderman, 2020). Consequently, AI deployment serves primarily as a complement to mitigate human and in data processing and task execution. Nonetheless, the advent of AI has indeed revolution multiple industries, reshaping work environments and while addressing industry challenges.

This industry not only includes hospitals, pharmaceutical companies, medical device makers, and digital health start-ups, but also experts dedicated to improving the health and overall well-being of people. Sunarti et al. (2021) mentioned that this is a multifaceted, dynamic industry. AI applied for the healthcare industry can provide policy-makers, doctors, and healthcare institutions with data-driven judgments. This will improve patient diagnoses of medical conditions as well as the effectiveness and cost efficiency for treating them. It will also make health services fairer to all (Business Standard, 2024.) The healthcare market with artificial intelligence (AI) is predicted to grow larger. According to a recent article in Business Standard, by 2024 it will be worth \$4.3 billion compared with only \$1.5 billion in 2019. The pandemic has further accelerated this trend. As businesses respond to new challenges caused by Covid-19, like those associated with digital health solutions and machine learning (ML), AI is increasingly incorporated into medical devices as well (for example, in remote monitoring and wearable). Within the healthcare sector, the market for AI health applications is given favorable forecast. The report by Collier et al. (2017) mentions that it will be worth \$6.6 billion by 2023, an annual growth rate of 40%. At the same time, around 20% of unmet clinical needs should be resolved with AI. Health costs will also be reduced by this move, and technology might help in preventing disease as well as treating it. Already, we have seen estimates that if 2026 sees AI being used in the health industry, then it will save approximately USD 150 billion annually for that sector. Khanijahani et al. (2022) In surveying the literature, Hoppe et al. (2023) restate that AI could potentially transform medical diagnosis and new drug finding. On the other hand, however, deploying AI is not entirely without challenges. There is the potential for injury to patients due system errors as well as risks of infringing individuals' personal privacy in collecting data and drawing inferences based upon them using AI systems.

12.2 The Role of HR in Sustainable Businesses

Many organizations are increasingly committed to integrating sustainability into their operations. This commitment is driven by various demands and challenges associated with sustainable development, such as climate change, resource scarcity, overpopulation, environmental degradation, and inequality. These pressing issues create a necessity for organizations to take action and implement sustainable management systems. In the realm of business sustainability, an approach known as the triple bottom line is commonly employed. This approach considers the simultaneous consideration of environmental, and economic performance, forming the foundation of an organization's sustainable business practices. Despite the extensive academic discourse on sustainability in organizations, there is a notable lack of research specifically focusing on the intersection of sustainability and human resource management (HR) (Heikkinen et al., 2021). Sustainable HRM aims to achieve business sustainability by implementing sustainable HRM practices within companies. These practices are designed to foster business sustainability and ensure the adoption of appropriate HRM strategies. Stahl et al. (2020) propose criteria evaluating sustainable HRM, suggesting that HRM practices can be deemed sustainable if they enhance social well-being, promote environmental stewardship, contribute to long-term economic viability. Conversely, HR practices are considered unsustainable when they lead to negative social, environmental, or economic outcomes.

The current business landscape is encountering a myriad of challenges stemming from social, environmental, and governance contexts (George et al., 2016), placing increasing pressure on companies to meet diverse requirements from internal and external stakeholders. Recognizing employees as key assets crucial for addressing these challenges and ensuring business sustainability, a different approach is taken by firms to embrace sustainable human resource management (SHRM). SHRM integrates social and environmental policies and regulations, governmental and societal pressures, as well as needs of customers and HR strategies and policies (Lucio & Stuart, 2011). Organizations are thus encouraged to vigilantly monitor their use of natural resources and infuse the ESG (environment, social, and governance) dimension across all operational aspects. SHRM entails engaging human capital in community involvement, promoting sustainable utilization of natural resources, and fostering awareness and accountability for ESG responsibilities among all employees to cultivate a sustainable business environment (Chams & García-Blandón, 2019). Consequently, ESG and traditional business practices are integrated strategies and policies for hiring, adherence, ethical practices, and overall sustainable business operations, as illustrated in Fig. 12.1.

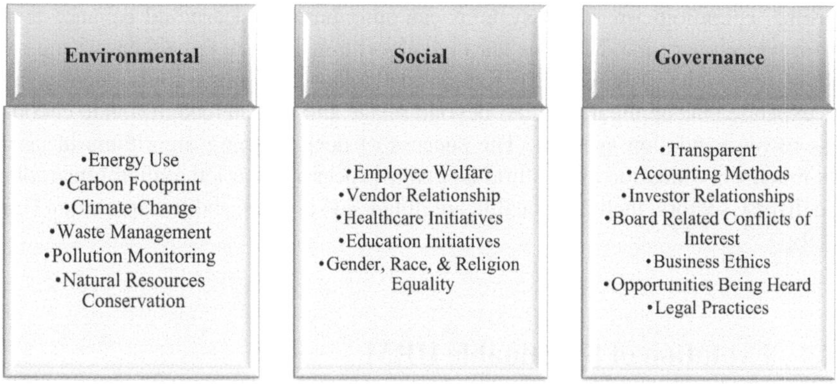

Fig. 12.1 Environment, Social, and Governance (ESG)

12.3 The Evolution of Artificial Intelligence

Since AI's inception in the mid-twentieth century, there have been great strides as it has shifted from early simple rule-based systems to complicated machine learning algorithms, which have changed the way we interact with technology and have the potential to impact every aspects of our lives. Our goal with this article is to provide a high-level overview of AI, covering the key milestones and achievements in the field. Research on AI dates back to the late 1950s and early 1960s, when researchers began to explore how computers could be used to perform tasks that required human-like intelligence. These early AI thinkers, such as John McCarthy, Marvin Minsky, and Claude Shannon, built the foundation for the field, developed the first AI systems, and conducted leading-edge research in natural language processing, knowledge representation, and reasoning (Bezboruah & Bora, 2020). A key milestone in the history of AI was the Dartmouth Conference in 1956. This conference initiated the field of AI research and is often held up as a key event in the history of AI. It brought together prominent computer scientists to explore the potential for computers to exhibit human-like intelligence.

In the past few decades, AI research has experienced remarkable growth, particularly in the fields of natural language processing, computer vision, and expert systems. The 1980s and early 1990s witnessed a resurgence of interest in AI due to advancements in hardware and software. These advancements enabled the development and utilization of more sophisticated AI systems, leading to increased research and innovation. As algorithms and hardware continued to advance, AI researchers made significant progress in constructing more advanced and powerful AI systems. Machine learning, such as neural networks, played a crucial role in driving this progress. These techniques allowed AI systems to learn from data, resulting in performance over time. Deep learning algorithms, a subset of machine learning, have been successfully applied in various domains, including image, speech recognition, and language translation. Recent years have witnessed notable breakthroughs in machine learning technology, in which systems can autonomously learn and enhance their performance over time. Machine learning algorithms analyze large datasets statistically, enabling the system identify patterns and make predictions. Consequently, this has expanded the application of AI beyond image and speech recognition to encompass recommendation systems. The success of deep learning algorithms in tasks like image classification and natural language processing has led to their integration into diverse sectors, including healthcare, financial services, and transportation (Lu, 2019).

12.4 Evolution of Sustainable HRM

In the last four years, the sustainable human resource management field has seen a significant increase in research. Sustainable human resource management (SHRM) is a concept created after 2000. Some researchers, such as Gollan (2000), Zaugg et al. (2001), Wagner and Marcus (2015), and Wilkinson et al. (2001), have started to highlight the significance of sustainable HRM, building from earlier works on environmental management, organization sustainability, and the HRM. An overview of the development on sustainability HRM literature, for example, as mentioned by Ehnert and Harry (2012), includes the Swiss, Australian, and German approaches. The Swiss approach emphasizes the economic, social and human side of sustainability, over the ecological side. It is so important; it emphasizes leveraging human resources, employability, individual responsibility, and also work-life balance play to manage the sustainability of the organization in question. Aside, the Australian case studies appear to be the human resource sustainability perspective and participative organizations that can do decision making with other components part of the network organizations. And the German case emphasizes all three types of sustainability—economical, ecological (nature), and human resources—are important resources for the organization to meet, or not meet, the organization targets and instructions on how to manage sustainability and survive in a competitive environment (Kramar & Robin, 2014).

Influential scholarly works have contributed to SHRM conceptualizations from several scholars (Ehnert et al., 2016, connecting sustainability and HRM practices (Hargreaves & Fink, 2012), elaborating on sustainable HR management and sustainable leadership; (Kramar & Robin, 2014), connecting SHRM with strategic HRM; (Mariappanadar & Sugumar, 2012), exploring SHRM; (De Prins, 2011), developing a full model of sustainable HRM, further developed by (Mazur & Barbara, 2015). Mazur and Barbara (2015) suggested that a SHRM framework should encompass a psychological perspective, a sociological perspective, a strategic perspective, and a green perspective. The theoretical underpinning of SHRM is about recognizing individuals in the organization, engaging them fully and connecting them institutionally to strategic policy and the environment.

12.5 Understanding the Advantages of AI in Human Resources

At present, human resource experts are required to be more productive, effective, and competitive at a global scale. Consequently, various human resource (HR) activities are now accompanied by artificial intelligence (AI). In the human resource (HR) field, artificial intelligence (AI) benefits the recruitment process in several ways, the performance management, and the onboarding phase, among others.

(a) Improving Onboarding Processes

The orientation process for new employees plays a significant role in their job satisfaction and performance, ultimately impacting their retention. Artificial intelligence (AI) now offers the opportunity to improve onboarding procedures by providing tailored learning experiences, development plans, and virtual support. This technology can streamline tasks, personalize the onboarding experience, and empower new hires with continuous access to resources. By automating administrative tasks and offering flexibility in terms of timing and location, the system lightens the workload of HR professionals. It is essential to carefully assess the risks of discrimination and security issues associated with AI algorithms. Continuous monitoring is necessary to ensure fairness, and stringent security measures must be in place to safeguard all data and prevent breaches. Implementing DMARC protection for email domains is crucial to prevent phishing and spoofing attempts. Artificial intelligence is reshaping the onboarding process, transforming it from a basic administrative task into strategic tool for boosting employee engagement and retention. Through automation of routine paperwork and customization of learning experiences, AI enhances the onboarding journey for new. This innovative technology ensures quick access to vital information and resources, empowering individuals with a sense of and belonging. Moreover, AI-driven analytics offer valuable feedback on onboarding strategies, enabling to identify areas for improvement and make well-informed, data-backed decisions. In essence, AI contributes significantly to streamlining, enhancing employee experiences, and providing actionable insights, laying a solid groundwork for new hires and nurturing ongoing success within companies.

(b) Developing Employee Training Strategies

AI improves training strategies by integrating personalized instruction, immediate feedback, and automating the execution of administrative processes, which increases productivity. By leveraging data capabilities, AI identifies areas of knowledge gaps and improves the effectiveness of the learning process in each HR department. AI is also capable of identifying a person's preferred way of learning and custom design education pathways, enabling trainers to dedicate attention on tasks of greater importance to the organization, such as priority skills, digital literacy, analytical thinking, strategic thinking, cultural understanding, collaboration, and emotional intelligence, and streamlining administrative tasks and allowing HR to focus on developing the talent of their asset pool. AI and the Changing Role of Training: AI will play a key role in revolutionizing training. Through extensive data, AI personalizes the learner's knowledge process. This personalized process combined with AI providing the information in many formats ensure that the material is accessible and engaging for learners. AI also automates administrative tasks which enables training experts to focus on creating robust educational resources and providing guidance. With the measurement of performance and the feedback loop, AI identifies areas needed for improvement and gaps in knowledge. Ultimately, AI will improve efficiency, effectiveness, enrich the learner's journey and the tools, and the journey will result in the up-skilling and reskilling of our global workforce.

(c) Data Quality and Restricted Quantity

The success of the AI in HR depends on the quality of data, because the reliability of the outcomes provided by the AI system is shaped by the data used to train the AI. One of the main challenges the HR industry faces includes the datasets are generally smaller in size compared to other industries, and the quality of the data is important to ensure the AI gives accurate, unbiased results. For example, Amazon pulled its recruitment software as it preferred male candidates, thus stressing the significance of utilizing good, unbiased data. Equally important, data security must be a key consideration when implementing AI in HR and organizations must follow data privacy regulations and use risk management measures to protect sensitive information. By adequately investing in data security and data privacy, organizations can design systems that are compliant and successful when implementing AI in HR, as the quality of data used is crucial. The effectiveness of artificial intelligence (AI) applications within human resources is heavily influenced by the quality and quantity of data. The reliability of insights and intelligence coming from AI will be based on high quality data, which supports informed decision-making. Yet, the HR sector often faces challenges in terms of data availability, in part due to its relatively smaller workforce in comparison to other sectors. Consequently, inadequate availability of data can slow down or make the development and use of AI systems less effective. In addition, privacy, safety, and security measures are concerns when handling HR data and organizations will need to ensure data security to meet employee privacy regulations and seek the necessary compliance. Given the high stakes organizational challenges discussed, the organizations role should be clear, and the organization must prioritize data quality, security and privacy when utilizing AI for HR. AI systems require a significant amount of high-quality, diverse to identify patterns and generate accurate predictions. However, biased or insufficient data can lead to erroneous outcomes and reinforce existing disparities. For instance an AI-based recruitment tool that primarily uses data on male applicants may unintentionally discriminate against female candidates. Additionally, safeguarding data privacy and is crucial in HR. To maintain trust and comply with legal standards, organizations must ensure the protection of sensitive employee data. By placing emphasis on data quality, quantity, and security, companies can establish a strong foundation for AI-enhanced HR strategies achieve optimal outcomes.

(d) Streamlining Recruitment and Hiring

AI is revolutionizing the talent acquisition and recruitment process, leveraging AI-powered chatbots, candidate screening, job matching algorithms, and predictive analytics. This paradigm shift moves information seamlessly from résumé to digital data, which in turn increases the efficiency throughout the hiring process. AI is transforming the traditional recruiting cycle, where algorithms can assess the candidates and make recommendations to lessen HR professional's workloads, automate repetitive tasks, and decrease bias. HR professionals can redirect their energy to more impactful work, such as focusing on the employee experience, all while ensuring the best candidate is awarded the opportunity. Candidates' expectations and behaviors have changed, and with the rise of AI, automation and administrative tasks will be executed to reduce costs, and talent management will be enabled on a broader scale. The advent of this sophisticated technology has modernized and automated time-consuming procedures and improved decision-making in hiring practices. Using AI technology to analyze résumé and capture the best candidates streamlines the hiring process, which in turn, helps HR teams shift from administrative duties to strategic initiatives and employer branding. Using AI to fairly assess and evaluate candidates removes bias and increases fairness in the hiring of top candidates. Through AI-enabled analytical platforms, organizations can gain valuable insights into the labor market and recruitment process, giving organizations provide higher quality and more efficient recruiting strategies. To sum up, AI is streamlining recruiting operations, and candidate selection processes, lowering costs, accelerating hiring processes, and most importantly, improving outcomes.

12.6 Sustainable HRM Practices in Healthcare Industries

Green HRM is a relatively new concept that has gained global attention, with definitions depending on the view of the individual. It is the common thread that businesses use to conserve energy and reduce industrial and daily pollution. Ultimately, the point of going green is to minimize the environmental impact related to energy consumption and pollution. Incorporating green processes with healthcare planning can boost firm-facing realities into the twenty-first century. It stands to reason that the organization would have a substantial role in combating environmental issues caused by global warming. Under the guidance of healthcare and management, HRM can provide the industry with resolutions that will improve the environment through eco-friendly innovations. Green HRM will establish easier targets to achieve than just looking at sustainability goals.

(a) Green Recruitment and Selection

The aim of green recruitment is to proactively identify and entice environmentally conscious candidates to apply for open positions in the organization either now or in the future (Shah, 2019). This first step is critical to putting Green HRM practices into practice which are critical to drawing in young talented people. Employer branding is enhanced when businesses highlight their environmental policies on their website. Web-based hiring frequently emphasizes the company's environmental efforts and standards for prospective candidates. After reading this information job seekers are more likely to apply to these environmentally conscious businesses if it fits with their values and aspirations. Choosing applicants who possess the required green expertise entails employing a range of methods in green selection. In this way, a talent pool is generated for the purpose of selection guaranteeing that applications for the position are limited to the most qualified individuals. The significance of green selection in accomplishing environmental goals was highlighted by Yusliza et al. (2019). Hiring environmentally conscious workers from the outset also saves money by preventing needless training costs. Selection of candidates with a positive attitude toward accepting positive change and familiarity with environmental issues is advised by Renwick et al. (2013). Ercantan and Eyupoglu (2022) emphasized that the application process used in the green selection ensures that candidates fulfill the requirements set forth by the organization. It is critical to assess candidates' green values and awareness using useful tools. Interviewers should thus gauge candidates' familiarity with the environment and pose pertinent questions.

(b) Green Training and Development

Training is an essential step given that it is through this process that a firm can reveal the value of hiring inexperienced staff members (Gardas et al., 2019). Therefore, it is important to conduct a training needs analysis before training is provided. The aim of green training is to enhance the knowledge and skills of the workforce so that they can implement environmental management practices effectively at work (Islam et al., 2023). Hence, whether or not an employee's position directly involves the environment, it is important for businesses to offer green training for all employees. Further, employees will need to be informed by the business, on how to manage waste, how to use energy more efficiently, and how to build green skills. In addition, businesses should consider introducing a job rotation scheme (Faisal, 2023). The three elements of green training identified by Tang et al. (2018) are knowledge management, raising green awareness and participating in environmental protection activities. All of these elements have the potential to enhance employee performance. The inclusion of green training is critical for modern organizations seeking to operate sustainably, ethically. Firms can reduce their ecological footprint, enhance operational efficiency, and enhance their reputation by equipping employees with essential environmental knowledge and skills. By creating shared environmental values, foster an organizational culture and facilitating green training helps companies to comply with environmental legislation, reducing risks, attracting environmentally friendly consumers. The commitment to green training is itself aspirational, envisaging a sustainable business and a broader society.

(c) Green Reward System

According to Tang et al. (2018), the green incentive program should be in line with the company's environmental goals, adding to the green incentives are necessary to enhance environment performance, to encourage employees to think of green concept and to influence daily behavior of the employee (Amrutha & Geetha, 2020). In addition, management constantly reinforces staff members to be environmentally friendly, according to the social exchange theory. The green incentive system includes monetary and non-monetary incentives given to staff members to ensure their commitment to the company's environmental goals. By setting up an effective green incentive program in the company, staff members are likely to participate in environmental projects and the working efficiency is expected to improve. Furthermore, apart from financial incentives, some companies link environmental performance to salary evaluations as a way to remind employees to work on the company's green initiatives (Rayner & Morgan, 2018). The green incentives can be financial or nonfinancial. Financial rewards involve cash awards bonuses and a performance-linked incentive while non-financial rewards consist of recognition of other forms of acknowledgment (Yusliza et al., 2019). In the end, incentives serve as motivation for employees to perform as good as they can.

(d) Green Performance

To keep an eye on workers actions and results in relation to the companies' green goals, green performance management is crucial (Mishra, 2017). Assessing long-term performance becomes difficult in the absence of this system. Policies and environmental responsibilities should be included in the companies' performance evaluation process to evaluate the efficacy of its employees. Employee accountability for environmental issues is promoted and environmental performance is improved by putting these policies into place. Setting attainable and transparent green targets for their departments or divisions communicating policies and guaranteeing accountability are all critical tasks performed by managers in this process (Mousa & Othman, 2020). It is the manager's responsibility to address any deviations from the targets related to environmental management. The incorporation of green performance is crucial to promoting sustainable business practices. By incorporating environmental criteria into performance evaluations, organizations can align employee goals with broader sustainability objectives. This approach not only assesses individual contributions to environmental conservation but also nurtures a culture of ecological responsibility. When employees understand that their impacts the company's environmental goals, they are more likely to embrace sustainable practices and seek ways to improve. Moreover, green performance management generates valuable data for tracking progress towards environmental targets and identifying areas that require attention. Insights can enhance programs, optimize resource distribution, and recognize exceptional employees who demonstrate outstanding environmental leadership. In essence, green performance management plays a key role in driving continuous progress, strengthening the organization's environmental stance, and ensuring long-term sustainability.

12.7 Benefits of AI in Healthcare

By integrating artificial intelligence (AI) into the healthcare field, there is potential for altering the treatment process in an entirely new way. There are many benefits associated with artificial intelligence (AI) including enhancing patient satisfaction, eliminating errors, and automating mundane or routine functions. In addition, given the scenario, AI could actually help healthcare organizations streamline how they work daily. When evidence is taken from the focus, AI has also proved to be very beneficial in helping healthcare professionals to be more effective overall.

(a) Cost Reductions

Artificial intelligence has the potential to completely change the healthcare system from a reactive to a proactive approach. AI systems can create individualized treatment plans by forecasting possible illnesses and enhancing diagnostic precision through the analysis of enormous volumes of medical data. Additionally, AI algorithms can predict errors and assess patient-specific risks, resulting in significant cost savings for hospitals and patients. A US study reported that approximately 27% of medical appointments in the US are canceled by patients, costing hospitals up to $200 per missed visit. The development of the Missed Visit Prediction Indicator has enabled the calculation of the risk of missing appointments, potentially saving around $150 billion annually by identifying individuals at risk of missing future appointments (Ong, 2019). By anticipating potential consequences, healthcare costs can be reduced and treatment outcomes improved. AI algorithms can calculate disease predispositions, assess the impact of various factors on health, predict reactions to treatments, and provide early warnings to patients, ultimately increasing the likelihood of successful disease management.

(b) Improved Accuracy and Efficiency

There are many benefits to using AI in healthcare but one of the most obvious is increased accuracy. Through the use of sophisticated algorithms, artificial intelligence (AI) is able to quickly and accurately analyze large volumes of data from a variety of medical documents including clinical notes electronic medical records (EMRs) and scientific publications. Moreover, artificial intelligence (AI) can simplify the coding process by finding connections patterns and important information in medical texts. Platforms like linguamatics leverage NLP-based text mining solutions to analyze medical documentation effectively (Linguamatics, 2024). This NLP platform can thoroughly examine all types of medical documents, ranging from patient records to research literature and regulatory compliance materials. The insights derived from extensive data analysis accelerate clinical procedures, enhance the quality of care, and ultimately lead to improved patient outcomes. Moreover, AI significantly enhances efficiency in administrative tasks within healthcare. Health professionals often engage in repetitive activities like appointment scheduling, billing, and claim processing. AI algorithms can automate these tasks, boosting overall efficiency in healthcare operations. The fusion of precision and effectiveness brought forth by artificial intelligence signifies a groundbreaking progression in the realm of healthcare. Through the optimization of mundane duties and the thorough examination of complex medical data, AI empowers healthcare professionals to devote additional time to patient interactions and critical decision-making. This refinement in operational efficacy leads to reduced wait, decreased administrative burdens, and increased patient contentment. Furthermore, early disease detection, more precise diagnoses, and customized treatment plans are made possible by AIs skill at finding anomalies in large datasets. The combination of human knowledge and AI power creates a powerful tool for bettering patient outcomes and the delivery of healthcare services as a whole.

(c) Enhanced Patient Experience

The benefits of artificial intelligence in healthcare extend to providing round-the-clock support to patients. Chatbots can offer potential diagnoses and guidance for next steps by leveraging the right knowledge base. By gathering patient data and cross-referencing symptoms, chatbots can offer valuable insights and evaluate the urgency of symptoms effectively. An excellent illustration of this is Babylon, a chatbot by BabylonHealth that assists patients remotely. Using NLP, this chatbot collects symptoms and offers recommendations for the next course of action. Additionally, it utilizes telemedicine to provide virtual consultations with healthcare professionals. Furthermore, AI and ML solutions enhance patients' experience with Remote Patient Monitoring (RPM). Biobeat, a med-tech company, has created an AI-powered remote monitoring platform that continuously gathers data from various wearable devices (Biobeat, 2022). These wearable monitors link to a cloud-based patient management system, providing real-time data to medical staff. The platform also features an integrated customizable early warning score system that analyzes patient data to detect deterioration and offer predictive analytics. The incorporation of artificial intelligence is reshaping the patient experience by enhancing accessibility, personalization, and proactive healthcare solutions. AI-powered chatbots provide round-the-clock assistance expediting initial evaluations and recommendations to cut down on wait times and alleviate anxiety. Furthermore, by continuously monitoring vital signs and early identifying possible health problems AI-facilitated remote monitoring revolutionizes the delivery of healthcare. This proactive approach allows healthcare providers to respond quickly when necessary while also helping patients better manage their well-being. AI is significantly enhancing the entire healthcare journey strengthening the bond between patients and caregivers through increased accessibility improved patient engagement and improved care quality.

(d) Effortless Exchange of Information

AI facilitates the exchange of medical data, knowledge, and insights among healthcare professionals through various platforms and formats. Additionally, AI applications can be seamlessly incorporated into Electronic Health Records (EHRs) and wearable devices used by patients (Kung et al., 2023). Patients also reap the benefits of this streamlined information sharing process. They are able to access their medical records, review lab results, manage appointment schedules, and receive recommendations. Through platforms like MyChart, patients can securely view and share their medical information. OpenAI GPT is linked to patients' cards, ensuring that their information is constantly updated (Wójcik et al., 2023). Additionally, patients are given the chance to share their opinions about hospitals and medical professionals. PatientPop uses artificial intelligence (AI)-powered software to gather and analyze reviews. Sentiment analysis and natural language processing (NLP) are used to assess user feedback. By assigning ratings to each visitor, this system assists others in selecting a hospital based on its overall estimation. The incorporation artificial intelligence is revolutionizing how information is shared and accessed in the healthcare industry. By facilitating effective data exchange among patients, healthcare providers, and institutions, AI greatly enhances care coordination and enables well-informed decision. AI-driven platforms empower patients to actively participate in managing their journey by access to medical records, scheduling appointments, and receiving personalized recommendations. Moreover, AI analyzes patient input to identify areas for improvement, leading to quality enhancements and heightened patient contentment. This harmonious flow of information, driven by AI, fosters a collaborative and patient-centric healthcare environment.

12.8 Adoption of AI for Sustainable HRM in Healthcare Industries

The integration of AI industrial technology and sustainable HRM practices will aid in achieving these goals and objectives and help organizations strive for excellence. Moreover, understanding the workforce's behaviors in a healthcare manner enables the hospital administration to act on the employees' behaviors. As a high-touch industry, employees' behaviors play a big role in the provision of quality healthcare and caring services to patients. We offer a conceptual framework to explain the prospective use of AI in the context of sustainable HRM practices in healthcare sectors (Refer to Fig. 12.2).

(a) Accuracy

Artificial intelligence technologies have been increasingly used in human resource management in the last two decades. The current findings suggest that HR professionals can now save time and can reduce the monotonous tasks that they usually engage in, due to AI (artificial intelligence). These activities may range from scanning job applicants at the start to sustaining the retention stage. Automation enhanced the efficacy of HR operations, plus it also supports HR professionals to remove barriers for diversity and inclusion, which will improve the HR procedures or operations holistically (Hmoud & Várallyai, 2020). The accuracy and efficiency of HR procedures have all increased, because of AI. HR professionals may now focus on strategic initiatives and employee relationship building, rather than reviewing resumes, or data entering. In addition, AI can be valuable for data-based decision making where the algorithms help HR professionals, understand the data by identifying trends or

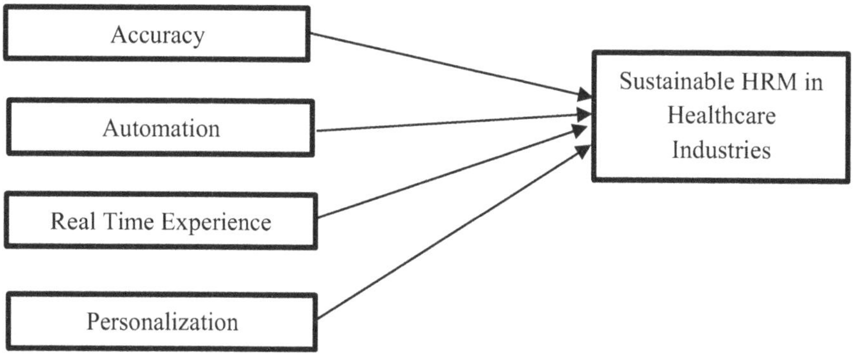

Fig. 12.2 Conceptual framework: Potential outcomes of AI adoption for sustainable HRM in healthcare industries

patterns. Specifically, AI can effectively predict an employee turning over to keep desired employees by minimizing employee turnover. Furthermore, AI algorithms can contribute to the reduction in bias when selecting candidates, and ensure, a fairer and more diverse hiring process. Ultimately, AI will empower HR professionals, by enhancing operations and contributing to company value with its speed, accurate, and objectivity.

(b) Automation

AI has shown benefits for HR professionals by automating repetitive and administrative tasks. A wide range of essential but tedious administrative tasks are frequently included in the duties of HR staff members. These tasks include posting job openings finding candidates screening setting up meetings and interviews creating schedules recording time and verifying accounts. Numerous research works have emphasized the importance of these assignments within the HR field (Nawaz & Gomes, 2019). AI could free HR professionals from tedious work and heavy paper usage in the healthcare setting which would be a huge benefit if it could be used to automate these tasks. Artificial intelligence (AI)-driven automation is transforming HR operations by increasing overall productivity and optimizing administrative tasks. AI frees up HR staff members to concentrate on strategic projects that propel company expansion by automating labor-intensive tasks like posting job ads finding candidates and setting up appointments. This change from routine tasks to more meaningful pursuits enhance job satisfaction and frees up HR departments to spend more time building rapport with employees and creating a positive work environment. Better HR performance and more informed decision-making result from automations ability to lower the possibility of human error and guarantee data accuracy. In the end, automation with AI enhancements enables HR departments to work more productively and add more value to the company.

(c) Real-Time Experience

AI chatbots offer instantaneous employee engagement and automate many HR tasks such as interview and candidate screening (Vedapradha et al. 2019). With AI, companies can access and analyze data immediately to support better decision making. AI also allows businesses to spot anomalies by giving managers real-time insights, and it enables control-decision making to address issues proactively. Real-time communication is crucial for maximizing resources, managing service delivery and driving down costs (Gopal et al. 2019). The latest technology such as IoT is allowing the digital world to be integrated seamlessly with the physical, thus providing a significant amount of real-time data. Companies can attract more applicants and speed up the hiring process by using AI-powered real-time video interviews with candidates from different locations (Thomas et al., 2020). AI-enhanced real-time chatbot analytics allows companies to respond quickly to problems and capitalize on opportunities. Proactively manage matters such as performance gaps and turnover identifying both trends and outliers with real-time AI chatbot analytics data. Device data, such as wearables and use of IoT devices in the workplace, is crucial for real-time analysis of workplace environment factors/appropriate well-being factors, which allows organizational decision making and implementing resources to be enhanced. Video interviews will expedite talent acquisition, expand the pool of potential candidates and simplify the hiring process. HR will be able to create a more nimble, responsive, and successful workplace by being able to collect, analyze, and act on real-time data.

(d) Personalization

Artificial intelligence (AI) can recognize, analyze, investigate, and perform in a customized way similar to a human (Khatri et al. by 2020). Chatbots are being used by businesses to assist candidates and employees by providing personalized advice and assistance based on their individual needs. The idea of mass personalization in staff development and training could be improved by AI. Organizations may use AI technology to determine the unique training needs of groups and individuals within their workforce and adjust their programs accordingly. AI can match employees with virtual personal mentors according to their needs. Even for complex issues AI-supported systems can provide personalized recommendations explanations or solutions to customer or employee queries in real-time by utilizing gathered data (Xu et al., 2020). HR may enable staff members to acquire the abilities and know-how required to effectively lead sustainability projects by providing them with focused learning opportunities. Artificial intelligence is bringing highly customized experiences to Human Resources revolutionizing the field. Artificial intelligence (AI) can recognize personal preferences and needs by utilizing data analysis. This allows for the creation of career development pathways and employee support. This degree of personalization encourages diversity and gives workers the tools they need to realize their greatest potential. Moreover, chatbots driven by artificial intelligence and virtual mentors offer real-time support improving communication and collaboration in the workplace. In the end, AI-enhanced personalized HR experiences promote retention engagement and satisfaction, which in turn promote organizational success.

12.9 Conclusion and Recommendation

In light of the COVID-19 outbreak, many HR processes have been disrupted, creating unknowns in human resource management. In response, have focused on studying more resilient and sustainable human resource management practices. While artificial intelligence has great potential to revolutionize healthcare, there are still limitations to its use. However, AI is already being in traditional HR practices such as talent acquisition and skills development. Embracing AI will become a key aspect of sustainable HR, requiring evaluation of HR's strategic role in healthcare and modification of existing practices to encourage innovation and change initiatives. HR teams will need to be adaptable to meet the challenges of evolving industry.

The integration of emerging technologies such as artificial intelligence (AI) with sustainable human resource management (HRM) poses significant challenges that require careful consideration. To effectively address these challenges, decision-makers must prioritize them according to their level of influence. Future research should use both qualitative and quantitative research methodologies to further analyze and prioritize these obstacles in relation to their impact on the adoption of AI for sustainable HRM in the healthcare sector. The Analytic Hierarchy Process (AHP) can be used to rank and evaluate the relative importance of each challenge by utilizing information from panels-of-matter experts, establishing an informed framework for decision-making. The results of this study can help decision-makers prioritize efforts and resources to improve the capacity of healthcare organizations to fully adopt AI. In addition, a quantitative approach can be employed to investigate the impact of AI-driven HR on turnover and success in the healthcare sector. Examining the viewpoints and experiences of healthcare workers with AI-based HR procedures can also address trust concerns surrounding AI.

References

Gollan, P. J. (2000). Human resources, capabilities and sustainability. In D. Dexter, J. Benveniste, A. Griffiths, P. Sutton (eds.), *Sustainability—The corporate challenge of the 21st century*. Allen and Unwin.

Ong & Adeline. (2019). Using randomforest to predict medical appointment no-shows. Available link <https://towardsdatascience.com/using-randomforest-to-predict-medical-appointment-no-shows-b33575e3ff42>

Amrutha, V. N., & Geetha, S. N. (2020). A systematic review on green human resource management: Implications for social sustainability. *Journal of Cleaner Production, 247*, 119131.

Bezboruah, T., & Bora, A. (2020). Artificial intelligence: The technology, challenges and applications. *Transactions on Machine Learning and Artificial Intelligence, 8*(5), 44–51. https://doi.org/10.14738/tmlai.85.8956

Biobeat. (2022). *The next generation of health-AI*. <https://www.bio-beat.com/>

Bohr, A., & Memarzadeh, K. (2020). *Artificial intelligence in healthcare* (pp. 25–60). Academic Press. The rise of artificial intelligence in healthcare applications.

Chams, N., Garcia-Blandon, J. (2019). On the Importance of Sustainable human resource management for the adoption of sustainable development goals. *Resources, Conservation and Recycling, 141*, 109–122.
Collier, M., Fu, R., Yin, L., Christiansen, P. (2017). Artificial intelligence: healthcare's new nervous system. Accenture. Available Link <https://www.accenture.com/content/dam/accenture/final/a-com-migration/manual/r3/pdf/pdf-49/Accenture-health-artificial-intelligence-j.pdf>
Dreyer, K. J., & Geis, J. R. (2017). When machines think: Radiology's next frontier. *Radiology, 285*(3), 713–718.
Ehnert, I., & Harry, W. (2012). Recent developments and future prospects on sustainable human resource management: Introduction to the special issue. *Management Revue, 23*, 221–238.
Ehnert, I., Parsa, S., Roper, I., Wagner, M., & Muller-Camen, M. (2016). Reporting on sustainability and HRM: A comparative study of sustainability reporting practices by the world's largest companies. *The International Journal of Human Resource Management, 27*, 88–108.
Ercantan, O & Eyupoglu, S. (2022). How do green human resource management practices encourage employees to engage in green behavior? Perceptions of university students as prospective employees. *Sustainability, 14*(3), 1718. https://doi.org/10.3390/su14031718
Esteva, A., Robicquet, A., Ramsundar, B., Kuleshov, V., DePristo, M., Chou, K., Cui, C., Corrado, G., Thrun, S., & Dean, J. (2019). A guide to deep learning in healthcare. *Nature Medicine, 25*(1), 24–29.
Faisal, S. (2023). Green human resource management—a synthesis. *Sustainability, 15*(3), 2259. https://doi.org/10.3390/su15032259
García-Arroyo, J., & Osca Segovia, A. (2021). Big data contributions to human resource management: A systematic review. *International Journal of Human Resource Management, 32*, 4337–4362.
Gardas, B. B., Mangla, S. K., Raut, R. D., Narkhede, B., & Luthra, S. (2019). Green talent management to unlock sustainability in the oil and gas sector. *Journal of Cleaner Production, 229*, 850–862.
George, G., Howard-Grenwille, J., Joshi, A., & Tihanyi, L. (2016). Understanding and tackling societal grand challenges through management research. *Academy of Management Journal, 59*(6), 1880.
Ghahramani, Z. (2015). Probabilistic machine learning and artificial intelligence. *Nature, 521*(7553), 452–459.
Goodfellow, I., Bengio, Y., & Courville, A. (2016). *Deep learning*. MA, MIT Press.
Gopal, G., Suter-Crazzolara, C., Toldo, L., & Eberhardt, W. (2019). Digital transformation in healthcare—architectures of present and future information technologies. *Clinical Chemistry and Laboratory Medicine (CCLM), 57*(3), 328–335. https://doi.org/10.1515/cclm-2018-0658
Hargreaves, A., & Fink, D. (2012). *Sustainable leadership*. John Wiley and Sons.
He, J., Baxter, S. L., Xu, J., Xu, J., Zhou, X., & Zhang, K. (2019). The practical implementation of artificial intelligence technologies in medicine. *Nature Medicine, 25*(1), 30–36.
Heikkinen, S., Lämsä, A.-M., & Niemistö, C. (2021). Work–family practices and complexity of their usage: A discourse analysis towards socially responsible human resource management. *Journal of Business Ethics, 171*(4), 815–831.
Hmoud, B. I., & Várallyai, L. (2020). Artificial intelligence in human resources information systems: Investigating its trust and adoption determinants. *International Journal of Engineering and Management Sciences, 5*(1), 749–765. https://doi.org/10.21791/IJEMS.2020.1.65
Hoppe, N., Härting, R. C., Rahmel, A. (2022). Potential benefits of artificial intelligence in healthcare. In C. P. Lim, A. Vaidya, Y. W. Chen, V. Jain, L. C. Jain (eds.), *Artificial intelligence and machine learning for healthcare: vol. 2: emerging methodologies and trends* (Vol. 229, pp. 225–249). Springer, Cham.
Hoppe, N., Härting, R. -C., & Rahmel, A. (2023). Potential benefits of artificial intelligence in healthcare. In *Artificial intelligence and machine learning for healthcare* (pp. 225–249). Springer.

Islam, M. A., Jantan, A. H., Yusoff, Y. M., Chong, C. W., & Hossain, M. S. (2023). Green Human Resource Management (GHRM) practices and millennial employees' turnover intentions in tourism industry in malaysia: Moderating role of work environment. *Global Business Review, 24*(4), 642–662. https://doi.org/10.1177/0972150920907000

Jainh, M. A. (2021). Comparison of nationalized and private banks' strategic human resource management practices. *J Contemp Issues Bus Gov, 27*, 709–713.

Khatri, S., Pandey, D. K., Penkar, D., & Ramani, J. (2020). *Impact of artificial intelligence on human resources bt—Data Management, analytics and innovation* (pp. 365–376). Springer Singapore.

Klein, G., Woods, D. D., Bradshaw, J. M., Hoffman, R. R., & Feltovich, P. J. (2004). Ten challenges for making automation a "Team Player" in joint human-agent activity. *IEEE Intelligent Systems, 19*(6), 91–95.

Klien, G., Woods, D. D., Bradshaw, J. M., Hoffman, R. R., & Feltovich, P. J. (2004). Ten challenges for making automation a "team player" in joint human-agent activity. *IEEE Intelligent Systems, 19*(6), 91–95.

Korteling, J. E., Brouwer, A.-M., & Toet, A. (2018). A neural network framework for cognitive bias. *Frontiers in Psychology*, 1561.

Korteling, J. H., van de Boer-Visschedijk, G. C., Blankendaal, R. A., Boonekamp, R. C., & Eikelboom, A. R. (2021). Human versus artificial intelligence. *Frontiers in Artificial Intelligence*, 4, Article ID: 622364. https://doi.org/10.3389/frai.2021.622364

Kramar & Robin. (2014). Beyond strategic human resource management: Is sustainable human resource management the next approach? *The International Journal of Human Resource Management, 25*, 1069–1089.

Kung, T. H., Cheatham, M., & Medenilla, A. (2023). Performance of ChatGPT on USMLE: Potential for AI-assisted medical education using large language models. *PLOS Digit Health, 2*(2), e0000198. https://doi.org/10.1371/journal.pdig.0000198

Linguamatics (2024). Accurate, scalable & trusted healthcare NLP. Available link <https://www.linguamatics.com/products/nlp-natural-language-processing>

Lu, Y. (2019). Artificial intelligence: A survey on evolution, models, applications and future trends. *Journal of Management Analytics, 6*(1), 1–29. https://doi.org/10.1080/23270012.2019.1570365

Lucio, M. M., & Stuart, M. (2011). The state, public policy and the renewal of HRM. *International Journal Human Resource Management, 22*(18), 3991–3671.

Mariappanadar & Sugumar. (2012). The harm indicators of negative externality of efficiency focused organizational practices. *International Journal of Social Economics, 39*, 209–220.

Marler, J. H., & Parry, E. (2016). Human resource management, strategic involvement and e-HRM technology. *International Journal of Human Resource Management, 27*, 2233–2253.

Mazur & Barbara. (2015). Sustainable human resource management. The attempt of holistic approach. *Ekonomia i Zarządzanie, 7*, 7–12.

Mishra, P. (2017). Green human resource management: A framework for sustainable organizational development in an emerging economy. *International Journal of Organizational Analysis, 25*(5), 762–788. https://doi.org/10.1108/IJOA-11-2016-1079

Moravec, H. (1998). When will computer hardware match the human brain. *Journal of Evolution and Technology, 1*(1), 10.

Mousa, S. K., & Othman, M. (2020). The impact of green human resource management practices on sustainable performance in healthcare organisations: A conceptual framework. *Journal of Cleaner Production, 243*, 118595. https://doi.org/10.1016/j.jclepro.2019.118595

Nawaz, N., Arunachalam, H., Pathi, B.K., & Gajenderan, V. (2024). *The adoption of artificial intelligence in human resources management practices.*

Nawaz, N., Gomes, A.M. (2019). Artificial intelligence chatbots are new recruiters. International *Journal of Advanced Computer Science and Applications (IJACSA), 10*(9). https://doi.org/10.14569/IJACSA.2019.0100901

Khanijahani, A., Iezadi, S., Dudley, S., Goettler, M., Kroetsch, P., & Wise, J. (2022). Organizational, professional, and patient characteristics associated with artificial intelligence adoption in

healthcare: A systematic review. *Health Policy and Technology, 11*(1). https://doi.org/10.1016/j.hlpt.2022.100602
De Prins, P. (2011). Duurzaam HRM: Synthetische academische introductie. J. Retrieved, Cited by: Rompa I. *Explorative Research on Sustainable Human Resource Management*. http://www.innovatiefinwerk.nl/sites/.../sustainable_hrm.pdf.downloaded
Rajaraman, V. (2014). JohnMcCarthy—Father of artificial intelligence. *Resonance, 19*, 198–207.
Rayner, J., & Morgan, D. (2018). An empirical study of 'green' workplace behaviours: Ability, motivation and opportunity. *Asia Pacific Journal of Human Resources, 56*(1), 56–78. https://doi.org/10.1111/1744-7941.12151
Renwick, D. W. S., Redman, T., & Maguire, S. (2013). Green human resource management: A review and research agenda. *International Journal of Management Reviews, 15*, 1–14.
Shah, M. (2019). Green human resource management: Development of a valid measurement scale. *Bus Strat Environ, 28*, 771–785.
Shneiderman, B. (2020). Design lessons from AI's two grand goals: Human emulation and useful applications. *IEEE Transactions on Technology and Society, 1*(2), 73–82.
Stahl, G. K., Brewster, C. J., Collings, D. G., & Hajro, A. (2020). Enhancing the role of human resource management in corporate sustainability and social responsibility: A multi-stakeholder, multidimensional approach to HRM. *Human Resource Management Review, 30*(3), 100708.
Business Standard. (2024). *AI platform market for healthcare industry to reach $4.3 bn in 2024: Report*. <https://www.business-standard.com/article/economy-policy/ai-platform-market-for-healthcare-industry-to-reach-4-3-bn-in-2024-report-123032000965_1.html>
Sunarti, S., Rahman, F. F., Naufal, M., Risky, M., Febriyanto, K., & Masnina, R. (2021). Artificial intelligence in healthcare: Opportunities and risk for future. *Gaceta Sanitaria, 35*, S67–S70.
Tang, G., et al. (2018). Green human resource management practices: Scale development and validity. *Asia Pacific Journal of Human Resources, 56*(1), 31–55. https://doi.org/10.1111/1744-7941.12147
Thomas, S., Kureshi, S., Suggala, S., Mendonca, V. (2020). HRM 4.0 and the Shifting Landscape of employer branding. In P. Kumar, A. Agrawal, P. Budhwar (Eds.) *Human & Technological Resource Management (HTRM): New insights into revolution 4.0* (pp. 37–51). Emerald Publishing Limited.
Van Hartskamp, M., Consoli, S., Verhaegh, W., Petkovic, M., & Van de Stolpe, A. (2019). Artificial intelligence in clinical health care applications. *Interactive Journal of Medical Research, 8*(2), e12100.
Vedapradha, R., Hariharan, R., & Shivakami, R. (2019). Artificial intelligence: A technological prototype in recruitment. *Journal of Service Science and Management, 12*(3)
Wagner & Marcus. (2015). A European perspective on country moderation effects: Environmental management systems and sustainability-related human resource benefits. *Journal of World Business, 50*, 379–388.
Wilkinson, A., Hill, M., & Gollan, P. (2001). The sustainability debate. *International Journal of Operations and Production Management, 21*, 1492–1502.
Wójcik, S., Rulkiewicz, A., Pruszczyk, P., Lisik, W., Poboży, M., & Domienik-Karłowicz, J. (2023). Beyond ChatGPT: What does GPT-4 add to healthcare? The dawn of a new era. *Cardiol J, 30*(6), 1018–1025. https://doi.org/10.5603/cj.97515. Epub
Xu, Y., Shieh, C.-H., van Esch, P., & Ling, I.-L. (2020). AI customer service: Task complexity, problem-solving ability, and usage intention. *Australasian Marketing Journal, 28*(4), 189–199. https://doi.org/10.1016/j.ausmj.2020.03.005

Yusliza, M.-Y., Norazmi, N. A., Jabbour, C. J. C., Fernando, Y., Fawehinmi, O., & Seles, B. M. R. P. (2019). Top management commitment, corporate social responsibility and green human resource management. *Benchmarking Int J, 26*, 2051–2078.

Zarringhalam, K., Enayetallah, A., Reddy, P., & Ziemek, D. (2014). Robust clinical outcome prediction based on bayesian analysis of transcriptional profiles and prior causal networks. *Bioninformatics, 30*(12), i69–i77.

Zaugg, R. J., Blum, A., & Thom, N. (2001). Sustainability in human resource management: Evaluation report. IOP-Press, Bern

Chapter 13
AI in Healthcare Education: A Systematic Review of Applications in Teaching and Learning

Venkat Bakthavatchaalam and Karthik Sivasankar

Abstract Artificial Intelligence (AI) has emerged as a transformative force in healthcare, showing a significant potential and requirement for advancements and adoption in healthcare education. This chapter conducts a systematic review of papers which have empirically analysed the how AI has been applied across the healthcare education. Owing to the fast pace of the field, the review considers papers written between 2018 and 2024. After screening 4,653 papers, 40 papers were finalised for the review. The key areas where AI is applied include teaching support, evaluation of training, assessments, feedback and personalising education. Studies also explored the effectiveness of AI through qualitative and quantitative methods and controlled experiments. Results show a growing belief in the use of AI in healthcare education. The chapter discusses the types of AI used, the fields, the methodology of implementation and the user experiences. Also, this chapter discusses the implications of AI adoption for future policymakers, showing the opportunities and challenges associated with the widespread implementation of AI technologies in healthcare education. Factors such as workforce training, data privacy, and regulatory frameworks are examined, considering the evolving field of AI-enabled healthcare education. By critically evaluating the results and insights from the reviewed papers, this chapter aims to inform policy decisions aimed at creating a responsible and effective curriculum, thus contributing to Sustainable Development Goals 3 and 4: Health and Good Health and Wellbeing and Quality Education.

Keywords Healthcare education · Artificial intelligence · Teaching · Learning · Industry 4.0

V. Bakthavatchaalam (✉)
Cardiff University, Cardiff, UK
e-mail: bakthavatchaalamv@cardiff.ac.uk

K. Sivasankar
University of Oslo, Oslo, Norway

R. Singh et al. (eds.), *Transforming Healthcare Sector Through Artificial Intelligence and Environmental Sustainability*, Approaches to Global Sustainability, Markets, and Governance, https://doi.org/10.1007/978-981-97-9555-0_13

13.1 Introduction

The integration of AI in healthcare education represents a significant change in teaching and learning methodologies within the medical domain (Garcia et al., 2024). AI, with its range of applications including Machine Learning (ML), Natural Language Processing (NLP), and robotics, provides a multitude of opportunities to enhance education outcomes, personalise learning experiences for individuals, and improve operational efficiencies across educational settings.

The significance of AI in the context of healthcare education is highlighted by the ever-growing demand for innovative training methods that are both accessible and effective. Applications of AI in this field are varied, ranging from virtual patient simulations that offer realistic clinical training scenarios to adaptive learning systems designed to adjust content delivery based on the learner's pace and level of understanding. Such tools are not only technologically advanced but also aim to substantially improve the quality of healthcare education by making it more interactive, engaging, and tailored to the needs and requirements of both students and professionals (Samaan et al., 2023).

The potential of AI to elevate healthcare education lies in its capacity to facilitate indivudualised learning experiences, enhance skill acquisition, and ensure that healthcare professionals are informed about the latest medical advancements (Narayanan et al., 2023; Gligorea et al, 2023). The healthcare students in several countries have expressed a strong interest in integrating AI into their curriculum and anticipate it to have a positive impact on various medical specialties, including radiology, surgery, and general medicine (Rainey et al., 2021; Sharma et al., 2023; Wong et al., 2021). All these emphasise the necessity for educational support to foster successful AI adoption in clinical practice.

While the integration of AI into healthcare education is promising in terms of improving educational outcomes and operational efficiencies, concerns have been raised about potential overreliance on AI technologies, loss of empathy, job displacement risks, and issues related to data privacy. Therefore, it is imperative to approach the integration of AI tools with caution, ensuring ethical and responsible use to address these challenges and limitations effectively (Al-Dhaen et al., 2023; Dave & Patel, 2023; Jeyaraman et al., 2023; Ueda et al., 2024).

Given these considerations, this systematic review examines the integration of AI technologies in healthcare education, focusing particularly on their influence on teaching and learning practices. The objectives are to identify the variety of AI technologies being incorporated in healthcare education, evaluate their effectiveness in improving learning outcomes and explore the challenges and limitations associated with their application. The review aims to contribute to a comprehensive understanding of how AI technologies are reshaping healthcare education, highlighting the benefits while also exploring the path towards effective and ethical integration of AI into educational strategies.

13.2 Background

In healthcare education, the adoption of AI technologies such as ML and NLP has marked a significant advancement (Wartman & Combs, 2018; Wiljer & Hakim, 2019). These technologies enable the development of adaptive learning systems and virtual patient simulations, which are crucial for personalising educational content, optimising learning pathways, enhancing the interaction between learners and automated assessment tools through natural human language understanding and generation (Dave & Patel, 2023; Nagi et al., 2023).

The integration of AI into healthcare education is grounded in established theoretical frameworks, constructivist theory and the theory of connectivism. The constructivist theory views learning as an active, constructive process wherein individuals build new knowledge based on existing knowledge (Dave & Patel, 2023; Komasawa & Yokohira, 2023). This principle is embodied in AI-enhanced educational tools that offer context-aware learning experiences, thus adapting to the learner's progress and understanding. Meanwhile, the theory of connectivism emphasises the importance of networks in the learning process (Goldie, 2016), aligning with AI's capability to analyse vast amounts of data from various sources to customise and improve educational experiences.

Together, these AI technologies and theoretical underpinnings provide a framework for future healthcare education. By leveraging the capabilities of ML and NLP, education institutions can offer more engaging, effective, and tailored learning experiences. This approach not only aligns with the dynamic nature of the medical field but also equips future healthcare professionals with the knowledge and skills necessary to meet the sector's evolving demands.

13.3 Methodology

The methodology of this systematic review was designed to collate and evaluate the current literature on the application of AI in medical and healthcare education. Central to the review's methodology was adherence to the Preferred Reporting Items for Systematic Reviews and Meta-Analyses (PRISMA) guidelines. This approach aligns with several notable studies in the healthcare field, which have similarly employed PRISMA to structure their systematic reviews and meta-analyses, covering a range of topics from cost-effectiveness in healthcare management to patient satisfaction and the impact of healthcare 4.0 digital technologies on the quality of care (Campra et al., 2021; Ferreira et al., 2023; Marbouh et al., 2023; Nagi et al., 2023; Yeow et al., 2023). A three-step approach was taken for the review: Identification, screening, and inclusion (Fig. 13.1).

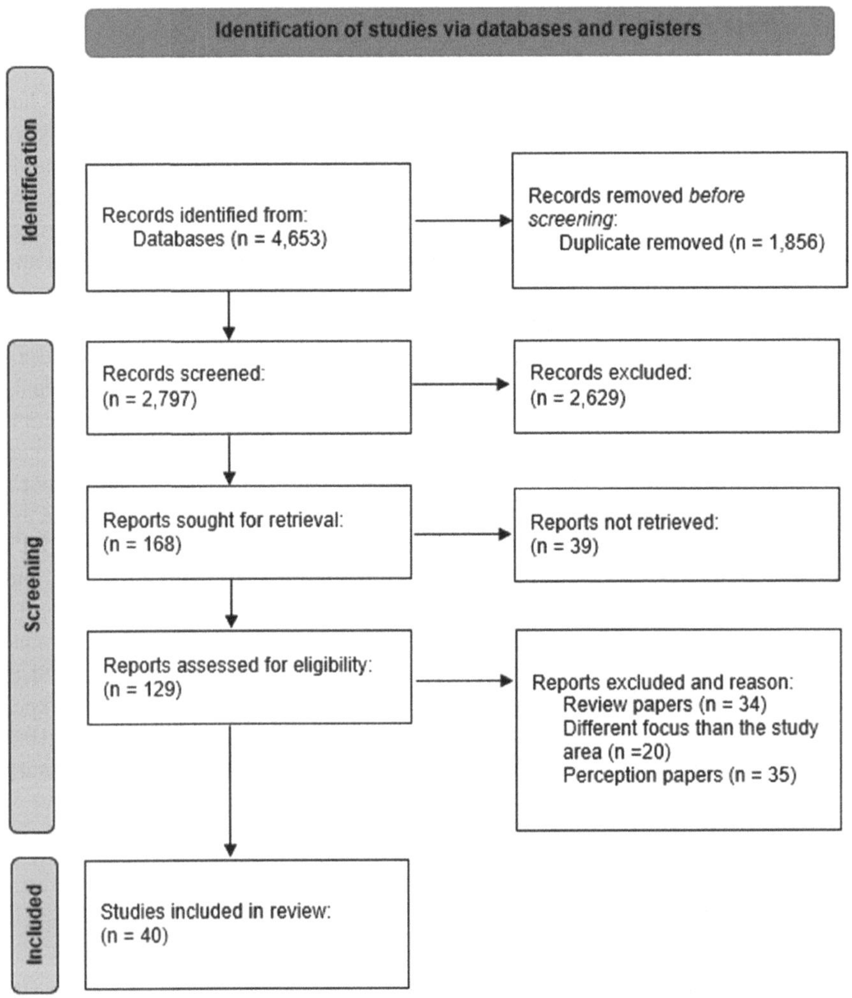

Fig. 13.1 Systematic literature review process steps. *Source* Authors

13.3.1 Identification Phase

The initial phase involved an extensive search of academic papers published in English from January 2018 to March 2024. This timeframe was selected to capture the most recent advancements and applications of AI within the context of healthcare education. The databases selected for the search included major repositories of scientific literature such as Scopus, IEEE Xplore, PubMed, and Web of Science. These databases were chosen due to their extensive coverage of both medical and AI/engineering disciplines, ensuring a comprehensive collection of literature. A targeted combination of keywords was employed to maximise the search's specificity and sensitivity. These keywords included "AI applications," "Higher Education," "Medical, Healthcare Education," "AI in medical Curriculum," and "Machine Learning," among others. The search strategy was tailored to each database to accommodate for different search algorithms and indexing terms.

13.3.2 Screening and Inclusion Phase

Following the identification phase, a two-step screening process was implemented. Initially, titles and abstracts were screened to exclude irrelevant papers, these include conference papers, posters, non-peer-reviewed articles, opinion pieces, surveys exploring the potential for AI implementation without actual application and studies not explicitly involving AI or ML as an intervention in healthcare education. This step ensured that only papers with a direct focus on the application of AI in healthcare education were considered for further review.

The inclusion criteria were rigorously defined to further refine the selection of papers. Papers were eligible for inclusion if they were published within the specified timeframe, indexed in Scopus, IEEE Xplore, PubMed, or Web of Science, written in English, peer reviewed and explicitly addressed the application of AI. Furthermore, only empirical research studies, including experimental, observational and case studies, were considered. This inclusion criterion was pivotal to ensure that the review focused on studies providing tangible evidence and insights into the application of AI in healthcare education.

13.3.3 Data Extraction and Analysis

For the papers meeting these inclusion criteria, the authors independently conducted data extraction to ensure accuracy and comprehensiveness. Extracted data included the study's objectives, methodology, AI technologies employed, educational settings, findings, and identified enablers and barriers to AI implementation. The analysis phase involved synthesising the extracted data to address the review's research questions. Quantitative data were summarised statistically, while qualitative findings were thematically analysed to identify patterns and insights regarding the application of AI in healthcare education.

Figure 13.1 pictorially represents the process and the number of papers. During the identification phase, a total of 4,653 records were identified from databases. Prior to screening, duplicates were removed, which accounted for 1,856 records. During screening, 2,797 records were examined. Subsequently, 2,629 records were excluded for eligibility reasons specified in this phase. In the eligibility assessment phase, 168 reports were sought for retrieval. Out of these, 39 reports were not retrieved as they were not able to be retrieved. The remaining 129 reports were assessed for eligibility. This assessment led to the exclusion of 89 reports, with the reasons for exclusion being: 34 were review papers, 20 had a different focus than the study area intended, and 35 were perception papers. Finally, 40 studies met the eligibility criteria and were included in the review.

13.4 Results and Analysis

The review provides an extensive examination of various facets, including publication year, geographical distribution of studies, research areas, AI technologies implemented, their educational applications, and the contexts within which AI is utilised in healthcare education. Thus the analysis enables a nuanced understanding of AI's integration into healthcare education, setting the stage for identifying trends, recognising gaps in the literature, and for deriving actionable insights for policymakers.

13.4.1 Distribution of the Studies

13.4.1.1 Geographical Distribution of the Studies

The distribution of studies presents a global snapshot of AI research in healthcare education (Fig. 13.2), showcasing the breadth of international engagement in this field. United States emerged as a significant contributor, with twelve studies, followed by Canada with four contributions. It shows the interests within North American countries in advancing AI applications within healthcare education. Europe's research landscape also featured prominently, with Germany contributing four studies and additional papers from Romania and the United Kingdom, cumulatively representing six European studies. The Asian research contribution was marked by China (three papers), Taiwan (two papers), and India (one paper), highlighting Asia's growing involvement in this area. Australia's contribution of two papers further indicates a wider global interest in AI's educational potential.

The inclusion of studies from other countries, each contributing a single paper, shows the budding interest and diversity in AI research within healthcare education. This geographical analysis not only delineates the research hotspots but also emphasises the importance of incorporating global perspectives and cross-cultural considerations in developing healthcare education curricula and research strategies. Such a broad overview is instrumental in developing pathways for future research collaborations and policy formulations that transcend regional boundaries. It is important to note that with the rapid adoption of AI within education system, more research needs to be conducted across the geographical boundaries to ensure that regional, cultural and varied regulations are considered.

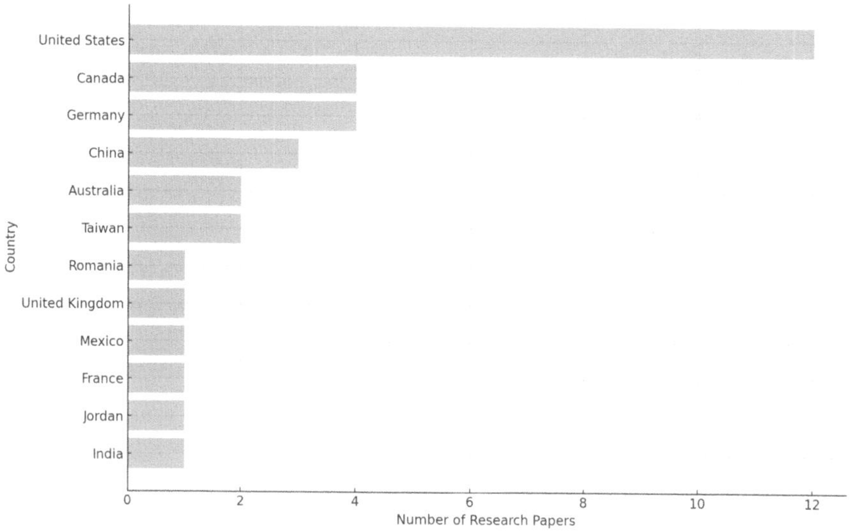

Fig. 13.2 Number of research papers by countries. *Source* Authors

13.4.1.2 Distribution of the Number of Papers by Year

The analysis of publication years traces the locus of interest and scholarly work in this new domain. Figure 13.3 illustrates the increasing interest and scholarly output in this area over the past seven years up to March 2024, evidencing a marked increase in the number of research. The data reveals an upward trajectory in the number of papers, with a total of 12 papers published in 2023 alone. The early months of 2024 have already witnessed the publication of four papers, indicating a continuing or even accelerating trend in research output.

This surge in scholarly interest shows the increasing recognition of AI's potential to change healthcare education-a trend further accelerated by the COVID-19 pandemic. The pandemic's impact, which necessitated a swift transition to online modes of education, has, as noted by Richardson and Ojeda (2022), Furlan et al. (2021), and Kaur et al. (2021), significantly propelled the adoption of AI within educational frameworks. The observed increase in research output, particularly in the wake of the pandemic, highlights the critical role of AI in facilitating adaptable and effective educational experiences. This growing body of literature reflects the academic community's response to contemporary challenges and suggests a robust and evolving discourse on integrating AI into healthcare education. Figure 13.3, therefore, not only documents this academic engagement but also signals the continued expansion of this research area, with implications for both current practices and future directions in the field.

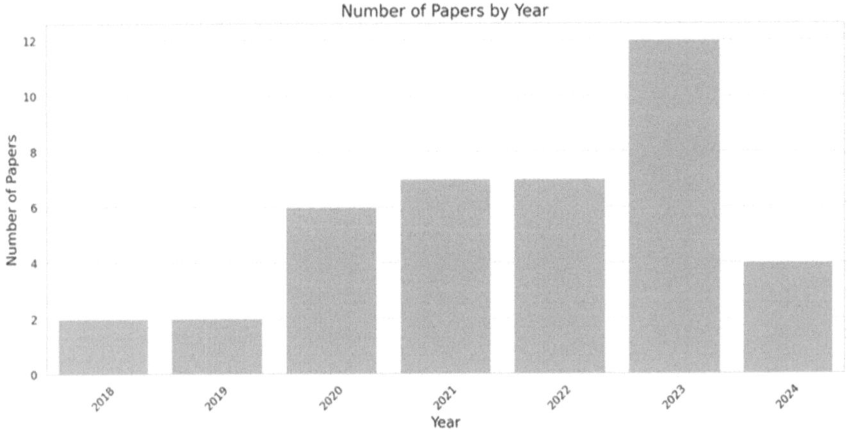

Fig. 13.3 Distribution of the number of papers by year. *Source* Authors

13.4.1.3 Distribution of Research Area

The review shows the varied scope of research areas where AI is being applied within healthcare education. From Fig. 13.4, the distribution of research categories reveals a predominant focus on medical education, with thirteen studies highlighting its central role in integrating AI technologies. Surgical practices and radiology each account for eight studies, indicating significant interest in applying AI to enhance procedural training and diagnostic skills. Clinical diagnostics using AI and ML represent emerging areas, with four and three studies respectively, suggesting an evolving interest in harnessing AI for more nuanced medical analyses and educational applications. Ophthalmology has one paper, and the 'other' categories have three papers. This distribution shows the breadth of AI's application in healthcare education, from broad-based medical training to specific clinical skills.

13.4.1.4 Distribution Based on the Types of Studies

This section provides an overview of the range of strategies employed by researchers to integrate AI into healthcare education. Winkler-Schwartz et al. (2019) use ML algorithms to distinguish levels of surgical expertise within virtual reality simulations, achieving notable accuracy. Bissonnette and Mirchi (2019) further validated the effectiveness of AI in assessing surgical skills through comparative analysis, successfully distinguishing between novice and experienced surgeons. Extending the application of AI to neurosurgical training, Ledwos et al. (2022), Lam et al. (2022) and Natheir et al. (2023) employed ML techniques alongside Electroencephalography (EEG) data to evaluate cognitive skills, showcasing the versatility of AI in medical education. Monlezun et al. (2018) demonstrated the integration of ML with

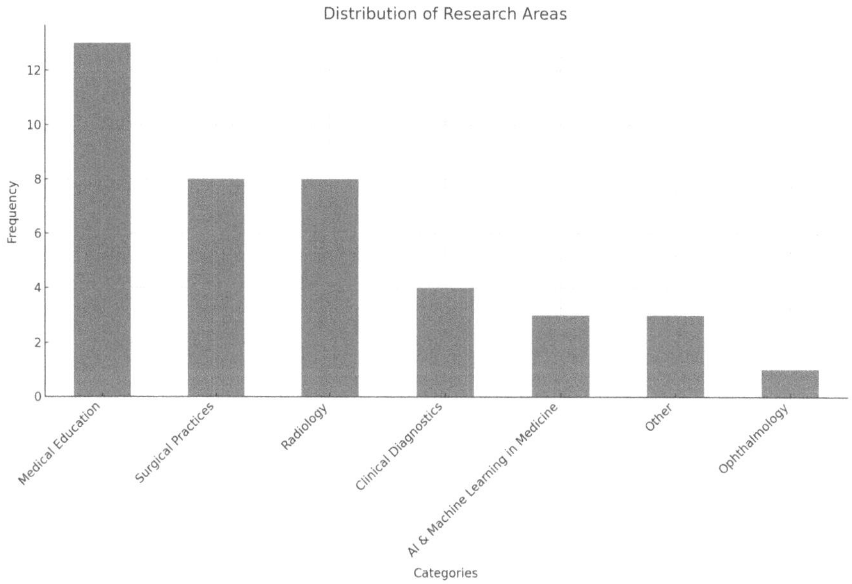

Fig. 13.4 Distribution of research areas of AI application. *Source* Authors

causal inference statistics to evaluate educational interventions in preventive cardiology, highlighting AI's potential in different medical domains. Jin et al. (2023) introduced an educational graphical user interface in Python aimed at bridging the gap between theoretical knowledge and practical application for radiology trainees.

Survey and perception studies by Sallam et al. (2023) and Kaur et al. (2021) explored attitudes towards AI in healthcare education, indicating an evolving role of AI in shaping medical training and assessment methodologies. Innovative development studies by Muntean et al. (2023), Mirchi et al. (2020), and Aldeman et al. (2021) explored tailored AI frameworks and platforms designed for personalised case-based learning, simulation-based learning feedback, and interactive learning environments, respectively, underscoring AI's potential within the different areas of medical education.

Curriculum integration studies by Hu et al. (2023), Lindqwister et al. (2021), Wiggins et al. (2021), and Richardson and Ojeda (2022) accentuate the integration of AI workshops and curricula to enhance understanding of AI fundamentals, AI literacy, and practical experience in developing and refining deep learning models for medical imaging, reflecting a shift towards more efficient, and objective educational methodologies.

Additionally, Otles et al. (2021) and Solano et al. (2021) trained and assessed NLP models to automatically classify the quality of surgical trainee feedback, while Nagaraj et al. (2023) explored the use of Convolutional Neural Network (CNN) models for video-based assessment and coaching of students in surgical skills. Ryder et al. (2024) utilised AI to gauge the competency of trainees in laparoscopic surgery, predicting student scores based on computer-generated metrics derived from video input, potentially aiding training for students and professionals in resource-limited settings. Lippitsch et al. (2024) developed and evaluated a software system for medical students to practice anamnestic interviews with virtual patient avatars, providing feedback on question completeness and essential items.

These studies show the progressive integration of AI in medical education, highlighting its role in enhancing training, assessment and teaching methodologies. From using ML algorithms for precise assessments to developing innovative educational tools and integrating AI into curricula, the evidence suggests a shift towards more adapted, efficient, and objective educational methodologies.

13.4.2 Applications of AI in Healthcare Education

Broadly categorising the selected papers, three areas emerged where AI has been applied in healthcare education. These include using AI for teaching and learning, training, and assessment. This section details these areas.

13.4.2.1 AI as a Tool for Curriculum and Teaching

The incorporation of AI in healthcare education represents a significant shift towards enhancing learning outcomes and assessment methodologies through technology-enhanced frameworks. This evolution is evidenced by various initiatives that collectively show AI's transformative potential in medical teaching and competency assessment.

On integrating AI into medical curricula, Lang and Repp (2020) highlight the benefits of interdisciplinary learning. Their work demonstrates the effectiveness of AI in fostering an enriched understanding among students from various fields, facilitating greater engagement and interdisciplinary dialogue. This initiative serves as an argument for the structured inclusion of AI in educational frameworks, emphasising its role in promoting a research-oriented learning environment. Building on this foundation, Hu et al. (2023) provide a concrete example of AI's application in diagnostic radiology residency training. By integrating an AI workshop into the radiology residency curriculum, they significantly enhanced residents' understanding of AI fundamentals. The initiative's success, measured through improvements in residents' confidence in AI topics post-workshop, showcases the dual potential of AI in teaching and assessing medical competencies.

Similarly, Lindqwister et al. (2021) developed an AI curriculum specifically for radiology residents at Dartmouth College, aimed at enhancing the understanding and application of AI in medical diagnostics. By combining foundational AI algorithms and ML techniques with practical engagement through lectures and journal clubs, they achieved an enhancement of residents' AI literacy. This approach not only enabled an intuitive understanding of AI among residents but also reportedly bolstered their confidence in critically evaluating relevant scientific literature, emphasising the necessity of AI literacy in radiology training programs. Krive et al. (2023) further extend the discourse on AI in medical education through their work at the University of Illinois. They introduced a competency-driven AI course, employing constructivist and backward design principles to craft online assignments that reflect real-world healthcare tasks. The significant enhancements in student knowledge and skills, evaluated through quizzes and project-based assessments, are reported to show the efficacy of AI as a progressive educational tool for assessment.

Park et al. (2024) investigate the integration of AI into medical education, focusing on enhancing educational assessment. Their study, conducted within a Digital Health Scholars, incorporates AI algorithms into clinical topics. Results indicate significant improvements in students' confidence regarding AI's applications and limitations, highlighting AI's critical role in preparing future healthcare professionals within a digital healthcare spectrum. Aldeman et al. (2021) contribute to AI incorporation in medical education with the development of SmartPathk, a ML-based platform tailored for renal pathology education. Their work demonstrates how SmartPathk increases the training capacity of new medical professionals in this field by enabling dynamic learning through student interactions, including quizzes, image analysis, and access to supplementary material. The notable success of SmartPathk highlights the significant potential of AI to serve as an invaluable asset in medical education. By enhancing the accessibility and efficiency of learning processes, AI-driven tools like this are poised to better the training and development of medical professionals, enabling a deeper understanding and proficiency in specialised areas of study.

Within radiology education, the introduction and application of deep learning (DL) is seen to have made a leap. Wiggins et al. (2021) investigate this by developing a 45-min hands-on introductory module. This module, designed for radiology trainees, is hosted on a web-based development environment and focuses on imparting a practical understanding of DL in medical imaging. Trainees are taken through the nuances of training a DL model to distinguish between chest and abdominal radiographs. The effectiveness of this module in closing the knowledge gap on DL among radiology trainees is notable, suggesting its potential applicability and benefit across a variety of medical specialities.

Building on this, Richardson and Ojeda (2022) further explore radiological education by creating the "Bumper-Car" curriculum. This innovative program is tailored to equip radiology residents with the necessary skills to develop, train, evaluate, and refine DL models, specifically focussing on radiology applications. By adopting a "no-code ML system" such as Lobe.ai, the curriculum prioritises practical experience over theoretical instruction. This methodological shift not only enhances the learners' ability to handle a variety of image datasets but also deepens their understanding of and interest in DL applications within the field of radiology. This coherent approach across different initiatives demonstrates the pivotal role of DL in enhancing radiology education, illustrating a seamless integration of theory and practice to foster a deeper comprehension and expertise in the field.

These initiatives collectively advocate for the strategic integration of AI into medical education. By enhancing both teaching methodologies and assessment practices, AI not only fosters an enriched learning environment but also equips future healthcare professionals with the essential competencies needed to understand and work with an increasingly digital healthcare industry.

13.4.2.2 AI as a Tool for Teaching and Learning

Integrating AI into medical training represents a significant evolution in educational methodologies, particularly highlighted by the challenges and opportunities presented during COVID-19 pandemic. The results show the sophisticated roles that AI can play in enhancing medical education across different specialisations.

Looking at these studies, the work by Kaur et al. (2021) explores the application of digital chatbots within medical education amidst the global health crisis. Their qualitative analysis, informed by the perceptions of medical students and academic staff, uncovered the potential of chatbots to mimic clinical scenarios, help revision processes, and deliver bespoke feedback. At the same time, this study also brought to light the challenges in adopting and integrating these digital tools into established education frameworks, laying the path for future research aimed at refining chatbots' functionality and integration in medical training. Looking into the digital tool exploration in medical education, Jin et al. (2023) introduced an educational graphical user interface (GUI) developed in Python, targeted at radiology trainees who lacked prior experience in AI coding. This GUI served as a bridge between theoretical knowledge and practical application, enabling trainees to construct and understand CNNs. The project not only enabled the understanding of complex ML and DL concepts but also highlighted the tangible applications of these technologies in radiology. The positive reception of the GUI, as indicated by high usability scores, reveals its effectiveness as a pedagogical tool, further showing AI's role in the future of radiology education.

Extending the application of AI into specialised medical training, Muntean et al. (2023) developed a tailored AI framework for ophthalmology residency programs. This framework, powered by a deep learning model trained on publicly accessible datasets of colour fundus photographs, enabling case-based learning through an expert-system algorithm. This algorithm optimises learning paths by assigning cases based on the residents' performance history and educational needs, thereby offering a customised training experience that aligns with individual learning objectives.

Looking at surgical education innovation, Mirchi et al. (2020) introduced the Virtual Operative Assistant, underpinned by explainable AI, to provide adapted feedback in simulation-based learning. This approach not only bridged the proficiency gap between novices and experts but also was reported to foster an engaging training environment. The significance of AI in surgical training is further evidenced by Alonso-Silverio et al. (2018), who developed an AI-enhanced laparoscopic box trainer to objectively assess surgical psychomotor skills with high accuracy. These studies collectively highlight AI's role in refining surgical education by offering precise feedback and realistic training simulations.

Expanding AI's application to radiology, Shiang et al. (2022) explore the implementation of AI-based Decision Support Systems in radiology education, revealing substantial support among radiology residents for integrating AI into their curriculum. Although the system's impact on diagnostic speed was modest, its utility in triaging and troubleshooting was significant, underscoring the need for early AI exposure in residency programs.

In the domain of diagnostic accuracy, Cheng et al. (2020) demonstrate the efficacy of AI in improving medical students' ability to interpret hip fractures using the AI-assisted "HipGuide" system. Similarly, in ophthalmology, Fang et al. (2022) and Han et al. (2022) investigate AI's effectiveness in enhancing the diagnostic skills of residents and students in identifying pathologic myopia and grading diabetic retinopathy, respectively. These studies collectively affirm the potential of AI to augment medical education by significantly improving diagnostic accuracy through targeted interventions. In addition, AI's applicability extends beyond clinical skills training to include educational content delivery and assessment. Hedderich et al. (2021) designed an online curriculum focused on AI applications in medical imaging, fostering a nuanced understanding of AI among medical professionals. Hsu et al. (2023) introduced "Termbot," a chatbot that employs gamification to enhance medical terminology learning, illustrating the innovative use of AI in engaging learners.

Innovations in teaching methodologies have also been highlighted by Joo et al. (2023) and Lippitsch et al. (2024), who assessed the impact of AI and clinical guidelines on medical students' proficiency in identifying heart failure and the effectiveness of ViPATalk, an interactive software for training in anamnestic interviews, respectively. Ryder et al. (2024) and Nagaraj et al. (2023) further demonstrate AI's utility in surgical skill assessment through AI-generated metrics and video-based evaluations, respectively. These approaches show the accuracy and effectiveness of AI in enhancing assessment and feedback mechanisms, offering a promising avenue for educational advancement in low and middle-income countries.

13.4.2.3 AI as a Tool for Assessment

The integration of AI as a tool for assessment and feedback was seen as a major development. Studies looked at training the algorithms with the existing data and using them as an assessment tool for healthcare students and trainees in several fields.

Beginning with Winkler-Schwartz et al. (2019), the utility of AI in differentiating surgical expertise through ML algorithms, such as K-nearest neighbour and support vector machine, was illustrated. Their study within virtual reality simulations highlighted a remarkable initial accuracy of up to 90% in classifying various levels of surgical experience. This research laid the groundwork for using AI in surgical education, emphasising its potential to offer targeted feedback based on nuanced performance metrics. Monlezun et al. (2018) expanded AI's application to evaluating educational interventions. By employing ML in conjunction with causal inference statistics, the study provides deep insights into the impacts of curriculum modifications on preventive cardiology, analysing data from a few thousand medical trainees. This showcased AI's versatility and its power in dissecting large datasets to evaluate educational outcomes. Bissonnette et al. (2019) further validated AI's efficacy in surgical skill assessment. Their work in virtual reality spinal surgery simulations managed to differentiate novice from experienced surgeons with a staggering 97.6% success rate. This research not only confirmed AI's accuracy in skill classification but also highlighted its role in standardising surgical training assessments.

The capabilities of AI in medical education were further evidenced by Huang et al. (2023), who demonstrated GPT-4's superior performance in a medical knowledge assessment when compared to its predecessor and human counterparts. This instance of AI outperforming human learners highlighted its robustness and reliability as an assessment tool. Furthermore, Ledwos et al. (2022) explored AI's application in assessing neurosurgical training learning curves through virtual reality simulations. The nuanced understanding of skill acquisition provided by AI-derived metrics pointed towards tailored educational strategies, marking a shift towards indivudualised learning experiences in surgical training. AI's applicability extended to grading and cognitive skills assessment, as shown by Lam. et al. (2022) and Natheir et al. (2023). These studies demonstrated ML's potential in objectively grading short answer questions and differentiating surgical expertise using EEG data, highlighting AI's versatile potential in medical education.

The evaluation of AI in the UK radiographer curriculum by Van de Venter et al. (2023) and the development of AI-enhanced learning tools like the Virtual Patient Simulator (Furlan et al., 2021) show AI's role in both professional development and improving reasoning skills. These examples underlined AI's capacity to enrich learning experiences and assessment methodologies across different domains of medical education. Also, the effectiveness of AI in providing a realistic platform for history-taking practice was illustrated by Holderried et al. (2024), emphasising AI's ability to simulate patient interactions and enhance communication skills without the need for actual or simulated patients. However, they caution that the AI not being reliable in providing accurate information. This, along with Sallam et al. (2023) exploration of attitudes towards ChatGPT in healthcare education, pointed towards AI's evolving role in medical training and assessment. The integration of NLP for assessing the quality of feedback, as investigated by Solano et al. (2021) and Otleş et al. (2021), highlighted AI's potential to automate and enhance feedback quality assessment. This application promises to refine educational feedback mechanisms, thus facilitating a more efficient learning process for surgical trainees. These studies illustrate a progressive integration of AI in medical education. From enhancing surgical training with considerably precise assessments to evaluating educational interventions and improving communication skills, AI's role is both transformative and expansive. It shows the shift towards more efficient, and objective educational methodologies, heralding a new era in medical education assessment.

13.5 Discussion

The potential of AI in healthcare education cannot be overstated. By leveraging AI, educators and institutions can evolve medical training, making it more aligned with the demands of contemporary and future healthcare milieu. AI's ability to analyse vast amounts of data and generate insights can result in more nuanced and compelling learning experiences, tailor educational content to individual learners, and provide simulation-based training that bridges the gap between theoretical knowledge and clinical practice. However, the integration of AI into healthcare education necessitates a thoughtful balance between technological innovation and the human aspects of learning. The essence of medical education not just to impart knowledge but to cultivate empathy, ethical reasoning, and interpersonal skills remains paramount. As such, future strategies should focus on refining AI technologies and addressing the technical challenges and ensuring that these tools complement traditional learning methods, support faculty in their roles as educators, and uphold the values central to healthcare.

It was interesting to note that none of the studies found that AI fared worse in their tests compared to humans, whether in reading images as a teaching aid, in tutoring, etc. This needs to be critically evaluated, as the amount of data available to train the AI is still in development, and the possibility of bias is rife throughout the system. AI is used primarily when there is a perceived shortage of specialists in the area. Either as a tool for learning for the students in their own time, as a continuous tutor, which is not possible physically, etc. It also creates the question: Who is training whom? Are the humans training the AI or is it the other way around?

The integration of AI has been shown to yield significant advantages characterised by improvements in training, teaching, and assessment methodologies across medical specialities. By offering tailored feedback and evaluations, AI has been shown to improve the learning experience, drawing on evidence from studies such as Winkler-Schwartz et al. (2019), Huang et al. (2023), and others. These studies illustrate AI's ability to analyse performance accurately, thus enabling targeted educational strategies. AI's capacity to evaluate educational interventions and curricula is demonstrated through research by Monlezun et al. (2018), highlighting its role in improving the effectiveness of teaching methods. Personalised learning experiences are further supported by AI, as shown by Ledwos et al. (2022), who emphasise the importance of tailored educational approaches. The pandemic accelerated AI's utility in adapting educational methodologies, with Jin et al. (2023) developing tools that bridge theoretical knowledge and practical application. In teaching, AI encourages interdisciplinary learning and fosters a research-oriented environment, as noted by Lang and Repp (2020) and Hu et al. (2023). In addition, AI contributes to developing specialised training frameworks and enhancing diagnostic accuracy, with Muntean et al. (2023), Cheng et al. (2020), and Fang et al. (2022) showcasing its efficacy. The development of educational tools like SmartPathk by Alderman et al. (2021) further attests to AI's role in enhancing the accessibility and efficiency of learning processes. Collectively, these papers show the potential of AI in medical education, highlighting its pivotal role in personalisation, assessment precision, and the modernisation of educational practices, ultimately equipping healthcare professionals with critical competencies for future challenges.

13.5.1 Issues of Using Artificial Intelligence in Medical Education

Despite AI's advantages in healthcare education, this evolution is not without its challenges, which according to the papers can be broadly categorise into technological integration, bias and equity, data privacy and security, clinical relevance and applicability, sustainability and scalability. Addressing these issues are essential to harnessing AI's full potential in an ethical and effective manner.

One of the primary hurdles in adopting AI within medical education is achieving seamless technological integration and fostering user acceptance. As Kaur et al. (2021) identify, integrating digital chatbots poses significant challenges, reflecting broader issues with incorporating advanced AI tools into existing educational frameworks. This challenge is not just about technological compatibility but also involves curriculum adaptation and preparing faculty to effectively utilise AI in their teaching and assessment practices (Kaur et al., 2021; Krive et al., 2023). Overcoming these hurdles requires a nuanced understanding of the technological and human elements involved in educational innovation.

Equity in medical education is another critical concern, with the potential for AI to either perpetuate or introduce biases posing significant risks. The reliance on historical data for training AI systems can inadvertently replicate existing disparities in healthcare education and delivery. This issue necessitates a deliberate approach to selecting training datasets, ensuring they are representative and devoid of biases to foster an inclusive educational environment (Lang & Repp, 2020; Monlezun et al., 2018). The handling of sensitive medical and personal information by AI systems in educational contexts also raises significant data privacy and security concerns. Rigorous data protection measures are essential to prevent unauthorised access and to comply with legal and ethical standards concerning data privacy.

Ensuring that AI applications in medical education are clinically relevant and applicable in real-world settings presents another layer of complexity. This is crucial in areas such as surgical education, where the practical application of skills learned through AI-enhanced simulations to actual clinical scenarios is of utmost importance (Huang et al., 2023; Ledwos et al., 2022). In addition, it should be noted that most of the papers reviewed speak of the limitations of AI within their specific applications, showing that it is not fool proof and not always reliable. Especially in a critical field as this, the limitations shown should be considered in detail before jumping on the AI bandwagon. Also, studies looking at mental health, elderly care, interdisciplinary fields such as medical engineering were scarce. The policy side of AI integration within healthcare should also be considered carefully (Shivashankar & Bakthavatchaalam, 2024).

Finally, the development of AI-based educational modules must consider their sustainability and ability to scale across different learning environments. This is especially important for initiatives like deep learning in radiology, where the effectiveness of educational content across different settings is a crucial concern (Jin et al., 2023; Richardson & Ojeda, 2022). Exploring the bredth of how AI can be integrated in medical education, it becomes clear that these challenges are complex and interlinked, requiring a comprehensive approach to resolution. By addressing the details of technological integration, bias and equity, data privacy, clinical relevance, and sustainability, the educational sector can unlock the rapidly increasing potential of AI, thus encouraging a future where medical training is more accessible, inclusive, and aligned with the evolving needs of healthcare.

13.5.2 Future Strategies

To explore and fully leverage the benefits of AI in medical education, the literature proposes a range of future strategies. These strategies aim to address the challenges identified, such as technological integration issues, bias, and data privacy concerns, ensuring that the deployment of AI in medical education is both practical and ethical. A pivotal strategy involves refining and expanding AI modules across various medical specialities, ensuring seamless incorporation into curricula and adaptability to meet a gamut of educational needs. This entails not only technological harmonisation with existing education structures but also customisation capabilities to cater to the specific requirements of different medical fields (Jin et al., 2023; Richardson & Ojeda, 2022).

Exploring an interdisciplinary approach that encompasses AI literacy within the medical curriculum is shown to be fundamental. Such an approach extends beyond technical knowledge to include ethical, legal, and social implications of AI usage. Lang and Repp (2020) and Hu et al. (2023) showcase the importance of integrating AI into educational frameworks, advocating for curricula that prepare students for a technologically advanced healthcare landscape while instilling a critical understanding of AI's ethical dimensions. Programmes such as medical engineering provided in some of the Universities could act as a bridge between medical and technological education.

Mitigating bias and ensuring fairness in AI applications have been shown to be major aspects that require work. They require transparent, inclusive development processes. Such a strategy would champion the involvement of a range of stakeholders in AI tool design and implementation, ensuring systems are equitable and representative. Continual evaluation of AI tools for bias and equity reveals a commitment to inclusivity in AI development (Kaur et al., 2021; Krive et al., 2023).

The formulation and enforcement of stringent data governance policies are critical for addressing data privacy and security concerns. Such policies ensure that AI systems in healthcare education adhere to the highest standards of data protection, safeguarding sensitive information and maintaining confidentiality (Lang & Repp, 2020; Monlezun et al., 2018). Utilising AI to enhance feedback and assessment methodologies represents another strategic direction. This includes employing AI for objective and standardised skill assessments, such as surgical competencies, and leveraging AI to improve the quality of feedback. The goal is to augment learning outcomes through precise, tailored feedback mechanisms, opening the path for more effective learning and skill development (Huang et al., 2023; Ledwos et al., 2022).

In conclusion, addressing the challenges posed by integrating AI necessitates a holistic and innovative approach. By focusing on strategies that emphasise enhanced technological integration, interdisciplinary and ethical education, inclusivity and bias mitigation, stringent data governance, and improved feedback mechanisms, the educational sector can effectively deal with the difficulties of AI deployment. These strategies, underpinned by a commitment to continuous innovation and research, offer a promising path forward for changing medical education to be more efficient, equitable, and attuned to the technological advancements shaping the healthcare education and sector.

13.6 Conclusion

This systematic review shows the dual impact of AI in healthcare education, highlighting its capacity to enhance learning through personalised approaches and innovative methodologies against challenges like technological integration, biases, and data privacy concerns. AI's breakthrough potential in healthcare education promises to align training with modern healthcare demands, emphasising the need for a balance between technological advancements and the human elements of learning. Addressing these challenges requires collaborative efforts to ensure AI integration complements traditional education values, ultimately improving healthcare education and patient care by bridging the gap between theoretical knowledge and clinical practices.

References

Abid, A., Murugan, A., Banerjee, I., Purkayastha, S., Trivedi, H., & Gichoya, J. (2024). AI education for fourth-year medical students: Two-year experience of a web-based, self-guided curriculum and mixed methods study. *JMIR Medical Education, 10*, e46500.

Aldeman, N. L. S., de Sá Urtiga Aita, K. M., Machado, V. P., da Mata Sousa, L. C. D., Coelho, A. G. B., da Silva, A. S., & do Monte, S. J. H. (2021). Smartpathk: A platform for teaching glomerulopathies using machine learning. *BMC Medical Education, 21*(1), 248.

Al-Dhaen, F., Hou, J., Rana, N. P., & Weerakkody, V. (2023). Advancing the understanding of the role of responsible AI in the continued use of IoMT in healthcare. *Information Systems Frontiers, 25*(6), 2159–2178.

Alonso-Silverio, G. A., Perez-Escamirosa, F., Bruno-Sanchez, R., Ortiz-Simon, J. L., Muñoz-Guerrero, R., Minor-Martinez, A., & Alarcón-Paredes, A. (2018). Development of a laparoscopic box trainer based on open-source hardware and artificial intelligence for objective assessment of surgical psychomotor skills. *Surgical Innovation, 25*(4), 380–388.

Bissonnette, V., Mirchi, N., Ledwos, N., Alsidieri, G., Winkler-Schwartz, A., & Del Maestro, R. F. (2019). Artificial intelligence distinguishes surgical training levels in a virtual reality spinal task. *JBJS, 101*(23), e127.

Campra, M., Orlandini, P., Amelio, S., & Brescia, V. (2021). Ordering cost-effectiveness management studies in healthcare: A PRISMA-compliant systematic literature review and bibliometric analysis. *International Journal of Business Administration*

Chassagnon, G., Billet, N., Rutten, C., Toussaint, T., Cassius de Linval, Q., Collin, M., Lemouchi, L., Homps, M., Hedjoudje, M., Ventre, J., Gregory, J., Canniff, E., Regnard, N. E., Bennani, S., & Revel, M. P. (2023). Learning from the machine: AI assistance is not an effective learning tool for resident education in chest x-ray interpretation. *European Radiology, 33*(11), 8241–8250. https://doi.org/10.1007/s00330-023-10043-1

Cheng, C. T., Chen, C. C., Fu, C. Y., Chaou, C. H., Wu, Y. T., Hsu, C. P., Chang, C. C., Chung, I. F., Hsieh, C. H., Hsieh, M. J., & Liao, C. H. (2020). Artificial intelligence-based education assists medical students' interpretation of hip fracture. *Insights into Imaging, 11*(1), 119. https://doi.org/10.1186/s13244-020-00932-0

Dave, M., & Patel, N. (2023). Artificial intelligence in healthcare and education. *British Dental Journal, 234*(10), 761–764. https://doi.org/10.1038/s41415-023-5845-2

Fang, Z., Xu, Z., He, X., & Han, W. (2022). Artificial intelligence-based pathologic myopia identification system in the ophthalmology residency training program. *Frontiers in Cell and Developmental Biology, 10*, 1053079.

Ferreira, D. C., Vieira, I., Pedro, M. I., Caldas, P., & Varela, M. (2023). Patient satisfaction with healthcare services and the techniques used for its assessment: A systematic literature review and a bibliometric analysis. *Healthcare (Basel, Switzerland), 11*(5), 639. https://doi.org/10.3390/healthcare11050639

Furlan, R., Gatti, M., Menè, R., Shiffer, D., Marchiori, C., Giaj Levra, A., & Dipaola, F. (2021). A natural language processing–based virtual patient simulator and intelligent tutoring system for the clinical diagnostic process: Simulator development and case study. *JMIR Medical Informatics, 9*(4), e24073.

Gao, Y., Kruger, U., Intes, X., Schwaitzberg, S., & De, S. (2020). A machine learning approach to predict surgical learning curves. *Surgery, 167*(2), 321–327.

Garcia, M. B., Arif, Y. M., Khlaif, Z. N., Zhu, M., de Almeida, R. P., de Almeida, R. S., & Masters, K. (2024). Effective integration of artificial intelligence in medical education: Practical tips and actionable insights. In M. Garcia & R. de Almeida (Eds.), *Transformative approaches to patient literacy and healthcare innovation* (pp. 1–19). IGI Global. https://doi.org/10.4018/979-8-3693-3661-8.ch001

Gligorea, I., Cioca, M., Oancea, R., Gorski, A. T., Gorski, H., & Tudorache, P. (2023). Adaptive learning using artificial intelligence in e-learning: A literature review. *Education Sciences, 13*(12), 1216.

Goldie, J. G. S. (2016). Connectivism: A knowledge learning theory for the digital age? *Medical Teacher, 38*(10), 1064–1069. https://doi.org/10.3109/0142159X.2016.1173661

Han, R., Yu, W., Chen, H., & Chen, Y. (2022). Using artificial intelligence reading label system in diabetic retinopathy grading training of junior ophthalmology residents and medical students. *BMC Medical Education, 22*(1), 258.

Hedderich, D. M., Keicher, M., Wiestler, B., Gruber, M. J., Burwinkel, H., Hinterwimmer, F., Czempiel, T., Spiro, J. E., Pinto Dos Santos, D., Heim, D., Zimmer, C., Rückert, D., Kirschke, J. S., & Navab, N. (2021). AI for doctors-a course to educate medical professionals in artificial intelligence for medical imaging. *Healthcare (Basel, Switzerland), 9*(10), 1278. https://doi.org/10.3390/healthcare9101278

Holderried, F., Stegemann-Philipps, C., Herschbach, L., Moldt, J. A., Nevins, A., Griewatz, J., Holderried, M., Herrmann-Werner, A., Festl-Wietek, T., & Mahling, M. (2024). A Generative Pretrained Transformer (GPT)-Powered chatbot as a simulated patient to practice history taking: Prospective. *Mixed Methods Study. JMIR Medical Education, 10*, e53961. https://doi.org/10.2196/53961

Hsu, M. H., Chan, T. M., & Yu, C. S. (2023). Termbot: A chatbot-based crossword game for gamified medical terminology learning. *International Journal of Environmental Research and Public Health, 20*(5), 4185.

Hu, R., Rizwan, A., Hu, Z., Li, T., Chung, A. D., & Kwan, B. Y. (2023). An artificial intelligence training workshop for diagnostic radiology residents. *Radiology: Artificial Intelligence, 5*(2), e220170.

Huang, R. S., Lu, K. J. Q., Meaney, C., Kemppainen, J., Punnett, A., & Leung, F. H. (2023). Assessment of resident and AI chatbot performance on the University of Toronto family medicine residency progress test: Comparative study. *JMIR Medical Education, 9*, e50514.

Jeyaraman, M., Balaji, S., Jeyaraman, N., Yadav, S. (2023). Unraveling the ethical enigma: Artificial intelligence in healthcare. *Cureus, 15*(8), e43262. https://doi.org/10.7759/cureus.43262

Jin, H., Wagner, M. W., Ertl-Wagner, B., & Khalvati, F. (2023). An educational graphical user interface to construct convolutional neural networks for teaching artificial intelligence in radiology. *Canadian Association of Radiologists Journal, 74*(3), 526–533.

Joo, H., Mathis, M. R., Tam, M., James, C., Han, P., Mangrulkar, R. S., Friedman, C. P., & Vydiswaran, V. G. V. (2023). Applying AI and guidelines to assist medical students in recognizing patients with heart failure: protocol for a randomized trial. *JMIR Research Protocols, 12*, e49842. https://doi.org/10.2196/49842

Kaur, A., Singh, S., Chandan, J. S., Robbins, T., & Patel, V. (2021). Qualitative exploration of digital chatbot use in medical education: A pilot study. *Digital Health, 7*, 20552076211038150.

Komasawa, N., Yokohira, M. (2023). Learner-centered experience-based medical education in an AI-driven society: A literature review. *Cureus. 15*(10), e46883. https://doi.org/10.7759/cureus.46883

Krive, J., Isola, M., Chang, L., Patel, T., Anderson, M., & Sreedhar, R. (2023). Grounded in reality: artificial intelligence in medical education. *JAMIA open, 6*(2), ooad037.

Lam, A., Lam, L., Blacketer, C., Parnis, R., Franke, K., Wagner, M., & Bacchi, S. (2022). Professionalism and clinical short answer question marking with machine learning. *Internal Medicine Journal, 52*(7), 1268–1271.

Lang, J., & Repp, H. (2020). Artificial intelligence in medical education and the meaning of interaction with natural intelligence–an interdisciplinary approach. *GMS Journal for Medical Education, 37*(6).

Ledwos, N., Mirchi, N., Yilmaz, R., Winkler-Schwartz, A., Sawni, A., Fazlollahi, A. M., & Del Maestro, R. F. (2022). Assessment of learning curves on a simulated neurosurgical task using metrics selected by artificial intelligence. *Journal of Neurosurgery, 137*(4), 1160–1171.

Lindqwister, A. L., Hassanpour, S., Lewis, P. J., & Sin, J. M. (2021). AI-RADS: An artificial intelligence curriculum for residents. *Academic Radiology, 28*(12), 1810–1816.

Lippitsch, A., Steglich, J., Ludwig, C., Kellner, J., Hempel, L., Stoevesandt, D., & Thews, O. (2024). Development and evaluation of a software system for medical students to teach and practice anamnestic interviews with virtual patient avatars. *Computer Methods and Programs in Biomedicine, 244*, 107964.

Marbouh, D., Swarnakar, V., Simsekler, M. C., Antony, J., Lizarelli, F. L., Jayaraman, R., Garza-Reyes, J. A., Shokri, A., Cudney, E. A., & Ellahham, S. (2023). Healthcare 4.0 digital technologies impact on quality of care: A systematic literature review. *Total Quality Management & Business Excellence, 34*, 2157–2182.

Mirchi, N., Bissonnette, V., Yilmaz, R., Ledwos, N., Winkler-Schwartz, A., & Del Maestro, R. F. (2020). The virtual operative assistant: An explainable artificial intelligence tool for simulation-based training in surgery and medicine. *PLoS ONE, 15*(2), e0229596.

Monlezun, D. J., Dart, L., Vanbeber, A., Smith-Barbaro, P., Costilla, V., Samuel, C., & Harlan, T. S. (2018). Machine learning-augmented propensity score-adjusted multilevel mixed effects panel analysis of hands-on cooking and nutrition education versus traditional curriculum for medical students as preventive cardiology: multisite cohort study of 3,248 trainees over 5 years. *BioMed Research International*.

Muntean, G. A., Groza, A., Marginean, A., Slavescu, R. R., Steiu, M. G., Muntean, V., & Nicoara, S. D. (2023). Artificial intelligence for personalised ophthalmology residency training. *Journal of Clinical Medicine, 12*(5), 1825.

Nagaraj, M. B., Namazi, B., Sankaranarayanan, G., & Scott, D. J. (2023). Developing artificial intelligence models for medical student suturing and knot-tying video-based assessment and coaching. *Surgical Endoscopy, 37*(1), 402–411.

Nagi, F., Salih, R., Alzubaidi, M., Shah, H., Alam, T., Shah, Z., & Househ, M. (2023). Applications of Artificial Intelligence (AI) in medical education: A scoping review. *Studies in Health Technology and Informatics, 305*, 648–651. https://doi.org/10.3233/SHTI230581

Narayanan, S., Ramakrishnan, R., Durairaj, E., & Das, A. (2023). Artificial intelligence revolutionizing the field of medical education. *Cureus., 15*(11), e49604. https://doi.org/10.7759/cureus.49604. PMID:38161821;PMCID:PMC10755136

Natheir, S., Christie, S., Yilmaz, R., Winkler-Schwartz, A., Bajunaid, K., Sabbagh, A. J., & Del Maestro, R. (2023). Utilizing artificial intelligence and electroencephalography to assess expertise on a simulated neurosurgical task. *Computers in Biology and Medicine, 152*, 106286.

Otles, E., Kendrick, D. E., Solano, Q. P., Schuller, M., Ahle, S. L., Eskender, M. H., Carnes, E., & George, B. C. (2021). Using natural language processing to automatically assess feedback quality: Findings from 3 surgical residencies. *Academic Medicine : Journal of the Association of American Medical Colleges, 96*(10), 1457–1460. https://doi.org/10.1097/ACM.0000000000004153

Park, S. H., Pinto-Powell, R., Thesen, T., Lindqwister, A., Levy, J., Chacko, R., Gonzalez, D., Bridges, C., Schwendt, A., Byrum, T., Fong, J., Shasavari, S., & Hassanpour, S. (2024). Preparing healthcare leaders of the digital age with an integrative artificial intelligence curriculum: A pilot study. *Medical Education Online, 29*(1), 2315684. https://doi.org/10.1080/10872981.2024.2315684

Rainey, C., O'Regan, T., Matthew, J., Skelton, E., Woznitza, N., Chu, K. Y., Goodman, S., McConnell, J., Hughes, C., Bond, R., McFadden, S., & Malamateniou, C. (2021). Beauty Is in the AI of the beholder: Are we ready for the clinical integration of artificial intelligence in radiography? An exploratory analysis of perceived ai knowledge, skills, confidence, and education perspectives of UK radiographers. *Frontiers in Digital Health, 3*, 739327. https://doi.org/10.3389/fdgth.2021.739327

Richardson, M. L., & Ojeda, P. I. (2022). A "Bumper-Car" curriculum for teaching deep learning to radiology residents☆. *Academic Radiology, 29*(5), 763–770.

Ryder, C. Y., Mott, N. M., Gross, C. L., Anidi, C., Shigut, L., Bidwell, S. S., Kim, E., Zhao, Y., Ngam, B. N., Snell, M. J., Yu, B. J., Forczmanski, P., Rooney, D. M., Jeffcoach, D. R., & Kim, G. J. (2024). Using artificial intelligence to gauge competency on a novel laparoscopic training system. *Journal of Surgical Education, 81*(2), 267–274. https://doi.org/10.1016/j.jsurg.2023.10.007

Sallam, M., Salim, N. A., Barakat, M., Al-Mahzoum, K., Al-Tammemi, A. B., Malaeb, D., Hallit, R., & Hallit, S. (2023). Assessing health students' attitudes and usage of ChatGPT in Jordan: Validation study. *JMIR Medical Education*, *9*, e48254. https://doi.org/10.2196/48254

Samaan, J. S., Yeo, Y. H., Rajeev, N., Ng, W. H., Srinivasan, N., & Samakar, K. (2023). Ensuring ethical and responsible integration of AI tools in healthcare. *Obesity Surgery, 33*(8), 2597. https://doi.org/10.1007/s11695-023-06688-y

Sharma, V., Saini, U., Pareek, V., Sharma, L., & Kumar, S. (2023). *Artificial intelligence (AI) integration in medical education: A pan-India cross-sectional observation of acceptance and understanding among students*. Scripta Medica.

Shiang, T., Garwood, E., & Debenedectis, C. M. (2022). Artificial intelligence-based decision support system (AI-DSS) implementation in radiology residency: Introducing residents to AI in the clinical setting. *Clinical Imaging, 92*, 32–37.

Shivashankar, K., & Bakthavatchaalam, V. (2024). Education policies through data driven decision making: Accelerating inclusive education for people with disabilities. In R. Singh, S. Khan, A. Kumar, & V. Kumar (Eds.), *Artificial Intelligence Enabled Management* (pp. 15–32). De Gruyter. https://doi.org/10.1515/9783111172408-002

Siyar, S., Azarnoush, H., Rashidi, S., Winkler-Schwartz, A., Bissonnette, V., Ponnudurai, N., & Del Maestro, R. F. (2020). Machine learning distinguishes neurosurgical skill levels in a virtual reality tumor resection task. *Medical & Biological Engineering & Computing, 58*, 1357–1367.

Solano, Q. P., Hayward, L., Chopra, Z., Quanstrom, K., Kendrick, D., Abbott, K. L., Kunzmann, M., Ahle, S., Schuller, M., Ötleş, E., & George, B. C. (2021). Natural language processing and assessment of resident feedback quality. *Journal of Surgical Education, 78*(6), e72–e77. https://doi.org/10.1016/j.jsurg.2021.05.012

Ueda, D., Kakinuma, T., Fujita, S., et al. (2024). Fairness of artificial intelligence in healthcare: Review and recommendations. *Japanese Journal of Radiology, 42*, 3–15. https://doi.org/10.1007/s11604-023-01474-3

van de Venter, R., Skelton, E., Matthew, J., Woznitza, N., Tarroni, G., Hirani, S. P., & Malamateniou, C. (2023). Artificial intelligence education for radiographers, an evaluation of a UK postgraduate educational intervention using participatory action research: A pilot study. *Insights into Imaging, 14*(1), 25.

Wartman, S. A., & Combs, C. D. (2018). Medical education must move from the information age to the age of artificial intelligence. *Academic Medicine, 93*(8), 1107–1109.

Wiggins, W. F., Caton, M. T., Jr., Magudia, K., Rosenthal, M. H., & Andriole, K. P. (2021). A conference-friendly, hands-on introduction to deep learning for radiology trainees. *Journal of Digital Imaging, 34*(4), 1026–1033.

Wiljer, D., & Hakim, Z. (2019). Developing an artificial intelligence–enabled health care practice: Rewiring health care professions for better care. *Journal of Medical Imaging and Radiation Sciences, 50*(4), S8–S14.

Winkler-Schwartz, A., Yilmaz, R., Mirchi, N., Bissonnette, V., Ledwos, N., Siyar, S., & Del Maestro, R. (2019). Machine learning identification of surgical and operative factors associated with surgical expertise in virtual reality simulation. *JAMA Network Open, 2*(8), e198363–e198363.

Wong, K., Gallant, F., & Szumacher, E. (2021). Perceptions of Canadian radiation oncologists, radiation physicists, radiation therapists and radiation trainees about the impact of artificial intelligence in radiation oncology–national survey. *Journal of Medical Imaging and Radiation Sciences, 52*(1), 44–48.

Yeow, H., Azman, A., & Singh, P.S. (2023). Perspectives of healthcare professionals regarding dignity and end-of-life care for the elderly: A systematic literature review. *Africa Journal of Nursing and Midwifery.*

Index

R. Singh et al. (eds.), *Transforming Healthcare Sector Through Artificial Intelligence and Environmental Sustainability*, Approaches to Global Sustainability, Markets, and Governance, https://doi.org/10.1007/978-981-97-9555-0

The manufacturer's authorised representative in the EU is Springer Nature Customer Service Centre GmbH, Europaplatz 3, 69115 Heidelberg, Germany. If you have any concerns regarding our products, please contact ProductSafety@springernature.com

Printed and bound by CPI Group (UK) Ltd, Croydon, CR0 4YY

01/10/2026

02225925-0014